Polyphenols in Health

Rajeev K. Singla · Vivek P. Chavda · Tapan Behl · Salvatore Parisi · Bairong Shen
Editors

# Polyphenols in Health

## Bioavailability, Mechanisms, and Therapeutic Innovations

Springer

*Editors*
Rajeev K. Singla
Institutes for Systems Genetics
West China Hospital of Sichuan University
Chengdu, Sichuan, China

Tapan Behl
Amity Institute of Pharmacy
Amity University Punjab
Mohali, Punjab, India

Bairong Shen
Institutes for Systems Genetics
West China Hospital of Sichuan University
Chengdu, Sichuan, China

Vivek P. Chavda
Department of Pharmaceutics
and Pharmaceutical Technology
L.M. College of Pharmacy
Ahmedabad, Gujarat, India

Salvatore Parisi
Lourdes Matha Institute of Hotel
Management and Catering Technology
Thiruvananthapuram, Kerala, India

ISBN 978-3-032-27323-9 ISBN 978-3-032-27324-6 (eBook)
https://doi.org/10.1007/978-3-032-27324-6



This Springer imprint is published by the registered company Springer Nature Switzerland AG
The registered company address is: Gewerbestrasse 11, 6330 Cham, Switzerland

**Acknowledgments** We thank Caproslaxy Media (https://www.caproslaxymedia.com/) for linguistic assistance during the preparation of this manuscript (Chaps. 1 and 13). To maintain uniformity and to standardise as per international nomenclature (Chap. 1), scientific terminologies for botanical sources have been adopted as per the information there on https://www.worldfloraonline.org/.

**Competing Interests** The authors declare that they have no competing interests.

# Contents

# Chapter 1
# Polyphenols: An Overview

**Ankit Kumar Dubey, Xiaoyu Li, Smriti Mishra, Bairong Shen, and Rajeev K. Singla**

**Abstract** Polyphenols are a diverse group of plant-derived secondary metabolites found in many common foods and beverages such as fruits, vegetables, whole grains, legumes, nuts, tea, coffee, cocoa, and wine. They are increasingly recognized as important bioactive components of plant-based diets. This Chapter offers an overview of polyphenols' chemistry and classification, covering major groups like flavonoids, phenolic acids, stilbenes, lignans, and tannins. It emphasizes how structural features, such as hydroxylation, glycosylation, and polymerization, affect their physical, chemical, and biological properties. The Chapter reviews dietary sources, intake variability, and methods for assessing intake through dietary tools, databases, and biomarkers. It explains how polyphenols act as antioxidants, modulating pathways like Nrf2, reducing inflammation via NF-κB and MAPK pathways, and providing cardiometabolic benefits like improved endothelial function, resistance to lipid oxidation, blood pressure regulation, and glucose control. It also discusses their interactions with gut microbiota, leading to bioactive metabolites and individual metabolic differences. Furthermore, it highlights links between polyphenol-rich diets

A. K. Dubey · X. Li · B. Shen · R. K. Singla (✉)
Department of Pharmacy and Institutes for Systems Genetics, Center for High Altitude Medicine, Frontiers Science Center for Disease-Related Molecular Network, West China Hospital, Sichuan University, Chengdu, Sichuan, China
e-mail: rajeevsingla26@gmail.com; rajeevkumar@scu.edu.cn

X. Li
e-mail: lixiaoyu@wchscu.cn

B. Shen
e-mail: bairong.shen@scu.edu.cn

A. K. Dubey
iGlobal Research and Publishing Foundation, New Delhi, India

S. Mishra
S.N. Pradhan Centre for Neuroscience, Ballygunge Science College, Calcutta University, College Square, India

R. K. Singla
School of Pharmaceutical Sciences, Lovely Professional University, Phagwara, Punjab, India

 
R. K. Singla et al. (eds.), *Polyphenols in Health*,
https://doi.org/10.1007/978-3-032-27324-6_1

and lower risk of cardiovascular disease, type-2 diabetes, some cancers, and neurodegenerative diseases. The chapter points out key research gaps, including issues with absorption, metabolism, lack of standardized reference materials, analytical challenges, and absence of universal intake guidelines. Finally, it suggests future directions such as better clinical trials, advanced analytical methods, delivery strategies to enhance bioavailability, sustainable sourcing of by-products, and personalized nutrition through omics and microbiome data, to improve understanding and application of polyphenols guidance.

**Keywords** Antioxidant mechanism · Bioavailability · Cardiometabolic health · Chronic disease prevention · Dietary assessment · Gut microbiota · Polyphenol

## Abbreviations

| | |
|---|---|
| 5-LOX | 5-Lipoxygenase |
| ABTS | 2,2'-Azinobis (3-ethylbenzothiazoline-6-sulfonic acid) |
| AD | Alzheimer's disease |
| APCI | Atmospheric pressure chemical ionization |
| BBB | Blood-brain barrier |
| CE | Capillary electrophoresis |
| COMT | Catechol-O-methyltransferase |
| COX | Cyclooxygenase |
| COX-2 | Cyclooxygenase-2 |
| CRP | C-reactive protein |
| CVD | Cardiovascular disease |
| DNA | Deoxyribonucleic acid |
| DAD | Diode-array detection |
| DPPH | 2,2-Diphenyl-1-picrylhydrazyl |
| DRI | Dietary reference intake |
| EPIC | European Prospective Investigation into Cancer and Nutrition |
| ERK | Extracellular signal-regulated kinase |
| ESI | Electrospray ionization |
| FRAP | Ferric reducing ability of plasma |
| FFQ | Food frequency questionnaire |
| GAE | Gallic acid equivalent |
| GC-MS | Gas chromatography-mass spectrometry |
| GLUT | Glucose transporter |
| GPx | Glutathione peroxidase |
| GST | Glutathione S-transferase |
| HDL | High-density lipoprotein |
| HPLC | High-performance liquid chromatography |
| HO-1 | Heme oxygenase-1 |
| OH | Hydroxyl group |

| | |
|---|---|
| iNOS | Inducible nitric oxide synthase |
| IL | Interleukin |
| IκB | Inhibitor κB |
| IKK | IκB kinase |
| JNK | C-Jun N-terminal kinase |
| LDL | Low-density lipoprotein |
| LC–MS/MS | Liquid chromatography coupled to mass spectrometry |
| MAPK | Mitogen-activated protein kinase |
| MW | Molecular weight |
| NF-κB | Nuclear factor kappa B |
| NO | Nitric oxide |
| Nrf2 | Nuclear factor erythroid 2-related factor 2 |
| NQO1 | NADPH quinone oxidoreductase |
| ORAC | Oxygen radical absorbance capacity |
| PD | Parkinson's disease |
| PPAR | Peroxisome proliferator-activated receptor |
| RNS | Reactive nitrogen species |
| ROS | Reactive oxygen species |
| SOD | Superoxide dismutase |
| SPE | Solid-phase extraction |
| SULT | Sulfotransferase |
| TOF | Time-of-flight |
| UDP | Uridine 5'-diphospho |
| UGT | UDP-glucuronosyltransferase |
| USDA | United States Department of Agriculture |
| UV-VIS | Ultraviolet-visible |

## 1.1 Introduction

### *1.1.1 Overview and Significance*

Polyphenols are a diverse and widespread group of naturally occurring organic compounds produced by plants as part of their secondary metabolism (Cory et al. 2018; Del Rio et al. 2013; Manach et al. 2004). These molecules are mainly characterized by containing one or more aromatic hydrocarbon rings with attached hydroxyl groups (Manach et al. 2004; Pandey and Rizvi 2009). Polyphenols are widely distributed across various plant species and are abundant in many common foods and beverages. They are present in fruits, vegetables, whole grains, pulses, nuts, seeds, herbs, spices, tea, coffee, chocolate, and wine, all of which are important sources of polyphenolic compounds in the diet (Cory et al. 2018; Grosso et al. 2017; Manach et al. 2004).

In recent decades, both the scientific and public health sectors have devoted increasing attention to polyphenols, especially as emerging evidence highlights their various health benefits. These biological effects are wide-ranging, including well-established antioxidative activity, anti-inflammation, cardiovascular protection, neurological preservation, metabolic regulation that supports normal blood sugar levels, and possible anticancer effects (Del Rio et al. 2013; Dubey et al. 2025b; Ozdal et al. 2016; Rudrapal et al. 2022; Singla and Bishayee 2025; Zamora-Ros et al. 2016). Epidemiological studies using cohort and case–control approaches consistently demonstrate that diets rich in polyphenols are associated with a significantly reduced risk of major noncommunicable diseases. These diseases include atherosclerotic cardiovascular disease, insulin resistance, type-2 diabetes, certain types of cancers, and neurodegenerative disorders (Del Rio et al. 2013; Ozdal et al. 2016; Perez-Jimenez et al. 2010; Zamora-Ros et al. 2016).

Historically, researchers studied polyphenols mainly through food chemistry and sensory science, emphasizing their role in sensory attributes like color, bitterness, astringency, and oxidative stability during storage and transport (Cory et al. 2018; Manach et al. 2004). However, modern nutritional science has shifted toward exploring the bioactive properties of these compounds and their ability to influence disease progression across the lifespan (Cory et al. 2018; Huang et al. 2020; Manach et al. 2004). Currently, polyphenols are increasingly recognized as vital bioactive components in evidence-based dietary patterns such as the Mediterranean diet, the Nordic diet, and various vegetarian and vegan diets (Huang et al. 2020; Saad et al. 2025). Despite this progress, exploring polyphenols remains challenging due to their structural and functional diversity, significant individual differences in absorption, metabolism, and use, and the multiple mechanisms through which they impact physiology (Hooper et al. 2012; Zamora-Ros et al. 2016). These challenges hinder efforts to establish causality, develop population-wide dietary guidelines, and translate mechanistic insights into clinical practice recommendations.

This comprehensive Chapter consolidates current knowledge of polyphenolic compounds, covering their chemical properties, structural classification, dietary and non-dietary sources, consumption patterns across populations, and analytical methods used for their study. It also details biological mechanisms with implications for human health, such as free radical scavenging, modulation of endogenous defenses, inhibition of inflammatory pathways, interaction with the gut microbiota, and prevention strategies for chronic diseases. The Chapter highlights common challenges, including issues with bioavailability studies, gaps in dietary guidance specific to polyphenols, and questions about individual differences in metabolism and therapeutic effects. Additionally, it explores future research directions, like the use of omics technologies, improved delivery and absorption strategies, and personalized nutrition approaches.

### 1.1.2 *Clinical Significance in Cardiovascular and Metabolic Health*

The epidemiological evidence base linking polyphenolic consumption to cardiovascular health outcomes has become increasingly robust. Large prospective cohort investigations conducted in European and North American populations consistently report quantifiable risk reductions when comparing individuals consuming polyphenol-rich diets with those with lower consumption patterns (Cardona et al. 2013; Grosso et al. 2018; Tresserra-Rimbau et al. 2014). Mechanistic investigations at the cellular and molecular level illuminate how these compounds influence key pathogenic processes underlying atherosclerotic disease, specifically, modification of endothelial function, regulation of lipoprotein oxidation, modulation of platelet aggregation behavior, and stabilization of arterial blood pressure within physiological ranges (Cardona et al. 2013; Grosso et al. 2017, 2018).

Metabolic effects of polyphenols extend beyond vascular protection to encompass improvements in glucose homeostasis and insulin signaling sensitivity. Epidemiological studies demonstrate associations between higher polyphenol consumption and reduced incidence of type-2 diabetes across multiple populations (Del Bo et al. 2019; Grosso et al. 2017; Jančić et al. 2022). Mechanistic studies reveal that polyphenolic compounds suppress intestinal carbohydrate-hydrolyzing enzymes, modulate glucose transporter expression and activity, enhance pancreatic beta cell function, and favorably modify adipose tissue biology (Del Bo et al. 2019; Quideau et al. 2011).

## 1.2 Chemical Definition, Classification, and Structural Properties

### 1.2.1 *Chemical Definition and Structural Fundamentals*

Polyphenols are organic compounds characterized by an aromatic ring, typically a benzene ring, with one or more hydroxyl groups (OH) attached covalently (Cory et al. 2018; Del Rio et al. 2013; Manach et al. 2004; Pandey and Rizvi 2009). In addition to this fundamental structure, polyphenols often feature further chemical modifications such as methoxy groups, conjugated carbohydrate units, esterified organic acids, or polymeric bonds linking multiple phenolic units (Grosso et al. 2017; Rudrapal et al. 2022). Their biosynthesis in plants primarily occurs through three metabolic pathways: the shikimate pathway, which produces simple phenolic acids; the phenylpropanoid pathway, responsible for flavonoids and similar compounds; and the polyketide synthesis pathway (Cory et al. 2018; Renaud and de Lorgeril 1992).

The extensive structural diversity of polyphenolic compounds is noteworthy. Current analytical and chemical databases list over 8,000 unique polyphenolic structures naturally found in plant tissues (Rudrapal et al. 2022; Tsao 2010). This number continues to grow with advances in chromatographic and spectrometric techniques, which offer greater resolution and sensitivity. Hydroxyl groups spread across the aromatic ring system impart distinctive physicochemical properties that greatly affect bioactivity. In particular, the number, stereochemical arrangement, and chemical reactivity of these hydroxyl groups influence their redox potential, ability to donate hydrogen in free radical neutralization, capacity to form chelation complexes with metals, and ability to engage in hydrogen bonding with biological macromolecules (Pandey and Rizvi 2009; Rudrapal et al. 2022; Tsao 2010).

### *1.2.2 Structural Characteristics Influencing Biological Activity*

The precise spatial arrangement of OH groups substantially influences antioxidative efficacy and susceptibility to enzymatic metabolism. Catechol moieties (two adjacent OH) and galloyl groups (three OH arranged on a benzene ring) demonstrate particularly pronounced antioxidative capacity if compared with monohydroxy phenolic compounds and macromolecules (Pandey and Rizvi 2009; Rudrapal et al. 2022; Tsao 2010). The hydroxylation pattern directly influences the thermodynamic feasibility of free radical scavenging reactions and the stability of intermediate radicals generated during redox cycling macromolecules (Pandey and Rizvi 2009; Rudrapal et al. 2022).

Glycosylation involves the covalent attachment of one or more sugar units to the polyphenolic aglycone, playing a key role in determining bioavailability and physiological effects (Bhagwat et al. 2014; Rudrapal et al. 2022; Saad et al. 2025). In typical plant foods, dietary polyphenols are mostly stored as O-glycosides (linked through hydroxyl oxygen) or C-glycosides (linked directly to the sugar via carbon), rather than as the free aglycones. This glycosidic linkage greatly enhances water solubility, which affects how they are absorbed in the gastrointestinal tract. It also requires enzymatic hydrolysis by intestinal enzymes or bacterial glycosidases in the gut to release the aglycone and influences the types of metabolites circulated after ingestion (Saad et al. 2025; Scalbert et al. 2005b).

The degree of polymerization (the formation of covalent bonds between multiple polyphenolic monomers) adds another layer of structural complexity that significantly affects bioavailability. Monomeric compounds like catechin can easily pass through the intestinal epithelium using active transport. In contrast, oligomeric and polymeric proanthocyanidins (condensed tannins), which may consist of three to over 100 linked monomer units, surpass the molecular weight (MW) limit for passive absorption in the intestine. These larger molecules largely stay intact during their passage through the small intestine and are extensively broken down and fermented by colonic microbiota, resulting in smaller phenolic metabolites that may have different bioactivities (Cardona et al. 2013; Rudrapal et al. 2022; Scalbert et al. 2005b; Willett et al. 2019).

### *1.2.3 Classification of Major Polyphenol Classes*

The extensive structural heterogeneity of polyphenolic compounds necessitates the development of systematic classification schemes. The most widely utilized classification approach categorizes polyphenols according to the fundamental carbon skeleton and the number of aromatic rings. Principal categories include flavonoids (C6-C3-C6 skeleton with three aromatic rings), phenolic acids (simple C6-C1 and C6-C3 compounds with single rings), stilbenes (C6-C2-C6 compounds), lignans (C6-C3-C3-C6 diphenolic structures), and various polymeric and oxidized forms (Dixon and Paiva 1995; Erlund 2004). Figure 1.1 illustrates the hierarchical classification of major polyphenol classes, including flavonoids (with subclasses flavonols, flavones, isoflavones, flavanones, flavanols, anthocyanins, and proanthocyanidins), phenolic acids (hydroxybenzoic and hydroxycinnamic acids), stilbenes (e.g., resveratrol), and lignans (e.g., secoisolariciresinol), accompanied by representative chemical structures.

#### 1.2.3.1 Flavonoids

Flavonoids constitute the largest and most extensively studied polyphenol class, encompassing more than 6,000 documented structures. The characteristic three-carbon heterocyclic central ring distinguishes flavonoids from other polyphenol categories (Manach et al. 2004; Rudrapal et al. 2022). Subdivisions within the flavonoid family include:

(a) Flavonols (such as quercetin, kaempferol, and isorhamnetin), widely distributed across onions, leafy greens, apples, berries, and tea preparations (Manach et al. 2004; Rudrapal et al. 2022)
(b) Flavones (including apigenin and luteolin), predominantly found in parsley, celery, and aromatic herbs (Cassidy et al. 2016; Li-Weber 2009)
(c) Flavanones (notably hesperidin and naringin), characteristic of citrus fruit varieties and their expressed juices (Li-Weber 2009; Messina 2014)
(d) Flavan-3-ols (catechins and their polymeric forms, proanthocyanidins), abundantly present in tea, cocoa, and grape tissues (Ludwig et al. 2014)
(e) Anthocyanidins and their glycosides (anthocyanins), conferring red, blue, and purple pigmentation in berry fruits, red grapes, and red vegetables (Baur and Sinclair 2006)
(f) And isoflavones (genistein, daidzein), essentially restricted to legumes, particularly soybeans (Adlercreutz 2007).

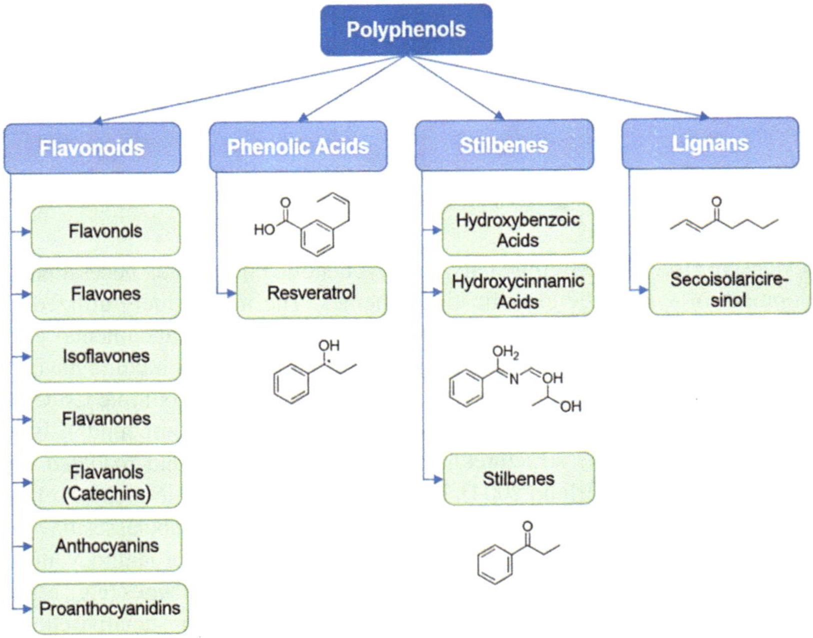


**Fig. 1.1** Hierarchical classification of major polyphenol classes, including flavonoids, phenolic acids, stilbenes, and lignans. This picture has been created by the authors

#### 1.2.3.2 Phenolic Acids

Phenolic acids constitute a quantitatively important secondary category, subdivided into hydroxybenzoic acids (C6-C1 structure) and hydroxycinnamic acids (C6-C3 structure) (Machado et al. 2023). Hydroxybenzoic acid derivatives, including gallic acid, protocatechuic acid, and vanillic acid, occur in tea, berries, and tree nuts, and additionally arise through microbial degradation of more complex polyphenols (Machado et al. 2023; Rudrapal et al. 2022). Hydroxycinnamic acid derivatives, encompassing caffeic acid, ferulic acid, *p*-coumaric acid, and sinapic acid, occur abundantly in coffee beans, cereal grains, fruits, and vegetables. Chlorogenic acids, which are esterified combinations of caffeic acid and quinic acid, predominate in coffee beverages and have received particular attention for their effects on glucose metabolism and vascular function (Ludwig et al. 2014; Tajik et al. 2017).

#### 1.2.3.3 Stilbenes

Stilbenes represent a smaller polyphenol class distinguished by a C6-C2-C6 backbone. Resveratrol, the most investigated stilbene compound, occurs in grape tissues, red wine, peanut legumes, and certain berries (Li-Weber 2009; Renaud and de Lorgeril 1992). The extensive investigation of resveratrol reflects evidence for cardioprotective, anticancer, and longevity-extending properties in experimental systems, mechanisms involving sirtuins, AMP-activated protein kinase signaling, and inflammatory pathway modulation (Sinha et al. 2002). However, the poor bioavailability of resveratrol following oral ingestion and the difficulty in translating promising in vitro findings to demonstrable clinical benefits in human populations remain a significant research challenge (Sinha et al. 2002; Walle 2011).

#### 1.2.3.4 Lignans

Lignans are diphenolic compounds formed through dimerization of phenylpropanoid metabolites, occurring principally in seeds (flaxseed, sesame), whole grains, legumes, and various fruits and vegetables (Adlercreutz 2007; Bentley et al. 2012). Plant-derived lignans, specifically secoisolariciresinol and matairesinol, undergo metabolism by intestinal symbiotic bacteria to generate enterolignans (enterodiol and enterolactone), which demonstrate weak estrogenic and anti-estrogenic activities (Smeds et al. 2007). Observational epidemiological studies indicate associations between elevated circulating enterolactone concentrations and reduced cardiovascular risk and certain cancer incidences, although evidence remains inconsistent across studies (Bentley et al. 2012; Smeds et al. 2007).

#### 1.2.3.5 Other Subclasses

Tannins constitute high-MW polymeric compounds subdivided into hydrolysable tannins (esters of gallic acid) and condensed tannins (polymeric flavan-3-ols) (Vaya and Mahmood 2006). Ellagitannins and related hydrolysable tannin compounds predominate in pomegranates, berries, and tree nuts (Cardona et al. 2013; Vaya and Mahmood 2006). Proanthocyanidins (condensed tannins) occur abundantly in grapes, cocoa preparations, and numerous berries. The high-MW and substantial protein-precipitating activity of tannins produces the characteristic astringent mouthfeel sensation, while their antimicrobial and antioxidative properties have been documented in multiple experimental systems (Cardona et al. 2013; Rudrapal et al. 2022).

## 1.3 Distinguishing Polyphenols from Other Phytochemical Classes

The broader category of plant-derived bioactive compounds (phytochemicals) encompasses multiple distinct classes beyond polyphenols, including carotenoids (lipophilic isoprenoid pigments), glucosinolates (sulfur-containing compounds prominent in cruciferous vegetables), phytosterols (plant-derived sterol lipids), and organosulfur compounds (derived from alliums) (Cory et al. 2018; Dixon and Paiva 1995). Despite some overlap in dietary sources and potential health effects, such as overlapping antioxidative capacity, polyphenols possess distinctive chemical structural characteristics, biosynthetic origins within phenylpropanoid and related metabolic pathways, and unique metabolic fates (Dixon and Paiva 1995; King and Young 1999). The polarity of most polyphenols contrasts with the lipophilicity of carotenoids; the sulfur-containing nature of glucosinolates differs from polyphenol chemistry; and the metabolism and mechanism of action of polyphenols exhibit distinct patterns if compared with other phytochemical classes (Dixon and Paiva 1995; King and Young 1999). Accurate classification is critical for interpreting population dietary studies, designing functional foods with specific bioactivity targets, and understanding potential synergistic or antagonistic interactions among phytochemical constituents within plant-based whole foods (Dixon and Paiva 1995; King and Young 1999).

## 1.4 Dietary and Non-dietary Sources of Polyphenols

### *1.4.1 Plant-Based Sources and Quantitative Content*

Plant-derived foods constitute the primary source of polyphenolic compounds for humans consuming conventional omnivorous diets (Perez-Jimenez et al. 2010; Tresserra-Rimbau et al. 2013). The specific polyphenol content, composition of constituent classes, and bioavailability profiles demonstrate substantial heterogeneity based on plant species and cultivar, tissue-specific localization, developmental ripeness stage, agronomic practices, environmental growth conditions, post-harvest handling, and food preparation techniques (Jurgonski et al. 2014; Tresserra-Rimbau et al. 2013). Broadly speaking, the major contributors to dietary polyphenol exposure encompass fruits, vegetables, whole grain cereals, legumes, tree nuts, seeds, medicinal and culinary herbs, spices, and beverages including tea, coffee, cocoa, and wine (Cory et al. 2018; Grosso et al. 2017).

#### 1.4.1.1 Fruits

Fruit sources provide the richest and most concentrated polyphenolic content of commonly consumed plant foods (Bhagwat et al. 2014; Kalt et al. 2020). Berries, including blueberries, blackberries, raspberries, and strawberries, concentrate particularly high levels of anthocyanin pigments (responsible for red, blue, and purple coloration), flavonol glycosides, and ellagitannin compounds (Khoo et al. 2017). Clinical and observational studies document associations between berry consumption and improvements in endothelial function, arterial pressure reduction, systemic lipid profile improvements, and enhanced cognitive performance (Wang et al. 2014). Grape fruits and grape-derived products, particularly red wine, provide flavan-3-ol catechins, anthocyanin pigments, and the stilbene resveratrol (Bain et al. 1998; Renaud and de Lorgeril 1992). Apple fruits contain quercetin glycosides (flavonols), phlorizin (a dihydrochalcone), and various phenolic acid derivatives (Ioannou et al. 2023). Stone fruits, including cherries, plums, and apricots, along with tropical fruits such as mangoes and pomegranates, contribute diverse flavonoid, tannin, and phenolic acid compounds (Ioannou et al. 2023). Importantly, polyphenol concentrations frequently demonstrate the highest levels in the fruit skin and peel tissues rather than the internal pulp, with significant implications for food processing practices and consumption patterns (Bhagwat et al. 2014; Jurgonski et al. 2014).

#### 1.4.1.2 Vegetables

Vegetable sources contribute more modest polyphenol concentrations relative to fruits and beverages, but their regular consumption contributes substantially to aggregate polyphenol exposure due to the substantial quantities typically consumed. Onion family vegetables (onions, leeks, and shallots) rank among the richest vegetable sources of flavonol compounds, particularly quercetin and kaempferol glycosides (Brown 2005). Brassica family vegetables, including kale, broccoli, and cabbage, contain flavonol glycosides, hydroxycinnamic acids, and additional phenolic compounds (Brown 2005; Esquivel-Ferrino et al. 2012). Leafy greens, including spinach, lettuce, and endive, contribute flavonoid and phenolic acid compounds (Perez-Jimenez et al. 2010). Root vegetables, including carrots, beets, and potatoes, particularly in skin tissues, contain chlorogenic acids and related caffeoylquinic acid derivatives. Fruits botanically classified as vegetables (tomatoes, peppers, and eggplants) provide flavonoid and phenolic acid compounds alongside carotenoid pigments (Brown 2005; Ludwig et al. 2014).

#### 1.4.1.3 Grains and Legumes

Cereal grains and legume crops represent quantitatively significant but often underappreciated sources of dietary polyphenolic compounds (Adom and Liu 2002; Chandrasekara and Shahidi 2011). The predominant polyphenol class in cereals comprises phenolic acids, notably ferulic acid, *p*-coumaric acid, and related hydroxycinnamic acid derivatives (Adom and Liu 2002). In cereal species including wheat, barley, oats, rye, and maize, these phenolic acids frequently exist as bound residues attached to cell wall polysaccharides (arabinoxylans and related hemicelluloses), affecting their chemical extractability and bioavailability following consumption (Adom and Liu 2002). The bran fractions removed during grain refining contain substantially elevated polyphenol concentrations if compared with the starchy endosperm tissue; consequently, whole grain products provide substantially more polyphenolic compounds than refined grain products (Adom and Liu 2002; Chandrasekara and Shahidi 2011). Legume crops, including soybeans, lentils, chickpeas, and various beans, contain isoflavone compounds (predominantly in soybeans), flavonol glycosides, and phenolic acids (Chandrasekara and Shahidi 2011; King and Bursill 1998). Soy isoflavones, particularly genistein and daidzein, have undergone extensive investigation for their phytoestrogenic properties and potential utility in hormonal regulation, cardiovascular protection, bone metabolism, and relief of menopausal vasomotor symptoms (King and Bursill 1998; Messina 2016). The regular consumption of whole grain cereals and legume crops not only increases polyphenol exposure but additionally provides dietary fiber, plant protein, and micronutrient density, reinforcing their inclusion in evidence-based healthy dietary patterns (Adom and Liu 2002; Chandrasekara and Shahidi 2011).

#### 1.4.1.4 Nuts and Seeds

Tree nuts and seed sources contain substantial polyphenol concentrations, often concentrated within the seed coat or outer skin tissues. Walnuts, hazelnuts, and almonds contain flavan-3-ol compounds, proanthocyanidin polymers, and phenolic acids (Sheng et al. 2021; Song et al. 2023). Pistachios and pecans provide substantial phenolic content. Flaxseeds and sesame seeds represent particularly rich sources of lignans, with documented associations between lignan consumption and potential health benefits, including reduced cardiovascular risk and hormonal modulation (Bentley et al. 2012; Smeds et al. 2007; Song et al. 2023). Regular consumption of tree nuts has been associated with reduced cardiovascular disease risk, all-cause mortality reduction, and improved cardiometabolic risk profiles, with polyphenols contributing to these benefits alongside unsaturated fatty acids, dietary fiber, and additional bioactive compounds (Hudthagosol et al. 2011).

#### 1.4.1.5 Herbs and Spices

Herbs, spices, and beverage sources contribute disproportionately to polyphenolic intake on a per-unit-weight basis, though absolute contribution to total intake depends on consumption quantity (Kadam and Lele 2017; Shelar et al. 2021). Culinary herbs, including rosemary, thyme, oregano, and sage, contain high polyphenol concentrations comprising flavonoids, phenolic acids, and specialized polyphenols such as rosmarinic and carnosic acids (Shelar et al. 2021). Spice compounds, including cloves, cinnamon, turmeric (curcuminoids), and black pepper, contain substantial phenolic content with potent antioxidative and antimicrobial activities. Even at the modest quantities typically added to meals, these ingredients substantially enhance the polyphenol density and antioxidative capacity of prepared foods (Kadam and Lele 2017; Shelar et al. 2021).

#### 1.4.1.6 Beverages

Beverage sources constitute major contributors to polyphenolic intake across many populations (Kadam and Lele 2017; Shelar et al. 2021; Westhuyzen et al. 2005). Tea preparations (both green and black varieties) provide abundant flavan-3-ol catechins and their oxidative polymerization products (theaflavins and thearubigins in black tea), representing major contributors to total polyphenol intake in tea-consuming populations (Sun et al. 2022). Coffee beverages provide chlorogenic acid, quinines, and related phenolic acids, frequently constituting a substantial proportion of total phenolic intake in Western dietary patterns (Belay et al. 2016; Sun et al. 2022). Cocoa and dark chocolate preparations provide flavan-3-ol compounds and proanthocyanidin polymers, with dark chocolate containing substantially higher polyphenol concentrations than milk chocolate preparations (Schramm et al. 2003). Wine, particularly red wine varieties, contains a complex mixture of anthocyanin pigments, flavan-3-ol compounds, and stilbene compounds, including resveratrol (German and Walzem 2000; Siemann and Creasy 1992). Tables 1.1, 1.2, 1.3, 1.4, 1.5, 1.6 and 1.7 present the representative polyphenol data for frequently consumed plant-based foods. These values reflect the heterogeneity across food categories and can guide dietary pattern optimization for enhanced polyphenolic exposure.

## 1.5 Effects of Processing and Culinary Preparation on Polyphenol Retention

Food processing techniques and culinary preparation methods substantially alter the concentration, composition, and bioavailability profile of dietary polyphenolic compounds. The mechanisms through which processing influences polyphenol content are multifactorial and include direct thermal degradation of heat-labile compounds, enzymatic modification including oxidation and polymerization, leaching into cooking media (particularly for water-soluble phenolic compounds), and alterations in food matrix structure affecting extractability (Choi et al. 2018; Wang et al. 2024; Zhu et al. 2013). Mechanical processing operations, including peeling, chopping, and milling, remove or destroy polyphenol-rich tissues. Peeling of fruits and vegetables eliminates the concentrated polyphenolic compounds localized in the outer peel and skin tissues, reducing substantially the total recoverable

**Table 1.1** Representative polyphenol content in commonly consumed plant-based foods (beverages) with retention optimization strategies

| Food category | Item | Major polyphenol classes | Optimal storage/Preparation | References |
|---|---|---|---|---|
| Beverage group | Green tea (freshly brewed) | Catechins (e.g., epigallocatechin gallate, EGCG), flavonol glycosides (e.g., quercetin, kaempferol); research highlights catechins as primary antioxidants with stability during short brewing | Steep 3–5 min at 80 °C to maximize catechin extraction without degradation; consume fresh to limit oxidation | Castillo-Munoz et al. (2007), Machado et al. (2023), Tresserra-Rimbau et al. (2013) |
| | Black tea (brewed) | Theaflavins (e.g., theaflavin-3-gallate), thearubigins (polymeric polyphenols formed via oxidation); studies emphasize oxidation products' role in bioactivity | Steep 3–5 min at 95–100 °C; moderate steeping preserves theaflavins while avoiding excess bitterness and polyphenol loss | Tresserra-Rimbau et al. (2013) |
| | Red wine | Anthocyanins (e.g., malvidin-3-glucoside), flavan-3-ols (e.g., catechin, epicatechin), resveratrol (stilbene); profiles vary by cultivar, with flavonols confirmed in grape sources | Store at 12–16 °C away from light to prevent anthocyanin degradation; decant briefly for moderate air exposure, enhancing bioavailability | Castillo-Munoz et al. (2007), Tresserra-Rimbau et al. (2013) |
| | Coffee (brewed, standard) | Chlorogenic acids (e.g., 5-caffeoylquinic acid), phenolic acids (e.g., caffeic, ferulic); by-product studies link these to high antioxidant capacity | Brew fresh with a medium grind at 90–96 °C, minimize air exposure post-brew to retain chlorogenic acids against oxidation | Machado et al. (2023), Tresserra-Rimbau et al. (2013) |

**Table 1.2** Representative polyphenol content in commonly consumed plant-based foods (berry fruits) with retention optimization strategies

| Food category | Item | Major polyphenol classes | Optimal storage/ Preparation | References |
|---|---|---|---|---|
| Berry fruits | Blueberries | Anthocyanins (e.g., delphinidin, malvidin glycosides), flavonols (e.g., quercetin); compositional data show high stability in frozen forms | Freeze immediately post-harvest or consume fresh; it avoids anthocyanin loss from enzymatic browning | Castillo-Munoz et al. (2007), Kalt et al. (2020), Jurgonski et al. (2014), Mihaylova et al. (2025), Skrovankova et al. (2015) |
| | Blackberries | Anthocyanins (cyanidin-based), ellagitannins (e.g., sanguiin H-6), and pomace extracts retain high levels | Fresh or freeze-dried to preserve ellagitannins; drying minimizes hydrolysis | Jurgonski et al. (2014), Mihaylova et al. (2025), Skrovankova et al. (2015) |
| | Raspberries | Anthocyanins (cyanidin-3-sophoroside), ellagitannins (e.g., lambertianin C); research supports ellagitannin stability in frozen storage | Fresh or frozen to retain intact ellagitannins for gut microbiota conversion to urolithins | Mihaylova et al. (2025), Skrovankova et al. (2015) |
| | Strawberries | Anthocyanins (pelargonidin-3-glucoside), ellagic acid (hydrolyzed from ellagitannins), are linked to high GAE in soft fruits | Consume fresh within 2 days; short storage limits pelargonidin degradation | Mihaylova et al. (2025), Skrovankova et al. (2015) |
| | Red grapes | Resveratrol (trans-resveratrol), anthocyanins (e.g., peonidin-3-glucoside); single-cultivar profiles detail flavonol co-occurrence | Consume with skin intact; cool, dark storage preserves resveratrol | Castillo-Munoz et al. (2007) |

polyphenol content. The industrial refining of cereal grains to produce white flour removes the fiber-rich and polyphenol-rich bran and germ fractions, resulting in dramatic reductions of phenolic acid content (Adom and Liu 2002; Wang et al. 2024).

Thermal processing effects on polyphenol content are complex and substance-specific. Thermal treatments can induce direct thermal degradation of certain heat-labile polyphenols (anthocyanins), demonstrating particular thermal sensitivity, with documented losses during prolonged boiling or baking. Conversely, moderate

**Table 1.3** Representative polyphenol content in commonly consumed plant-based foods (citrus and stone fruits) with retention optimization strategies

| Food category | Item | Major polyphenol classes | Optimal storage/ Preparation | References |
|---|---|---|---|---|
| Citrus and stone fruits | Oranges | Flavanones (e.g., hesperidin, naringin); population studies confirm citrus as the top dietary source | Fresh or minimal processing (e.g., juicing) to retain flavanone glycosides | Tresserra-Rimbau et al. (2013) |
| | Lemons/ Limes | Flavanones (e.g., eriocitrin), phenolic acids (e.g., ferulic); general fruit matrices support classification | Fresh juice preparation maximizes eriocitrin extraction without heat-induced loss | Jurgonski et al. (2014), Tresserra-Rimbau et al. (2013) |
| | Cherries | Anthocyanins (e.g., keracyanin), flavonols; soft fruit data underpin contents | Fresh or tart varieties; quick freezing preserves anthocyanins | Skrovankova et al. (2015) |
| | Pomegranate | Ellagitannins (e.g., punicalagins), punicalagins (unique hydrolysable tannins); high phenolic yields | Fresh fruit, seeds, or juice; avoid prolonged storage to prevent punicalagin breakdown | Mihaylova et al. (2025) |

thermal processing can disrupt cell wall structures and release phenolic compounds sequestered within cellular compartments, potentially increasing extractability and bioavailability (Tsao 2010; Wang et al. 2024; Zhu et al. 2013). Prolonged boiling of vegetables frequently results in substantial losses of water-soluble phenolic acids through leaching into the cooking water medium (Tsao 2010; Wang et al. 2024). Fermentation processes can substantially modify polyphenol profiles. During fermentation of tea (oxidation to produce black and oolong varieties), theaflavins and thearubigins form through enzymatic condensation of catechin monomers, producing orange-red colored compounds with distinct sensory and potentially bioactive properties distinct from green tea (Tsao 2010; Wang et al. 2024; Zhu et al. 2013). Fermentation in cocoa processing similarly transforms catechin composition. Coffee roasting produces novel phenolic compounds through Maillard reactions and non-enzymatic browning reactions, potentially creating compounds with distinct biological activities (Choi et al. 2018; Tsao 2010; Wang et al. 2024). Optimization of processing conditions to maximize retention of polyphenolic compounds while maintaining food safety, sensory appeal, and technical functionality represents an active and important area of food science and technology research (Tsao 2010; Wang et al. 2024; Zhu et al. 2013).

**Table 1.4** Representative polyphenol content in commonly consumed plant-based foods (vegetables) with retention optimization strategies

| Food category | Item | Major polyphenol classes | Optimal storage/ Preparation | References |
|---|---|---|---|---|
| Vegetable sources | Kale | Flavonols (e.g., kaempferol glycosides), phenolic acids (e.g., ferulic); leafy greens contribute significantly to intake | Raw or brief steaming (<5 min) to retain heat-sensitive flavonols | Bento-Silva et al. (2021), Robbins (2003), Tresserra-Rimbau et al. (2013) |
| | Spinach | Flavonols (quercetin derivatives), caffeic acid; analytical methods confirm ranges | Fresh or lightly steamed to minimize caffeic acid oxidation | Robbins (2003) Tresserra-Rimbau et al. (2013) |
| | Red onions | Quercetin glycosides (e.g., quercetin-4'-glucoside), anthocyanins; preferred raw for stability | Raw preparation to preserve quercetin bioavailability | Robbins (2003), Tresserra-Rimbau et al. (2013) |
| | Broccoli | Flavonols (kaempferol), phenolic acids (sinapic); GAE data from vegetables | Steamed or raw; steaming enhances extractability without major loss | Robbins (2003) |
| | Red peppers | Flavonols (quercetin), phenolic acids; stable across methods | Any preparation; minimal impact on flavonols | Robbins (2003), Tresserra-Rimbau et al. (2013) |
| | Artichokes | Chlorogenic acid, flavonols (luteolin); high in structured tissues | Steamed or roasted to optimize chlorogenic acid release | Bento-Silva et al. (2021), Robbins (2003) |

## 1.6 Geographic, Agronomic, and Environmental Variability

The polyphenol content within plant tissues exhibits substantial heterogeneity across geographic locations, growing conditions, and agronomic practices (Dixon and Paiva 1995; Jurgonski et al. 2014; Stalikas 2007). Cultivar selection dramatically influences polyphenol accumulation: certain apple cultivars, for example, contain substantially higher quercetin concentrations than others (Jurgonski et al. 2014; Stalikas 2007). Soil mineral composition, water availability, temperature patterns, solar radiation intensity, altitude, and exposure to biotic stress (pathogen pressure and herbivory) all exert significant influences on phenolic compound accumulation (Dixon and Paiva 1995; Jurgonski et al. 2014; Stalikas 2007). Environmental stressors, including ultraviolet radiation, pathogenic attack, and herbivorous insect feeding, often stimulate the synthesis of phenolic compounds as part of the plant's innate defense mechanism,

**Table 1.5** Representative polyphenol content in commonly consumed plant-based foods (nuts and seeds) with retention optimization strategies

| Food category | Item | Major polyphenol classes | Optimal storage/ Preparation | References |
|---|---|---|---|---|
| Nuts and seeds | Artichokes | Chlorogenic acid, flavonols (luteolin); high in structured tissues | Steamed or roasted to optimize chlorogenic acid release | Bento-Silva et al. (2021), Robbins (2003) |
| | Almonds | Flavan-3-ols (catechins), phenolic acids; nut matrices yield high extracts | Raw or light roasting (<140 °C) to retain flavan-3-ols | Jurgonski et al. (2014), Robbins (2003), Tresserra-Rimbau et al. (2013) |
| | Hazelnuts | Flavan-3-ols, proanthocyanidins (condensed tannins); dietary sources confirmed | Natural form, vacuum-sealed storage prevents oxidation | Tresserra-Rimbau et al. (2013) |
| | Walnuts | Phenolic acids (ellagic precursors), ellagic acid; applicable to seeds | Raw, cool storage (<4 °C) to stabilize ellagic acid | Robbins (2003) |
| | Flaxseeds | Lignans (secoisolariciresinol diglucoside), phenolic acids; grinding aids bioavailability | Grind fresh immediately before use for lignan absorption | Jurgonski et al. (2014), Robbins (2003) |

potentially increasing concentrations of protective flavonoids and other phenolics (Dixon and Paiva 1995; Stalikas 2007; Winkel-Shirley 2002).

Geographic differences in dietary polyphenol exposure patterns across populations reflect differential consumption of major polyphenol-contributing food categories. Populations in Asia often derive substantial polyphenol exposure from green tea, soy products, and vegetables, producing distinct polyphenol intake profiles if compared with European and American populations deriving major polyphenol contributions from coffee, wine, chocolate, and assorted fruits (Huang et al. 2020; Grosso et al. 2018; Martinez-Castaneiras et al. 2025). Within Countries, polyphenol intake varies by demographic characteristics, including age, biological sex, socioeconomic status, and lifestyle behavior, including smoking, physical activity levels, and alcohol consumption (Huang et al. 2020; Martinez-Castaneiras et al. 2025; Zamora-Ros et al. 2016).

**Table 1.6** Representative polyphenol content in commonly consumed plant-based foods (cereals and legumes) with retention optimization strategies

| Food category | Item | Major polyphenol classes | Optimal storage/ Preparation | References |
|---|---|---|---|---|
| Cereals and legumes | Whole wheat grain | Phenolic acids (ferulic acid-bound forms); cereal grains as key sources | Whole grain milling/ consumption releases bound ferulic acid | Bento-Silva et al. (2021), Robbins (2003), Tresserra-Rimbau et al. (2013) |
| | Oats (steel-cut) | Phenolic acids, avenanthramides (unique to oats); minimal processing retains them | Minimal processing, store dry to avoid avenanthramide loss | Bento-Silva et al. (2021) |
| | Brown rice | Phenolic acids (ferulic, tricin); less refinement preserves bounds | Minimal refinement, cook with bran intact | Robbins (2003) |
| | Lentils (cooked) | Flavonols (kaempferol), phenolic acids; pulses contribute to intake | Soak and cook to enhance flavonol digestibility | Tresserra-Rimbau et al. (2013) |
| | Chickpeas (cooked) | Flavonols, phenolic acids, similar to those in lentils | Soak overnight, cook thoroughly for polyphenol release | Tresserra-Rimbau et al. (2013) |
| | Oregano (dried) | Phenolic acids (e.g., rosmarinic acid), flavonoids (e.g., quercetin, apigenin); exceptional antioxidant rankings due to high phenolic density | Fresh or dried in shaded, airtight storage; minimal cooking to retain volatile flavonoids | Floegel et al. (2011), Robbins (2003) |
| | Ginger (fresh) | Gingerols, phenolic acids, flavonoids | Fresh grating, minimal cooking | Ghasemzadeh et al. (2010) |

## 1.7 Bioavailability from Diverse Dietary Sources

The term "bioavailability" encompasses the proportion of an orally ingested compound that is absorbed through the gastrointestinal epithelium, enters systemic circulation, and reaches target tissues in active chemical forms capable of producing biological effects (Cardona et al. 2013; Scalbert et al. 2005b). Polyphenol bioavailability demonstrates remarkable heterogeneity across different chemical structures, food matrix compositions, individual digestive and metabolic capacities, and microbiota composition (Le et al. 2023; Scalbert et al. 2005b).

The chemical structure of polyphenolic compounds fundamentally determines bioavailability patterns. Small, hydrophilic compounds with relatively simple structures, including certain phenolic acids such as caffeic acid, exhibit relatively high absorption efficiency within the small intestine, with bioavailability percentages approaching 30–50% in some studies (Saad et al. 2025; Scalbert et al. 2005b).

**Table 1.7** Representative polyphenol content in commonly consumed plant-based foods (herbs and spices) with retention optimization strategies

| Food category | Item | Major polyphenol classes | Optimal storage/ Preparation | References |
|---|---|---|---|---|
| Herbs and spices | Cloves | Phenolic acids (gallic), flavonoids (eugenol derivatives); exceptionally high GAE | Ground fresh, use sparingly; dry storage maintains potency | Floegel et al. (2011), Robbins (2003) |
| | Rosemary (dried) | Phenolic acids (e.g., rosmarinic acid), flavonoids (e.g., luteolin, apigenin), and high antioxidant capacity linked to these compounds in spice analyses | Dried leaves in a cool, dark place; use fresh sprigs for maximal rosmarinic acid retention, avoiding prolonged heat exposure | Floegel et al. (2011), Robbins (2003) |
| | Turmeric | Curcuminoids (e.g., curcumin), phenolic acids, and essential oil studies confirm that associated phenolics are bioavailability-enhanced by piperine | Store whole rhizomes in cool, dry conditions; grind fresh and pair with black pepper for improved curcumin absorption | Dosoky and Setzer (2018) |
| | Cinnamon (ground) | Phenolic acids (e.g., protocatechuic acid), proanthocyanidins; high GAE values from bark polyphenols, with the Ceylon variety preferred for lower coumarin | Store ground or sticks in airtight containers; use in moderate heat applications to preserve proanthocyanidins | Floegel et al. (2011), Robbins (2003) |
| | Oregano (dried) | Phenolic acids (e.g., rosmarinic acid), flavonoids (e.g., quercetin, apigenin); exceptional antioxidant rankings due to high phenolic density | Fresh or dried in shaded, airtight storage; minimal cooking to retain volatile flavonoids | Floegel et al. (2011), Robbins (2003) |
| | Ginger (fresh) | Gingerols, phenolic acids, flavonoids | Fresh grating, minimal cooking | Ghasemzadeh et al. (2010) |

Conversely, large, complex, poorly water-soluble compounds, including proanthocyanidin polymers, ellagitannins, and other high-MW polyphenols, exhibit minimal absorption within the small intestine and reach the colon largely intact, where they undergo extensive microbial enzymatic transformation (Ozdal et al. 2016; Saad et al. 2025; Scalbert et al. 2005b).

Glycosylation status substantially influences polyphenol bioavailability. Dietary polyphenols predominantly occur as glycosides (sugars covalently attached to the aglycone), requiring enzymatic hydrolysis to liberate the free aglycone form. Intestinal glycosidases (particularly β-glucosidase, lactase-phlorizin hydrolase) and symbiotic bacterial glycosidases catalyze this liberation process. Variable expression of these enzymes across individuals and within microbial communities contributes to inter-individual differences in bioavailability (Le et al. 2023; Saad et al. 2025; Scalbert et al. 2005b).

The food matrix, encompassing co-ingested macronutrients, fiber, additional phytochemicals, and food structure, substantially modulates polyphenol bioavailability. Fat intake can enhance the absorption of lipophilic polyphenols through enhanced micellization. High dietary fiber can both enhance bioavailability of certain polyphenols through increased gastrointestinal transit time and residence time in absorptive regions, and potentially reduce absorption through competitive interactions (Le et al. 2023; Saad et al. 2025). Protein and polysaccharide interactions may enhance or diminish bioavailability depending on specific compounds and food context (Le et al. 2023; Saad et al. 2025).

Individual host factors, including variations in metabolic enzyme expression—particularly uridine 5'-diphospho (UDP)–glucuronosyltransferases, sulfotransferases, and catechol-O-methyltransferase—intestinal permeability, hepatic function, and renal excretion capacity, contribute substantially to inter-individual variability in polyphenol bioavailability and metabolic fate (Le et al. 2023; Cassidy et al. 2011; Scalbert et al. 2005b). Genetic polymorphisms in genes encoding metabolic enzymes create population-level heterogeneity in bioavailability and biological responsiveness (Cassidy et al. 2011; Le et al. 2023).

The composition and functional capacity of an individual's resident gut microbiota represent a critical determinant of polyphenol bioavailability and metabolism, particularly for polyphenols reaching the colon intact (Le et al. 2023). Microbial enzymes can cleave glycosidic bonds, open aromatic rings, and reduce double bonds, producing bioavailable metabolites such as urolithins from ellagitannins and enterolignans from plant lignans (Le et al. 2023). Microbial composition differs substantially across individuals based on dietary history, antibiotic exposure, disease states, and genetic factors, generating distinct "metabotypes" with divergent polyphenol metabolic capacity (Le et al. 2023).

## 1.8 Dietary Intake Patterns and Assessment

### *1.8.1 Global Variations in Polyphenol Consumption*

Estimated polyphenol intake demonstrates remarkable geographic heterogeneity reflecting regional dietary traditions, agricultural availability, cultural preferences, and socioeconomic factors (Martinez-Castaneiras et al. 2025; Rothwell et al. 2013). Comprehensive dietary studies conducted within the European Prospective Investigation into Cancer and Nutrition (EPIC) cohort indicated median total polyphenol intakes within ranges of 600–1,600 mg/day across participating European populations (Grosso et al. 2018; Rothwell et al. 2013). Phenolic acids, predominantly derived from coffee and whole grain sources, frequently constitute approximately

50% of total polyphenol intake in Northern European populations. Flavonoid compounds from fruits, vegetables, tea, and wine constitute the major remaining fraction of total intake (Grosso et al. 2018; Rothwell et al. 2013).

In Northern European regions, coffee emerges as the predominant polyphenol contributor, frequently providing 35–40% of total intake due to widespread coffee beverage consumption. Tea, fruits, and wine contribute substantially to remaining intake (Rothwell et al. 2013). Southern European populations demonstrate more balanced contributions across coffee, wine, fruits, and vegetables, with greater wine contribution reflecting traditional Mediterranean dietary patterns (He et al. 2013; Rothwell et al. 2013). East Asian populations derive major polyphenol contributions from green tea, soy and legume products, and various vegetables, with tea frequently providing 30–45% of total intake (Martinez-Castaneiras et al. 2025; Yates et al. 1998). North American populations demonstrate contribution patterns intermediate between Northern European and Southern European patterns, with coffee, fruits, tea, and chocolate providing substantial proportions (Martinez-Castaneiras et al. 2025; Rothwell et al. 2013). Within-Country variation in polyphenol intake according to demographic characteristics, including age, biological sex, socioeconomic status, and lifestyle factors, remains substantial (Martinez-Castaneiras et al. 2025).

### *1.8.2 Methodological Approaches to Assessing Polyphenol Intake*

Quantification of polyphenol intake in population-level epidemiological investigations and clinical studies confronts substantial methodological challenges rooted in the complexity of polyphenol diversity, limitations inherent to food composition databases, and inaccuracies in dietary assessment instruments (Baldrick et al. 2011; Pedisic and Bauman 2015; Rothwell et al. 2013). The most commonly utilized approaches encompass food frequency questionnaires (FFQ), 24 h dietary recalls, and prospective dietary records, each combined with polyphenol composition data from specialized databases such as Phenol-Explorer and United States Department of Agriculture (USDA) databases (Baldrick et al. 2011; Pedisic and Bauman 2015; Rothwell et al. 2013).

FFQ asks participants to estimate their consumption frequency of specific foods over a defined time period (typically the past year), combined with portion size estimates (Baldrick et al. 2011; Rothwell et al. 2013). FFQ approaches offer practical advantages for large epidemiological studies through relatively low participant burden and feasibility for studying long-term dietary patterns, but confront limitations including imprecise quantification of portion sizes, memory-based errors, particularly regarding lower consumption foods, and systematic tendency toward underestimation of intake magnitude (Baldrick et al. 2011; Rothwell et al. 2013).

24 h dietary recalls involve detailed questioning of a trained nutritionist regarding foods and beverages consumed during the preceding 24 h period, with subsequent estimation of nutrient and phytochemical content through database linkage (Baldrick et al. 2011; Pedisic and Bauman 2015; Rothwell et al. 2013). This approach provides detailed food-level data for a specific day but requires multiple recalls (typically two to four recalls) to characterize habitual intake, as single-day information demonstrates substantial within-person variability. Prospective dietary records require participants to record all foods and beverages consumed during a specified period (typically 3–7 days), offering detailed information with reduced memory-based errors if compared with recalls, but require higher participant literacy and motivation (Baldrick et al. 2011; Pedisic and Bauman 2015; Rothwell et al. 2013).

Biomarker-based assessment approaches, utilizing plasma or urinary concentrations of specific polyphenols or their metabolites, provide objective indices of polyphenol exposure, offering an alternative to dietary questionnaire-based assessment (Baldrick et al. 2011; Cassidy et al. 2011). However, biomarker approaches confront limitations including variable polyphenol absorption and metabolism across individuals, rapid elimination of many polyphenols generating short half-lives, difficulty in linking specific biomarker concentrations to specific food sources, and limited standardization of bioanalytical methods (Baldrick et al. 2011; Cassidy et al. 2011).

The integration of improved and expanded food composition databases (particularly Phenol-Explorer), standardization of analytical methodologies for polyphenol quantification, and validation of biomarker approaches have advanced the accuracy of polyphenol intake assessment, though substantial methodological challenges remain (Baldrick et al. 2011; Pedisic and Bauman 2015; Rothwell et al. 2013). Future advances in non-targeted metabolomics technologies may identify novel biomarkers providing improved characterization of polyphenol exposure and metabolism (Pedisic and Bauman 2015; Scalbert et al. 2005b).

### *1.8.3 Absence of Formalized Dietary Reference Intakes*

In contrast to established dietary reference intakes (DRI) for essential nutrients and several bioactive compound classes, no widely accepted quantitative dietary recommendations specific to polyphenols have been formally established by major regulatory or scientific bodies (Willett et al. 2019; Yang et al. 2020). Multiple factors contribute to this absence of formalized recommendations. The extraordinary structural and functional heterogeneity of polyphenolic compounds, encompassing >8,000 known structures with divergent biological activities and bioavailability profiles, creates challenges in establishing a unified intake recommendation (Willett et al. 2019; Yang et al. 2020). Uncertainties regarding dose–response relationships for specific polyphenol classes and individual compounds, variable inter-individual responsiveness to polyphenol consumption, and incomplete understanding of optimal intake levels remain substantial (Hooper et al. 2012; Willett et al. 2019; Yang et al. 2020).

Consequently, contemporary public health dietary recommendations emphasize consumption of diverse plant-based foods, including fruits, vegetables, whole grains, legumes, nuts, seeds, and selected beverages, recommendations that implicitly increase polyphenol exposure while accommodating broader nutritional objectives (Hooper et al. 2012; Willett et al. 2019; Yang et al. 2020). This food-based approach avoids the challenges of compound-specific recommendations while providing practical guidance for achieving optimal polyphenol intake within the context of comprehensive dietary quality (Hooper et al. 2012; Willett et al. 2019; Yang et al. 2020).

Some investigative teams have proposed tentative polyphenol intake targets based on epidemiological evidence and intervention trial results, generally in the range of 500–1,000 mg/day for total polyphenols, though such proposals remain provisional and have not achieved regulatory adoption (Del Bo et al. 2019; Del Rio et al. 2013; Scalbert et al. 2005a, b). Further research clarifying dose–response relationships, identifying polyphenol classes or individual compounds with the greatest health impact, and characterizing responder and non-responder populations will likely advance the development of evidence-based recommendations in future years (Del Bo et al. 2019; Del Rio et al. 2013; Scalbert et al. 2005a, b).

### *1.8.4 Individual Factors Influencing Absorption and Metabolic Utilization*

The absorption, metabolic transformation, tissue distribution, and biological activity of dietary polyphenols demonstrate substantial heterogeneity across individuals, reflecting divergent genetic backgrounds, physiological states, microbiota composition, and dietary contexts (Cassidy et al. 2011; Le et al. 2023). Recognition of these sources of inter-individual variability is critical for interpreting epidemiological associations and clinical trial findings.

The chemical structure of specific polyphenolic compounds exerts a foundational influence on bioavailability. Glycosylation patterns, conjugation with organic acids, polymerization degree, and presence of additional chemical substituents all influence water solubility, gastrointestinal epithelial permeability, and need for enzymatic hydrolysis (Le et al. 2023). More complex, polymeric compounds generally demonstrate lower small intestine bioavailability but undergo more extensive bacterial fermentation with the generation of alternative bioavailable metabolites (Scalbert et al. 2005a, b).

Host genetic variation in metabolic enzyme expression creates population-level heterogeneity in polyphenol transformation. The UDP-glucuronosyltransferase (UGT) family of phase II metabolic enzymes catalyze glucuronidation (a major polyphenol conjugation mechanism). Common genetic polymorphisms in UGT genes create variable enzyme expression and activity across individuals (Cassidy et al. 2011; Le et al. 2023). Similarly, catechol-O-methyltransferase (COMT) polymorphisms affect the methylation and metabolism of catecholic polyphenols, while sulfotransferase (SULT) variants affect sulfate conjugation (Cassidy et al. 2011; Le et al. 2023). These genetic differences contribute to "phenotypic responder" and "non-responder" classifications in the context of clinical polyphenol interventions (Cassidy et al. 2011; Le et al. 2023).

The composition and metabolic capacity of resident gut microbiota demonstrate remarkable inter-individual heterogeneity influenced by dietary patterns, antibiotic exposures, disease states, and host genetics (Le et al. 2023). Specific bacterial taxa possess enzymatic capacity to metabolize distinct polyphenol subclasses, some bacteria degrade ellagitannins to urolithins, others metabolize plant lignans to enterolignans, and still others cleave polymeric proanthocyanidins (Le et al. 2023). Substantial differences in the proportion and relative abundance of these bacterial taxa across individuals create divergent "metabotypes" with distinct polyphenol metabolic profiles and generation of bioactive metabolites (Le et al. 2023; Scalbert et al. 2005a, b).

Health status and physiological state influence polyphenol absorption. Gastrointestinal inflammatory conditions, altered intestinal epithelial permeability, hepatic dysfunction, and renal impairment can all modify polyphenol absorption and metabolism (Le et al. 2023; Scalbert et al. 2005a, b). Age-related changes in gastrointestinal function and declining microbiota diversity in aging populations may influence polyphenol metabolism in older adults (Cassidy et al. 2011; Le et al. 2023).

## 1.9 Biological Functions and Health-Related Mechanisms

### *1.9.1 Antioxidant Mechanisms and Cellular Protection*

The antioxidative properties of polyphenolic compounds have constituted the most intensively studied and best-characterized biological activity across the entire polyphenol research field (Prochazkova et al. 2011; Rice-Evans et al. 1996). Antioxidative mechanisms encompass both direct chemical reactions with reactive oxygen species (ROS) and reactive nitrogen species (RNS), and indirect activation of endogenous antioxidative defense systems (Prochazkova et al. 2011; Rice-Evans et al. 1996).

Direct antioxidative mechanisms involve the chemical reactivity of the aromatic polyphenol structure with oxygen-centered and nitrogen-centered free radicals. The hydrogen atoms attached to OH on the aromatic ring can be donated to free radicals, stabilizing and neutralizing them (Rice-Evans et al. 1996). The aromatic ring structure itself is resonance-stabilized, allowing delocalization of unpaired electron density across the ring, creating a stabilized radical intermediate of reduced reactivity compared with the original free radical species (Rice-Evans et al. 1996; Tsao 2010). The particular antioxidative reactivity of polyphenols with OH positioning (catechol and galloyl arrangements) reflects the enhanced resonance stabilization in these structural configurations (Rice-Evans et al. 1996; Rudrapal et al. 2022).

Beyond direct free radical scavenging, polyphenols chelate redox-active transition metal ions, including ferrous and cuprous forms, preventing their participation in Fenton and Haber–Weiss reactions that generate highly reactive hydroxyl radicals through hydrogen peroxide decomposition (Prochazkova et al. 2011). The catechol moiety is particularly effective in metal chelation, allowing coordination of metal-centered through the two adjacent hydroxyl oxygen atoms (Rudrapal et al. 2022; Tsao 2010).

Polyphenols prevent lipid peroxidation in biological membranes and circulating lipoproteins through multiple mechanisms. Direct scavenging of lipid peroxyl radicals and preventing the propagation phase of lipid autoxidation chains protects membrane and lipoprotein lipids from oxidative damage (Rice-Evans et al. 1996). This protection against low-density lipoprotein (LDL) oxidation is particularly relevant to atherosclerosis pathogenesis, as oxidatively modified LDL drives endothelial dysfunction and atherosclerotic lesion progression (Prochazkova et al. 2011; Rice-Evans et al. 1996).

Indirect antioxidative mechanisms involve modulation of endogenous cellular antioxidative defense enzyme systems. Many polyphenols activate the transcription factor nuclear factor erythroid 2-related factor 2 (Nrf2), a master regulator of cytoprotective gene expression. Nrf2 activation induces transcription of genes encoding phase II detoxification and antioxidative defense enzymes, including glutathione S-transferases (GST), NADPH quinone oxidoreductase (NQO1), and heme oxygenase-1 (HO-1) (Rice-Evans et al. 1996; Tsao 2010). This adaptive cellular response enhances the cell's capacity to manage oxidative stress and electrophilic challenges. Other polyphenols modulate mitochondrial ROS production or enhance expression of the superoxide dismutase (SOD), catalase, and glutathione peroxidase (GPx) antioxidative enzyme families (Rudrapal et al. 2022; Tsao 2010).

The physiological relevance of polyphenol antioxidative capacity has been somewhat debated in recent years, as the direct free radical scavenging capacity of polyphenols at circulating concentrations following dietary consumption is modest compared with endogenous antioxidative defense systems (Rudrapal et al. 2022; Saad et al. 2025). However, the indirect mechanisms of enzyme induction and signaling effects occur at physiologically relevant polyphenol concentrations, indicating that indirect cellular signaling mechanisms likely predominate over direct free radical scavenging in mediating health benefits (Pandey and Rizvi 2009; Tsao 2010).

### 1.9.2 Anti-inflammatory Effects and Immune Modulation

Chronic systemic low-grade inflammation constitutes a central pathogenic mechanism in the etiology of noncommunicable diseases, including atherosclerotic cardiovascular disease, insulin resistance, metabolic syndrome, obesity, neurodegeneration, and malignant neoplasia. Polyphenols exert anti-inflammatory effects through suppression of multiple inflammatory signaling pathways and modulation of pro- and anti-inflammatory mediator production (Gonzalez-Castejon and Rodriguez-Casado 2011; Perez-Vizcaino and Fraga 2018). The transcription factor nuclear factor kappa B (NF-κB) represents a central regulator of pro-inflammatory gene expression, controlling transcription of numerous inflammatory cytokines (tumor necrosis factor-α, interleukin-6(IL-6), and IL-1β), chemokines, adhesion molecules (intercellular adhesion molecule-1, vascular cell adhesion molecule-1), and enzymes including inducible nitric oxide synthase (iNOS) and cyclooxygenase-2 (COX-2). Many polyphenols suppress NF-κB activation through multiple mechanisms, including prevention of inhibitor κB (IκB) phosphorylation and degradation, inhibition of IκB kinase (IKK) enzyme complex, and direct inhibition of the p65 subunit deoxyribonucleic acid (DNA)-binding activity. These mechanisms prevent the nuclear translocation and transcriptional activation of NF-κB, thereby suppressing pro-inflammatory gene expression (Del Rio et al. 2013; Gonzalez-Castejon and Rodriguez-Casado 2011; Perez-Vizcaino and Fraga 2018).

The mitogen-activated protein kinase (MAPK) family of signaling enzymes, including extracellular signal-regulated kinases (ERK), c-Jun N-terminal kinases (JNK), and p38 MAPK, regulates inflammatory responses, cell proliferation, differentiation, and apoptosis. Polyphenols modulate MAPK signaling cascades, often through inhibition of upstream kinases or phosphatase activation, resulting in suppression of pro-inflammatory responses (Del Rio et al. 2013; Gonzalez-Castejon and Rodriguez-Casado 2011; Perez-Vizcaino and Fraga 2018). Polyphenolic compounds inhibit key enzymes in eicosanoid synthesis pathways. Inhibition of cyclooxygenase (COX) and 5-lipoxygenase (5-LOX) enzymes reduces the production of pro-inflammatory eicosanoids, including prostaglandins and leukotrienes, mediators critical in inflammatory and immune responses (Del Rio et al. 2013; Gonzalez-Castejon and Rodriguez-Casado 2011; Perez-Vizcaino and Fraga 2018). Such enzymatic inhibition mechanisms are mechanistically related to the biological effects of non-steroidal anti-inflammatory drugs, though achieved through a distinct chemical mechanism specific to polyphenols (Del Rio et al. 2013; Rudrapal et al. 2022).

Human intervention studies with polyphenol-rich foods, including berry preparations, tea infusions, cocoa beverages, and polyphenol-enriched supplements, frequently document reductions in circulating inflammatory markers, including C-reactive protein (CRP), IL-6, IL-1β, and cell adhesion molecules. Meta-analyses of randomized controlled trials consistently document small but statistically significant reductions in CRP following polyphenol interventions, although the magnitude of effect varies substantially based on baseline inflammatory status, polyphenol compound and dose, intervention duration, and participant population characteristics (Duenas et al. 2015; Perez-Vizcaino and Fraga 2018).

### 1.9.3 Role in Chronic Disease Prevention

The multifactorial biological effects of polyphenols, encompassing antioxidant, anti-inflammatory, metabolic modulatory, and cellular signaling effects, collectively contribute to their potential in preventing or delaying the onset of major chronic diseases. Tables 1.8 and 1.9 summarize epidemiological evidence linking polyphenolic compound consumption to chronic disease prevention, including estimated effect sizes and identified mechanistic pathways.

### 1.9.4 Cardiovascular Disease Prevention

Polyphenolic compounds exert cardioprotective effects through multiple converging mechanisms (Cardona et al. 2013; Cassidy et al. 2016). Enhancement of endothelial function represents a central mechanism—polyphenols stimulate nitric oxide (NO) synthesis and bioavailability—which improves vasodilation capacity, reduces vascular inflammation, and inhibits platelet aggregation. Reduced oxidative modification of LDL particles diminishes their atherogenic potential and uptake by macrophages, inhibiting foam cell formation and atherosclerotic lesion development. Anti-inflammatory effects suppress endothelial activation and inflammatory cell recruitment into arterial walls. Modulation of systemic blood pressure through multiple mechanisms, including endothelial vasodilation, reduced sodium reabsorption, and potassium channel modulation, contributes to cardiovascular protection. Favorable modification of lipoprotein cholesterol profiles, including LDL reduction, triglyceride reduction, and high-density lipoprotein (HDL) elevation, further reduces cardiovascular risk (Hartley et al. 2013; Hooper et al. 2015).

Large prospective cohort studies from European and American populations consistently document 7–15% reductions in cardiovascular event risk (myocardial infarction, stroke) associated with elevated polyphenol consumption. Meta-analyses of randomized controlled intervention trials, while typically of shorter duration and smaller size than cohort studies, demonstrate consistent cardioprotective effects on intermediate risk biomarkers, including blood pressure, endothelial function indices, and inflammatory markers (Grosso et al. 2018; Hartley et al. 2013; Hooper et al. 2015).

**Table 1.8** Evidence linking polyphenol consumption to chronic disease prevention with estimated effect magnitudes and mechanistic pathways. Considered diseases do not include AD, PD, osteoporosis, and metabolic syndrome

| Disease entity | Polyphenol class most studied | Observed risk reduction (95% CI) | Proposed mechanistic pathways | References |
|---|---|---|---|---|
| Cardiovascular disease (incident events) | Flavan-3-ols, anthocyanins, phenolic acids | 7–15% per 200 mg/day intake increase | Endothelial NO bioavailability, LDL oxidation resistance, blood pressure reduction, and platelet aggregation inhibition | Cassidy et al. (2016), Heiss et al. (2010), Tresserra-Rimbau et al. (2014) |
| Coronary heart disease | Flavonoids (all subclasses) | 5–10% per 100 mg/day intake increase | Antioxidant, anti-inflammatory, platelet function modulation, lipid profile improvement | Del Bo et al. (2019), Heiss et al. (2010) |
| Stroke (ischemic) | Anthocyanins, flavonols | 8–12% reduced risk (highest vs. lowest intake quintiles) | Endothelial function, blood pressure reduction, and anticoagulation effects | Cassidy et al. (2016), Del Bo et al. (2019) |
| Type-2 diabetes mellitus | Flavonoids, phenolic acids | 5–12% per 100 mg/day intake increase | Pancreatic beta cell preservation, glucose transporter modulation, PPAR activation, insulin secretion enhancement | Del Bo et al. (2019), Gonzalez-Castejon and Rodriguez-Casado (2011), Perez-Vizcaino and Fraga (2018) |
| All-cause mortality | Total polyphenols | 7–10% reduced risk (highest vs. lowest intake quintiles) | Multifactorial: CVD, metabolic disease, cancer prevention | Del Bo et al. (2019), Hartley et al. (2013), Wang et al. (2014, 2021) |
| Colorectal cancer | Phenolic acids, total polyphenols | 8–15% inverse association | Anti-proliferative, pro-apoptotic, phase II enzyme induction, microbiota modulation | Duenas et al. (2015), Grosso et al. (2017) |

(continued)

**Table 1.8** (continued)

| Disease entity | Polyphenol class most studied | Observed risk reduction (95% CI) | Proposed mechanistic pathways | References |
|---|---|---|---|---|
| Breast cancer | Isoflavones, anthocyanins | Variable; context-dependent (hormone receptor status) | Estrogen receptor modulation, anti-proliferative effects | Grosso et al. (2017), Khoo et al. (2017) |
| Prostate cancer | Lignans, isoflavones | Inconsistent evidence across studies | Phytoestrogenic effects, PPAR activation | Grosso et al. (2017) |

**Table 1.9** Evidence linking polyphenol consumption to chronic disease prevention with estimated effect magnitudes and mechanistic pathways. Considered diseases are: AD, PD, osteoporosis, and metabolic syndrome

| Disease entity | Polyphenol class most studied | Observed risk reduction (95% CI) | Proposed mechanistic pathways | References |
|---|---|---|---|---|
| Neurodegenerative diseases (Alzheimer's) | Mixed polyphenols, phenolic acids | 10–20% reduced risk (highest vs. lowest intake) | Antioxidant, anti-inflammatory, amyloid aggregation inhibition, tau phosphorylation reduction | Perez-Vizcaino and Fraga (2018), Scheltens et al. (2016), Wang et al. (2024) |
| Parkinson's disease | Phenolic acids (particularly from coffee) | 25–40% reduced risk in high consumers | Antioxidant, anti-inflammatory, and neuroprotection mechanisms | Del Bo et al. (2019), Perez-Vizcaino and Fraga (2018) |
| Metabolic syndrome | Flavonoids, phenolic acids | 5–8% reduced risk | Lipid profile improvement, adipocyte function, and glucose control enhancement | Gonzalez-Castejon and Rodriguez-Casado (2011), Hooper et al. (2015) |
| Osteoporosis | Isoflavones, phenolic acids | 3–6% increased bone density | Osteoblast activation, osteoclast suppression, estrogen receptor modulation | Messina (2014) |

### 1.9.5 Cancer Prevention

Polyphenolic compounds may influence multiple stages of cancer pathogenesis: initiation (prevention of mutagenic damage), promotion (suppression of proliferation), progression (inhibition of invasion and metastasis), and angiogenesis (prevention of new vessel formation) (Perez-Vizcaino and Fraga 2018; Wang et al. 2021). Anti-proliferative mechanisms include cell cycle arrest through modulation of cyclin-dependent kinases and checkpoint proteins, induction of apoptosis through both intrinsic (mitochondrial) and extrinsic (death receptor) pathways, and inhibition of survival signaling (Perez-Vizcaino and Fraga 2018; Wang et al. 2021). Anti-inflammatory effects suppress tumor-associated chronic inflammation, a key cancer-promoting microenvironment (Perez-Vizcaino and Fraga 2018; Wang et al. 2021). Modulation of phase I and phase II xenobiotic-metabolizing enzymes influences DNA adduct formation and elimination of mutagenic compounds. Antioxidant effects reduce oxidative DNA damage and enhance DNA repair mechanisms (Perez-Vizcaino and Fraga 2018; Wang et al. 2021).

Epidemiological evidence indicates inverse associations between polyphenol consumption and colorectal cancer incidence, with effect magnitudes of approximately 8–15% reduced risk in high-consumption vs. low-consumption populations (Hooper et al. 2015; Wang et al. 2021). Evidence for breast and prostate cancer prevention is less conclusive, with variable findings potentially related to context-dependent effects involving hormone receptor status (Perez-Vizcaino and Fraga 2018; Wang et al. 2021).

### 1.9.6 Type-2 Diabetes and Metabolic Health

Polyphenolic compounds improve glucose homeostasis through multiple metabolic mechanisms. Inhibition of carbohydrate-digesting enzymes (α-glucosidaseand α-amylase) slows monosaccharide absorption and reduces postprandial glycemic excursions. Modulation of glucose transporter (GLUT) expression and activity affects cellular glucose uptake. Enhancement of insulin secretion from pancreatic beta cells and improvement of insulin sensitivity in target tissues (liver, muscle, adipose) promote normoglycemic homeostasis (Ammar et al. 2020; Magrone and Jirillo 2010). Modulation of peroxisome proliferator-activated receptors (PPARs) (transcription factors regulating metabolic gene expression) favorably affects lipid and glucose metabolism. Favorable modification of adipocyte metabolism and reduced systemic inflammation both contribute to improved insulin sensitivity (Ammar et al. 2020; Magrone and Jirillo 2010). Prospective epidemiological studies indicate a 5–12% reduced type-2 diabetes risk associated with elevated polyphenol consumption, with particular benefits from flavonoid and phenolic acid classes (Ammar et al. 2020; Magrone and Jirillo 2010).

### *1.9.7 Neurodegenerative Disease Prevention*

Polyphenolic compounds may reduce the risk of Alzheimer's disease (AD), Parkinson's disease (PD), and other neurodegenerative conditions through antioxidative, anti-inflammatory, and neuroprotective mechanisms (Harman 1992; Sies et al. 2005; Vauzour et al. 2008). The brain, despite its substantial metabolic activity and related ROS generation, has a relatively modest endogenous antioxidative defense capacity compared with other tissues. Dietary polyphenols, some of which penetrate the blood–brain barrier (BBB) to variable extents, may enhance neuroprotection (Harman 1992; Renaud and de Lorgeril 1992; Sies et al. 2005; Vauzour et al. 2008). Polyphenols inhibit amyloid-β aggregation and fibrillogenesis, central pathological processes in AD. Modulation of tau protein phosphorylation, another key pathological process in Alzheimer's pathology, has been documented in multiple investigations (Harman 1992; Sies et al. 2005; Vauzour et al. 2008). Suppression of neuroinflammation (activation of microglia and astrocytes producing pro-inflammatory mediators) contributes to neuroprotection. Enhancement of neuronal signaling through modulation of MAPK and other survival signal pathways supports neuronal function (Harman 1992; Sies et al. 2005; Vauzour et al. 2008).

Epidemiological evidence suggests a 10–20% reduced AD risk in populations with elevated polyphenol consumption, with particularly strong evidence for the Mediterranean dietary pattern. Coffee polyphenol consumption has been associated with a 25–40% reduced PD risk in large observational studies (Harman 1992; Ludwig et al. 2014; Sies et al. 2005).

## 1.10 Polyphenol-Gut Microbiota Interactions

The relationship between polyphenolic compounds and the resident bacterial microbiota of the gastrointestinal tract represents a bidirectional interactive system with profound implications for health and disease (Queipo-Ortuno et al. 2012; Rastmanesh et al. 2008). Many dietary polyphenols escape absorption within the small intestine and transit to the colon, where they encounter the resident bacterial community comprising thousands of species and billions of individual organisms (Rastmanesh et al. 2008).

Polyphenolic compounds exert selective antimicrobial effects, inhibiting the growth of potentially pathogenic bacteria while sparing or promoting the growth of commensal beneficial bacteria. These selective effects create alterations in microbial community composition, favoring the expansion of beneficial taxa including *Bifidobacterium*, *Lactobacillus*, and *Faecalibacterium* species (Queipo-Ortuno et al. 2012; Rastmanesh et al. 2008). The enhancement of beneficial microbial populations supports multiple facets of health, including synthesis of short-chain fatty acids (acetate, propionate, and butyrate types), production of vitamin K, enhanced intestinal barrier function through tight junction stabilization, and modulation of local and systemic immunity (Queipo-Ortuno et al. 2012; Rathod et al. 2023).

Conversely, the resident microbiota extensively metabolizes polyphenolic compounds through enzymatic pathways not available to the host organism. Bacterial β-glucosidases cleave glycosidic bonds, liberating aglycones. Ring-fission enzymes open aromatic rings, reducing polyphenols to simpler phenolic acids. Decarboxylation, demethylation, and other modifications generate numerous metabolites (Queipo-Ortuno et al. 2012; Rastmanesh et al. 2008). These microbially-derived metabolites, like urolithins from ellagitannins, enterolignans from plant lignans, and simple phenolic acids from complex polyphenols, may possess bioactivities distinct from their parent compounds. Evidence indicates that these microbial metabolites, rather than parent polyphenols, may constitute the bioactive species circulating in the bloodstream and exerting systemic effects (Queipo-Ortuno et al. 2012; Rastmanesh et al. 2008).

Inter-individual heterogeneity in microbiota composition determines distinct metabolic capacities for polyphenol transformation. Individuals harboring particular bacterial taxa capable of metabolizing specific polyphenol subclasses generate distinct patterns of circulating metabolites (Queipo-Ortuno et al. 2012; Rastmanesh et al. 2008). This microbial metabolic heterogeneity contributes substantially to inter-individual variations in biological response to polyphenol consumption, identified as "responder" and "non-responder" phenotypes in clinical intervention trials. Genetic sequencing and metabolomic profiling may eventually enable the prediction of individual microbiota-based response capacity to polyphenol interventions (Queipo-Ortuno et al. 2012; Rastmanesh et al. 2008). Figure 1.2 presents a schematic overview of the primary mechanisms through which dietary polyphenols and their metabolites exert protective effects against chronic diseases.

## 1.11 Additional Bioactive Properties

Beyond the well-characterized antioxidative and anti-inflammatory properties, polyphenolic compounds demonstrate additional diverse biological activities with potential health implications:

(a) Antimicrobial activity: multiple polyphenols demonstrate direct antimicrobial activity against pathogenic bacteria, viruses, and fungi, mechanisms involving cell membrane disruption, enzyme inhibition, and nucleic acid damage (Sies et al. 2005; Tsao 2010)
(b) Phytoestrogenic effects: isoflavones and certain other polyphenols interact with estrogen receptors, exerting weak estrogenic or anti-estrogenic effects depending on receptor isoform and cellular context (Adlercreutz 2007; Sies et al. 2005)

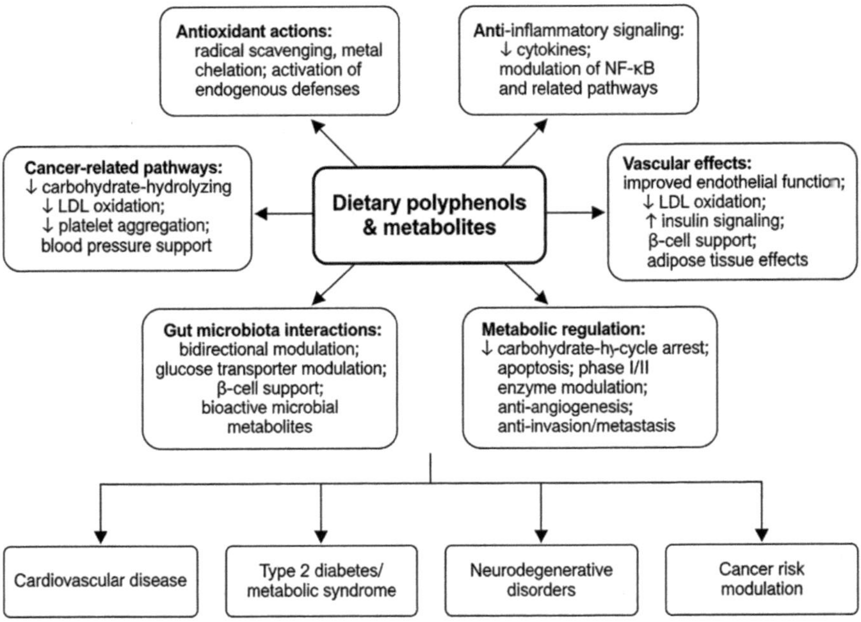


**Fig. 1.2** Major mechanisms of polyphenols in chronic disease prevention: schematic diagram of multifactorial actions of polyphenols, linking intake to health outcomes via antioxidant defense, inflammation control, vascular protection, metabolic regulation, gut microbiota modulation, and anticancer pathways. This picture has been created by the authors

(c) Metabolic enzyme modulation: polyphenols modulate the expression of xenobiotic-metabolizing enzymes, affecting drug metabolism, potentially influencing therapeutic efficacy or toxicity of pharmaceutical agents (Rudrapal et al. 2022; Tsao 2010)
(d) Receptor and kinase modulation: polyphenols interact with diverse cellular receptors (aryl hydrocarbon receptor, Toll-like receptors) and kinases (Src family kinases, Bruton's tyrosine kinase), with implications for signaling pathway modulation (Sies et al. 2005; Tsao 2010)
(e) Skin health and photoprotection: topical and dietary polyphenols protect against ultraviolet radiation-induced damage, oxidative stress, and photoaging through antioxidative and anti-inflammatory mechanisms (Rudrapal et al. 2022; Tsao 2010).

## 1.12 Analytical Methodologies for Polyphenol Characterization

### 1.12.1 Extraction and Sample Preparation

Accurate quantification and characterization of polyphenolic compounds in food and biological samples require optimization of extraction procedures. The extractability of polyphenols from solid matrices depends on solvent composition, pH, temperature, extraction duration, and solid-to-solvent ratio (Rudrapal et al. 2022; Tsao 2010; Wang et al. 2024; Zhu et al. 2013). Commonly utilized solvents include aqueous mixtures of methanol, ethanol, acetone, or acetonitrile, frequently acidified with formic acid or acetic acid to enhance the stability of certain polyphenol classes and improve extraction efficiency. Solid-phase extraction (SPE) procedures enable sample cleanup, fractionation of polyphenol classes, and concentration of dilute samples before chromatographic analysis (Rudrapal et al. 2022; Tsao 2010; Wang et al. 2024; Zhu et al. 2013).

Enzymatic treatments, including cellulase, pectinase, and hemicellulose digestion, can enhance the extraction of bound polyphenols associated with cell wall polysaccharides, enabling measurement of both free and bound fractions (Chandra et al. 2014; Wang et al. 2024; Zhu et al. 2013). The selection of extraction methodology critically influences measured polyphenol content and composition. Standardization and detailed documentation of extraction conditions are essential for result comparability across studies (Chandra et al. 2014).

### 1.12.2 Chromatographic and Spectrometric Analysis Techniques

Polyphenol identification and quantification rely primarily on high-performance liquid chromatography (HPLC) combined with ultraviolet–visible (UV–VIS) spectrophotometric detection or diode-array detection (DAD) (Chandra et al. 2014; Sivakumar et al. 2020). HPLC separation based on differential partitioning between stationary and mobile phases, followed by UV–VIS detection exploiting the characteristic light-absorbing properties of aromatic polyphenol structures, enables identification of individual compounds through comparison with authentic standards (Chandra et al. 2014; Sivakumar et al. 2020).

Liquid chromatography coupled to mass spectrometry (LC–MS/MS) represents a more sophisticated analytical approach providing MW and structural information through fragmentation patterns, enabling identification of polyphenols without requiring authentic reference standards (Chandra et al. 2014; Sivakumar et al. 2020). Electrospray ionization (ESI) and atmospheric pressure chemical ionization

(APCI) represent common ionization techniques in LC–MS/MS analyses. The sensitivity of mass spectrometry detection enables quantification of polyphenols at lower concentration levels compared with UV–VIS detection (Sivakumar et al. 2020).

High-resolution mass spectrometry techniques, including time-of-flight (TOF) and Orbitrap mass spectrometry, enable ultra-high mass accuracy measurements, facilitating the discovery of novel polyphenolic compounds and their metabolites in complex biological samples. Untargeted metabolomics approaches utilize such technologies to comprehensively characterize polyphenol-derived metabolite profiles (Scalbert et al. 2005a, b; Sivakumar et al. 2020).

Gas chromatography-mass spectrometry (GC–MS) addresses volatile or thermally stable polyphenols, frequently following chemical derivatization to enhance volatility. Capillary electrophoresis (CE) provides alternative separation mechanisms for charged polyphenolic compounds and remains useful in specialized applications (Chandra et al. 2014; Sivakumar et al. 2020).

### *1.12.3 Colorimetric and Spectrophotometric Methods for Global Assessment*

Colorimetric assays provide rapid estimation of global polyphenolic content without the requirement for chromatographic separation or expensive instrumentation, facilitating high-throughput screening applications. The Folin-Ciocalteu assay measures the reducing capacity of polyphenol samples through absorbance change at 765 nm following reaction with the Folin-Ciocalteu reagent. Results are expressed as gallic acid equivalents (GAE), a reference compound standard. The assay demonstrates good reproducibility but lacks compound specificity, measuring total reducing capacity without distinguishing among polyphenol classes (Chandra et al. 2014; Rudrapal et al. 2022; Tsao 2010).

Flavonoid-specific colorimetric assays utilizing aluminum chloride or aluminum nitrate form-colored complexes with flavonoid-specific structural motifs, enabling quantification of flavonoid subsets. Proanthocyanidin-specific assays utilizing vanillin or other chemical reagents enable quantification of condensed tannins (Chandra et al. 2014; Rudrapal et al. 2022; Tsao 2010).

Antioxidant capacity assays, including 2,2-diphenyl-1-picrylhydrazyl (DPPH), 2,2'-azinobis (3-ethylbenzothiazoline-6-sulfonic acid) (ABTS), ferric reducing ability of plasma (FRAP), and oxygen radical absorbance capacity (ORAC), measure polyphenol electron/hydrogen donating capacity through mechanistic variations. Results provide indices of antioxidative potential but do not specifically quantify individual polyphenol compounds. These methods have utility for quality control and research applications, but require recognition that chemical antioxidative capacity does not necessarily correlate with in vivo biological antioxidative effects (Chandra et al. 2014; Rudrapal et al. 2022; Tsao 2010).

### 1.12.4 Limitations and Future Analytical Directions

Contemporary challenges in polyphenol analysis include the extraordinary structural diversity encompassing >8,000 compounds, rendering comprehensive analysis of all compounds infeasible in typical studies. Limited commercial availability of authentic reference standards for numerous polyphenolic compounds restricts the capacity for confident compound identification and quantification. Matrix interference effects in complex food and biological samples can obscure analytes of interest or generate false identifications. Dynamic polyphenol metabolism, including rapid conjugation by phase II enzymes, microbial transformation, and tissue-specific metabolism, complicates biomarker identification and temporal kinetic characterization (Chandra et al. 2014; Saad et al. 2025; Scalbert et al. 2005a, b).

Future analytical advancement directions encompassing untargeted metabolomics with high-resolution mass spectrometry, development of novel selective standards and reference materials, standardization of analytical methodologies and reporting units, improved bioinformatic platforms for managing complex datasets, and integration of chromatographic and MS technologies with emerging techniques including ion mobility spectrometry, will advance the sophistication and accuracy of polyphenol research (Chandra et al. 2014; Sivakumar et al. 2020; Urquiaga and Leighton 2000).

## 1.13 Challenges and Future Research Directions

### 1.13.1 Bioavailability Limitations and Efficacy Translation Gaps

A fundamental challenge in translating polyphenol research from laboratory and animal investigations into clinical benefit for human populations centers on the frequently low and variable bioavailability of these compounds. Many polyphenolic compounds exhibit poor absorption, extensive hepatic first-pass metabolism generating conjugated forms with altered or diminished bioactivity compared with parent compounds, and rapid urinary elimination producing short circulating half-lives. These bioavailability limitations frequently result in systemic concentrations substantially lower than the concentrations used in vitro experimental investigations demonstrating bioactivity (Manach et al. 2005; Mateos et al. 2001).

Intensive investigation into enhancement of polyphenol bioavailability through food matrix modifications, co-administration with absorption-enhancing compounds, utilization of nanotechnology-based delivery vehicles, and microencapsulation approaches represents an active research area (Machado et al. 2021; Mateos et al. 2001; Dubey et al. 2025a). Improved understanding of which circulating metabolites (parent compounds, conjugates, or microbial metabolites) mediate health benefits will inform optimization strategies (Manach et al. 2005; Mateos et al. 2001).

### 1.13.2 Gaps in Contemporary Dietary Guidance

Despite mounting epidemiological evidence supporting the health benefits of polyphenol-rich diets, formal quantitative dietary reference intakes specific to polyphenols remain absent from regulatory agency guidelines (Wierzbicki 2008; Yang et al. 2020). The absence reflects acknowledged challenges, including structural diversity of polyphenol compounds, uncertainties regarding dose–response relationships and optimal intake levels, incomplete understanding of inter-individual variation in response, and incomplete mechanistic understanding of health benefit mechanisms (Wierzbicki 2008; Yang et al. 2020).

The current strategy of emphasizing plant food consumption through qualitative dietary recommendations (e.g., "increase fruit and vegetable consumption") implicitly increases polyphenol exposure while accommodating broader nutritional objectives (Yang et al. 2020). Whether more specific polyphenol-targeted quantitative recommendations would offer additional practical value beyond existing food-based guidance remains uncertain (Wierzbicki 2008).

### 1.13.3 Emerging Technologies and Personalized Nutrition

Rapid advancement in omics technologies, including genomics, transcriptomics, proteomics, and metabolomics, combined with mechanistic microbiome investigations, provides unprecedented capacity to understand polyphenol-human biology interactions at molecular systems levels. Metabolomic profiling enables comprehensive characterizations of polyphenol-derived metabolite patterns following consumption, potentially identifying novel biomarkers of exposure and biological effect. Microbiome sequencing and functional metagenomic approaches elucidate the microbial enzymatic capacity for polyphenol metabolism and identify bacterial taxa associated with favorable metabolite generation (Aleman-Jimenez et al. 2021; Hidalgo et al. 2012; Urquiaga and Leighton 2000).

Gene expression profiling (transcriptomics) in peripheral blood mononuclear cells or tissue biopsies enables characterization of polyphenol-induced changes in molecular pathways and identification of "responder" and "non-responder" phenotypes to polyphenol interventions. Integration of individual genomic data with metabolomic and microbiome information may enable the prediction of individual-level response to polyphenol interventions and optimization of personalized nutritional recommendations. Systems biology approaches incorporating computational modeling and machine learning analysis of complex multidimensional datasets may identify novel polyphenol-gene-environment interactions with implications for precision nutrition implementation (Aleman-Jimenez et al. 2021; Hidalgo et al. 2012; Urquiaga and Leighton 2000).

### *1.13.4 Future Research Priorities*

Consensus regarding future polyphenol research priorities encompasses:

(1) Execution of large-scale, adequately powered, long-duration randomized controlled trials with clinically relevant health outcomes (cardiovascular events, cancer incidence, neurodegenerative disease progression) as primary endpoints, combined with careful characterization of polyphenol exposure and metabolite generation (Atherton et al. 2006; Manach et al. 2005; Mateos et al. 2001)
(2) Integration of dietary assessment, biomarker measurements, and omics data to characterize inter-individual variability and identify predictive biomarkers of response to polyphenol interventions (Aleman-Jimenez et al. 2021; Hidalgo et al. 2012)
(3) Translation of mechanistic discoveries into evidence-based, implementable dietary recommendations and substantiation of health-related marketing claims through rigorous clinical and epidemiological evidence (Wierzbicki 2008)
(4) Advancement of sustainable sourcing strategies and valorization of polyphenol-rich agricultural by-products (fruit peels, vegetable discards, seed coats) to enhance nutritional value while reducing food system waste (Borah et al. 2023; Ravindran and Jaiswal 2016)
(5) Development and validation of novel delivery and bioavailability enhancement technologies enabling optimization of polyphenol administration and biological activity (Machado et al. 2021; Manach et al. 2005; Mateos et al. 2001).

## 1.14 Conclusion

Polyphenolic compounds represent structurally diverse, ubiquitously distributed plant-derived secondary metabolites with substantial influence on human nutrition and health. The major recognized polyphenol classes, including flavonoids (> 6,000 known structures), phenolic acids, stilbenes, lignans, and polymeric tannins, demonstrate distinct chemical structural characteristics, disparate dietary sources, divergent bioavailability profiles, and differentiated biological activities. Substantial intake heterogeneity across global populations reflects regional dietary traditions, agricultural practices, and cultural food preferences, with estimated median intakes ranging from 600 to 1,600 mg/day across studied populations.

Convergent evidence from in vitro mechanistic studies, animal models, human intervention trials, and large prospective epidemiological investigations supports important roles for polyphenol-rich dietary patterns in the prevention and mitigation of major chronic noncommunicable diseases, including cardiovascular disease, type-2 diabetes mellitus, select malignant neoplasms, and neurodegenerative diseases. Health-promoting biological mechanisms encompass direct antioxidant free radical scavenging and indirect activation of endogenous cellular defense systems, suppression of pro-inflammatory signaling cascades and cytokine production, favorable modulation of endothelial function and vascular biology, enhancement of insulin sensitivity and glucose homeostasis, and complex bidirectional interactions with the resident gut microbiota generating bioactive microbial metabolites.

Contemporary challenges in polyphenol research include the frequently limited bioavailability and variable absorption and metabolism across individuals, uncertainty regarding optimal intake levels and dose–response relationships, incomplete characterizations of which specific polyphenol classes or individual compounds drive health benefits, heterogeneity in individual metabolic and microbiota-based response capacity, and gaps between mechanistic discoveries and translation into clinical practice. Rapid advancement in analytical chemistry, omics technologies, and systems biology approaches offers promising future directions for enhanced understanding of polyphenol exposure, metabolism, biological effects, and identification of personalized nutrition strategies optimized to individual genetic and microbiota characteristics.

Within the context of contemporary public health nutrition, polyphenol-rich plant-based dietary patterns represent an evidence-supported, practical, and sustainable strategy for chronic disease prevention and health promotion. Future research integrating long-term randomized controlled trials, advanced analytical and computational technologies, and precision nutrition frameworks will further refine the understanding of how polyphenols can be optimally leveraged for disease prevention and health maintenance within the broader context of sustainable, culturally appropriate, and nutritionally adequate dietary patterns.

**Acknowledgements** We thank Caproslaxy Media (https://www.caproslaxymedia.com/) for linguistic assistance during the preparation of this manuscript. To maintain uniformity and to standardize as per international nomenclature, scientific terminologies for botanical sources have been adopted as per the information available on https://www.worldfloraonline.org/.

**Funding Source** This work was supported by the Sichuan Science and Technology Program (Grant No. 2025YFHZ0213).

## References

Adlercreutz H (2007) Lignans and human health. Crit Rev Clin Lab Sci 44(5–6):483–525. https://doi.org/10.1080/10408360701612942

Adom KK, Liu RH (2002) Antioxidant activity of grains. J Agric Food Chem 50(21):6182–6187. https://doi.org/10.1021/jf0205099

Aleman-Jimenez C, Dominguez-Perles R, Medina S, Prgomet I, Lopez-Gonzalez I, Simonelli-Munoz A, Campillo-Cano M, Aunon D, Ferreres F, Gil-Izquierdo A (2021) Pharmacokinetics and bioavailability of hydroxytyrosol are dependent on the food matrix in humans. Eur J Nutr 60(2):905–915. https://doi.org/10.1007/s00394-020-02295-0

Ammar A, MounaTurki TK, Bragazzi NL, Boukhris O, Bouaziz M, Ayadi F, El Abed K, Driss T, Souissi N, Chtourou H, Bailey SJ, Hoekelmann A (2020) Effects of natural polyphenol-rich pomegranate juice on the acute and delayed response of Homocysteine and steroidal hormones following weightlifting exercises: a double-blind, placebo-controlled trial. J Int Soc Sports Nutr 17(1):15. https://doi.org/10.1186/s12970-020-00345-w

Atherton HJ, Bailey NJ, Zhang W, Taylor J, Major H, Shockcor J, Clarke K, Griffin JL (2006) A combined 1H-NMR spectroscopy- and mass spectrometry-based metabolomic study of the PPAR-alpha null mutant mouse defines profound systemic changes in metabolism linked to the metabolic syndrome. Physiol Genomics 27(2):178–186. https://doi.org/10.1152/physiolgenomics.00060.2006

Bain BJ, Amos RJ, Bareford D, Chapman C, Davies SC, Old JM, Wild BJ, Cavill I, Hyde K, Matutes E, Reilly J, Wood JK (1998) The laboratory diagnosis of haemoglobinopathies. Br J Haematol 101(4):783–792. https://doi.org/10.1046/j.1365-2141.1998.00809.x

Baldrick FR, Woodside JV, Elborn JS, Young IS, McKinley MC (2011) Biomarkers of fruit and vegetable intake in human intervention studies: a systematic review. Crit Rev Food Sci Nutr 51(9):795–815. https://doi.org/10.1080/10408398.2010.482217

Baur JA, Sinclair DA (2006) Therapeutic potential of resveratrol: the in vivo evidence. Nat Rev Drug Discov 5(6):493–506. https://doi.org/10.1038/nrd2060

Belay A, Kim HK, Hwang YH (2016) Binding of caffeine with caffeic acid and chlorogenic acid using fluorescence quenching, UV/vis and FTIR spectroscopic techniques. Luminescence 31(2):565–572. https://doi.org/10.1002/bio.2996

Bentley AR, Kritchevsky SB, Harris TB, Newman AB, Bauer DC, Meibohm B, Clark AG, Cassano PA, Health ABCS (2012) Genetic variation in antioxidant enzymes and lung function. Free Radic Biol Med 52(9):1577–1583. https://doi.org/10.1016/j.freeradbiomed.2012.02.025

Bento-Silva A, Duarte N, Mecha E, Belo M, Serra AT, Vaz Patto MC, Bronze MR (2021) Broa, an ethnic maize bread, as a source of phenolic compounds. Antioxidants (Basel) 10(5). https://doi.org/10.3390/antiox10050672

Bhagwat S, Haytowitz DB, Holden JM (2014) USDA database for the flavonoid content of selected foods, Release 3.1. U.S. Department of Agriculture, Agricultural Research Service

Borah MS, Tiwari A, Sridhar K, Narsaiah K, Nayak PK, Stephen Inbaraj B (2023) Recent trends in valorization of food industry waste and by-products: encapsulation and in vitro release of bioactive compounds. Foods 12(20). https://doi.org/10.3390/foods12203823

Brown CR (2005) Antioxidants in potato. Am J Potato Res 82(2):163–172. https://doi.org/10.1007/bf02853654

Cardona F, Andres-Lacueva C, Tulipani S, Tinahones FJ, Queipo-Ortuno MI (2013) Benefits of polyphenols on gut microbiota and implications in human health. J Nutr Biochem 24(8):1415–1422. https://doi.org/10.1016/j.jnutbio.2013.05.001

Cassidy A, O'Reilly EJ, Kay C, Sampson L, Franz M, Forman JP, Curhan G, Rimm EB (2011) Habitual intake of flavonoid subclasses and incident hypertension in adults. Am J Clin Nutr 93(2):338–347. https://doi.org/10.3945/ajcn.110.006783

Cassidy A, Bertoia M, Chiuve S, Flint A, Forman J, Rimm EB (2016) Habitual intake of anthocyanins and flavanones and risk of cardiovascular disease in men. Am J Clin Nutr 104(3):587–594. https://doi.org/10.3945/ajcn.116.133132

Castillo-Munoz N, Gomez-Alonso S, Garcia-Romero E, Hermosin-Gutierrez I (2007) Flavonol profiles of Vitis vinifera red grapes and their single-cultivar wines. J Agric Food Chem 55(3):992–1002. https://doi.org/10.1021/jf062800k

Chandra S, Khan S, Avula B, Lata H, Yang MH, Elsohly MA, Khan IA (2014) Assessment of total phenolic and flavonoid content, antioxidant properties, and yield of aeroponically and conventionally grown leafy vegetables and fruit crops: a comparative study. Evid Based Complement Alternat Med 2014:253875. https://doi.org/10.1155/2014/253875

Chandrasekara A, Shahidi F (2011) Bioactivities and antiradical properties of millet grains and hulls. J Agric Food Chem 59(17):9563–9571. https://doi.org/10.1021/jf201849d

Choi S, Jung S, Ko KS (2018) Effects of coffee extracts with different roasting degrees on antioxidant and anti-inflammatory systems in mice. Nutrients 10(3). https://doi.org/10.3390/nu10030363

Cory H, Passarelli S, Szeto J, Tamez M, Mattei J (2018) The role of polyphenols in human health and food systems: a mini-review. Front Nutr 5:87. https://doi.org/10.3389/fnut.2018.00087

Del Bo C, Bernardi S, Marino M, Porrini M, Tucci M, Guglielmetti S, Cherubini A, Carrieri B, Kirkup B, Kroon P, Zamora-Ros R, Liberona NH, Andres-Lacueva C, Riso P (2019) Systematic review on polyphenol intake and health outcomes: is there sufficient evidence to define a health-promoting polyphenol-rich dietary pattern? Nutrients 11(6). https://doi.org/10.3390/nu11061355

Del Rio D, Rodriguez-Mateos A, Spencer JP, Tognolini M, Borges G, Crozier A (2013) Dietary (poly)phenolics in human health: structures, bioavailability, and evidence of protective effects against chronic diseases. Antioxid Redox Signal 18(14):1818–1892. https://doi.org/10.1089/ars.2012.4581

Dixon RA, Paiva NL (1995) Stress-induced phenylpropanoid metabolism. Plant Cell 7(7):1085–1097. https://doi.org/10.1105/tpc.7.7.1085

Dosoky NS, Setzer WN (2018) Chemical composition and biological activities of essential oils of curcuma species. Nutrients 10(9). https://doi.org/10.3390/nu10091196

Dubey AK, Brijesh H, Zhang Y, Shen B, Calina D, Sharifi-Rad J, Singla RK (2025a) Tissue engineering with bionanomaterials: a new frontier. Phytomedicine 149:157380. https://doi.org/10.1016/j.phymed.2025.157380

Dubey AK, Chandragiri SS, Geevarghese AV, Kapoor B, Gulati M, Rani P, Singh G, Chavda VP, Gundamaraju R, Bansal H, Gautam RK, Goyal R, Okoh MP, Shen B, Singla RK (2025b) Exploring the pharmacokinetics, drug-likeness, and toxicological features of anticancer flavonoids: a Boulevard to explore their clinical translational potential. Front Pharmacol 16:1648395. https://doi.org/10.3389/fphar.2025.1648395

Duenas M, Munoz-Gonzalez I, Cueva C, Jimenez-Giron A, Sanchez-Patan F, Santos-Buelga C, Moreno-Arribas MV, Bartolome B (2015) A survey of modulation of gut microbiota by dietary polyphenols. Biomed Res Int 2015:850902. https://doi.org/10.1155/2015/850902

Erlund I (2004) Review of the flavonoids quercetin, hesperetin, and naringenin. Dietary sources, bioactivities, bioavailability, and epidemiology. Nutrition Research 24 (10):851–874. https://doi.org/10.1016/j.nutres.2004.07.005

Esquivel-Ferrino PC, Favela-Hernandez JM, Garza-Gonzalez E, Waksman N, Rios MY, del Rayo Camacho-Corona M (2012) Antimycobacterial activity of constituents from Foeniculum vulgare var. dulce grown in Mexico. Molecules 17(7):8471–8482. https://doi.org/10.3390/molecules17078471

Floegel A, Kim D-O, Chung S-J, Koo SI, Chun OK (2011) Comparison of ABTS/DPPH assays to measure antioxidant capacity in popular antioxidant-rich US foods. J Food Compos Anal 24(7):1043–1048. https://doi.org/10.1016/j.jfca.2011.01.008

German JB, Walzem RL (2000) The health benefits of wine. Annu Rev Nutr 20:561–593. https://doi.org/10.1146/annurev.nutr.20.1.561

Ghasemzadeh A, Jaafar HZ, Rahmat A (2010) Antioxidant activities, total phenolics and flavonoids content in two varieties of Malaysia young ginger (Zingiber officinale Roscoe). Molecules 15(6):4324–4333. https://doi.org/10.3390/molecules15064324

Gonzalez-Castejon M, Rodriguez-Casado A (2011) Dietary phytochemicals and their potential effects on obesity: a review. Pharmacol Res 64(5):438–455. https://doi.org/10.1016/j.phrs.2011.07.004

Grosso G, Stepaniak U, Micek A, Kozela M, Stefler D, Bobak M, Pajak A (2018) Dietary polyphenol intake and risk of hypertension in the Polish arm of the HAPIEE study. Eur J Nutr 57(4):1535–1544. https://doi.org/10.1007/s00394-017-1438-7

Grosso G, Godos J, Lamuela-Raventos R, Ray S, Micek A, Pajak A, Sciacca S, D'Orazio N, Del Rio D, Galvano F (2017) A comprehensive meta-analysis on dietary flavonoid and lignan intake and cancer risk: Level of evidence and limitations. Mol Nutr Food Res 61(4). https://doi.org/10.1002/mnfr.201600930

Harman D (1992) Free radical theory of aging. Mutat Res 275(3–6):257–266. https://doi.org/10.1016/0921-8734(92)90030-s

Hartley L, Igbinedion E, Holmes J, Flowers N, Thorogood M, Clarke A, Stranges S, Hooper L, Rees K (2013) Increased consumption of fruit and vegetables for the primary prevention of

cardiovascular diseases. Cochrane Database Syst Rev 2013(6):CD009874. https://doi.org/10.1002/14651858.CD009874.pub2
He FJ, Li J, Macgregor GA (2013) Effect of longer term modest salt reduction on blood pressure: cochrane systematic review and meta-analysis of randomised trials. BMJ 346:f1325. https://doi.org/10.1136/bmj.f1325
Heiss C, Keen CL, Kelm M (2010) Flavanols and cardiovascular disease prevention. Eur Heart J 31(21):2583–2592. https://doi.org/10.1093/eurheartj/ehq332
Hidalgo M, Oruna-Concha MJ, Kolida S, Walton GE, Kallithraka S, Spencer JP, de Pascual-Teresa S (2012) Metabolism of anthocyanins by human gut microflora and their influence on gut bacterial growth. J Agric Food Chem 60(15):3882–3890. https://doi.org/10.1021/jf3002153
Hooper L, Kay C, Abdelhamid A, Kroon PA, Cohn JS, Rimm EB, Cassidy A (2012) Effects of chocolate, cocoa, and flavan-3-ols on cardiovascular health: a systematic review and meta-analysis of randomized trials. Am J Clin Nutr 95(3):740–751. https://doi.org/10.3945/ajcn.111.023457
Hooper L, Abdelhamid A, Bunn D, Brown T, Summerbell CD, Skeaff CM (2015) Effects of total fat intake on body weight. Cochrane Database Syst Rev 2015(8):CD011834. https://doi.org/10.1002/14651858.CD011834
Huang Q, Braffett BH, Simmens SJ, Young HA, Ogden CL (2020) Dietary polyphenol intake in US Adults and 10-Year trends: 2007–2016. J Acad Nutr Diet 120(11):1821–1833. https://doi.org/10.1016/j.jand.2020.06.016
Hudthagosol C, Haddad EH, McCarthy K, Wang P, Oda K, Sabate J (2011) Pecans acutely increase plasma postprandial antioxidant capacity and catechins and decrease LDL oxidation in humans. J Nutr 141(1):56–62. https://doi.org/10.3945/jn.110.121269
Ioannou GD, Savva IK, Christou A, Stavrou IJ, Kapnissi-Christodoulou CP (2023) Phenolic profile, antioxidant activity, and chemometric classification of carob pulp and products. Molecules 28(5). https://doi.org/10.3390/molecules28052269
Jančić D, Šuković D, Rešetar J, Delić L, Nikolić M (2022) Nutritional composition, biologically active substances and antioxidant activity of young spelt grass extract. JSFA Rep 2(8):385–397. https://doi.org/10.1002/jsf2.73
Jurgonski A, Juskiewicz J, Zdunczyk Z, Matusevicius P, Kolodziejczyk K (2014) Polyphenol-rich extract from blackcurrant pomace attenuates the intestinal tract and serum lipid changes induced by a high-fat diet in rabbits. Eur J Nutr 53(8):1603–1613. https://doi.org/10.1007/s00394-014-0665-4
Kadam D, Lele SS (2017) Extraction, characterization and bioactive properties of Nigella sativa seedcake. J Food Sci Technol 54(12):3936–3947. https://doi.org/10.1007/s13197-017-2853-8
Kalt W, Cassidy A, Howard LR, Krikorian R, Stull AJ, Tremblay F, Zamora-Ros R (2020) Recent research on the health benefits of blueberries and their anthocyanins. Adv Nutr 11(2):224–236. https://doi.org/10.1093/advances/nmz065
Khoo HE, Azlan A, Tang ST, Lim SM (2017) Anthocyanidins and anthocyanins: colored pigments as food, pharmaceutical ingredients, and the potential health benefits. Food Nutr Res 61(1):1361779. https://doi.org/10.1080/16546628.2017.1361779
King RA, Bursill DB (1998) Plasma and urinary kinetics of the isoflavones daidzein and genistein after a single soy meal in humans. Am J Clin Nutr 67(5):867–872. https://doi.org/10.1093/ajcn/67.5.867
King A, Young G (1999) Characteristics and occurrence of phenolic phytochemicals. J Am Diet Assoc 99(2):213–218. https://doi.org/10.1016/S0002-8223(99)00051-6
Le PH, Ho LTT, Le DHT, Nguyen V (2023) Purification of coffee polyphenols extracted from coffee pulps (Coffee arabica L.) using aqueous two-phase system. Molecules 28(15). https://doi.org/10.3390/molecules28155922
Li-Weber M (2009) New therapeutic aspects of flavones: the anticancer properties of Scutellaria and its main active constituents Wogonin, Baicalein and Baicalin. Cancer Treat Rev 35(1):57–68. https://doi.org/10.1016/j.ctrv.2008.09.005

Ludwig IA, Clifford MN, Lean ME, Ashihara H, Crozier A (2014) Coffee: biochemistry and potential impact on health. Food Funct 5(8):1695–1717. https://doi.org/10.1039/c4fo00042k

Machado ND, Gutierrez G, Matos M, Fernandez MA (2021) Preservation of the antioxidant capacity of resveratrol via encapsulation in niosomes. Foods 10(5). https://doi.org/10.3390/foods10050988

Machado M, Espirito Santo L, Machado S, Lobo JC, Costa ASG, Oliveira M, Ferreira H, Alves RC (2023) Bioactive potential and chemical composition of coffee by-products: from pulp to silverskin. Foods 12(12). https://doi.org/10.3390/foods12122354

Magrone T, Jirillo E (2010) Polyphenols from red wine are potent modulators of innate and adaptive immune responsiveness. Proc Nutr Soc 69(3):279–285. https://doi.org/10.1017/S0029665110000121

Manach C, Scalbert A, Morand C, Remesy C, Jimenez L (2004) Polyphenols: food sources and bioavailability. Am J Clin Nutr 79(5):727–747. https://doi.org/10.1093/ajcn/79.5.727

Manach C, Williamson G, Morand C, Scalbert A, Remesy C (2005) Bioavailability and bioefficacy of polyphenols in humans. I. Review of 97 bioavailability studies. Am J Clin Nutr 81(1 Suppl):230S–242S. https://doi.org/10.1093/ajcn/81.1.230S

Martinez-Castaneiras P, Ortiz C, Fernandez de Larrea-Baz N, Lope V, Sanchez-Gordon G, Ruiz-Moreno E, Alonso I, Garcia-Esquinas E, Perez-Gomez B, Pastor-Barriuso R, Galan I, Castello A (2025) Intake of fruit, vegetables and pulses, and all-cause, cardiovascular and cancer mortality: results from a population-based prospective study. Public Health 239:169–178. https://doi.org/10.1016/j.puhe.2024.12.014

Mateos R, Espartero JL, Trujillo M, Rios JJ, Leon-Camacho M, Alcudia F, Cert A (2001) Determination of phenols, flavones, and lignans in virgin olive oils by solid-phase extraction and high-performance liquid chromatography with diode array ultraviolet detection. J Agric Food Chem 49(5):2185–2192. https://doi.org/10.1021/jf0013205

Messina M (2014) Soy foods, isoflavones, and the health of postmenopausal women. Am J Clin Nutr 100(Suppl 1):423S-430S. https://doi.org/10.3945/ajcn.113.071464

Messina M (2016) Soy and health update: evaluation of the clinical and epidemiologic literature. Nutrients 8(12). https://doi.org/10.3390/nu8120754

Mihaylova R, Elincheva V, Momekov G, Simeonova R (2025) unlocking the therapeutic potential of ellagitannins: a comprehensive review of key representatives. Molecules 30(22). https://doi.org/10.3390/molecules30224328

Ozdal T, Sela DA, Xiao J, Boyacioglu D, Chen F, Capanoglu E (2016) The reciprocal interactions between polyphenols and gut microbiota and effects on bioaccessibility. Nutrients 8(2):78. https://doi.org/10.3390/nu8020078

Pandey KB, Rizvi SI (2009) Plant polyphenols as dietary antioxidants in human health and disease. Oxid Med Cell Longev 2(5):270–278. https://doi.org/10.4161/oxim.2.5.9498

Pedisic Z, Bauman A (2015) Accelerometer-based measures in physical activity surveillance: current practices and issues. Br J Sports Med 49(4):219–223. https://doi.org/10.1136/bjsports-2013-093407

Perez-Jimenez J, Neveu V, Vos F, Scalbert A (2010) Systematic analysis of the content of 502 polyphenols in 452 foods and beverages: an application of the phenol-explorer database. J Agric Food Chem 58(8):4959–4969. https://doi.org/10.1021/jf100128b

Perez-Vizcaino F, Fraga CG (2018) Research trends in flavonoids and health. Arch Biochem Biophys 646:107–112. https://doi.org/10.1016/j.abb.2018.03.022

Prochazkova D, Bousova I, Wilhelmova N (2011) Antioxidant and prooxidant properties of flavonoids. Fitoterapia 82(4):513–523. https://doi.org/10.1016/j.fitote.2011.01.018

Queipo-Ortuno MI, Boto-Ordonez M, Murri M, Gomez-Zumaquero JM, Clemente-Postigo M, Estruch R, Cardona Diaz F, Andres-Lacueva C, Tinahones FJ (2012) Influence of red wine polyphenols and ethanol on the gut microbiota ecology and biochemical biomarkers. Am J Clin Nutr 95(6):1323–1334. https://doi.org/10.3945/ajcn.111.027847

Quideau S, Deffieux D, Douat-Casassus C, Pouysegu L (2011) Plant polyphenols: chemical properties, biological activities, and synthesis. Angew Chem Int Ed Engl 50(3):586–621. https://doi.org/10.1002/anie.201000044

Rastmanesh R, Abargouei AS, Shadman Z, Ebrahimi AA, Weber CE (2008) A pilot study of potassium supplementation in the treatment of hypokalemic patients with rheumatoid arthritis: a randomized, double-blinded, placebo-controlled trial. J Pain 9(8):722–731. https://doi.org/10.1016/j.jpain.2008.03.006

Rathod NB, Elabed N, Punia S, Ozogul F, Kim SK, Rocha JM (2023) Recent developments in polyphenol applications on human health: a review with current knowledge. Plants (Basel) 12(6). https://doi.org/10.3390/plants12061217

Ravindran R, Jaiswal AK (2016) Exploitation of food industry waste for high-value products. Trends Biotechnol 34(1):58–69. https://doi.org/10.1016/j.tibtech.2015.10.008

Renaud S, de Lorgeril M (1992) Wine, alcohol, platelets, and the French paradox for coronary heart disease. Lancet 339(8808):1523–1526. https://doi.org/10.1016/0140-6736(92)91277-f

Rice-Evans CA, Miller NJ, Paganga G (1996) Structure-antioxidant activity relationships of flavonoids and phenolic acids. Free Radic Biol Med 20(7):933–956. https://doi.org/10.1016/0891-5849(95)02227-9

Robbins RJ (2003) Phenolic acids in foods: an overview of analytical methodology. J Agric Food Chem 51(10):2866–2887. https://doi.org/10.1021/jf026182t

Rothwell JA, Perez-Jimenez J, Neveu V, Medina-Remon A, M'Hiri N, Garcia-Lobato P, Manach C, Knox C, Eisner R, Wishart DS, Scalbert A (2013) Phenol-Explorer 3.0: a major update of the Phenol-Explorer database to incorporate data on the effects of food processing on polyphenol content. Database (Oxford) 2013:bat070. https://doi.org/10.1093/database/bat070

Rudrapal M, Khairnar SJ, Khan J, Dukhyil AB, Ansari MA, Alomary MN, Alshabrmi FM, Palai S, Deb PK, Devi R (2022) Dietary polyphenols and their role in oxidative stress-induced human diseases: insights into protective effects, antioxidant potentials and mechanism(s) of action. Front Pharmacol 13:806470. https://doi.org/10.3389/fphar.2022.806470

Saad AM, Mohammed DM, Alkafaas SS, Ghosh S, Negm SH, Salem HM, Fahmy MA, Semary HE, Ibrahim EH, AbuQamar SF, El-Tarabily KA, El-Saadony MT (2025) Dietary polyphenols and human health: sources, biological activities, nutritional and immunological aspects, and bioavailability- a comprehensive review. Front Immunol 16:1653378. https://doi.org/10.3389/fimmu.2025.1653378

Scalbert A, Johnson IT, Saltmarsh M (2005a) Polyphenols: antioxidants and beyond. Am J Clin Nutr 81(1 Suppl):215S-217S. https://doi.org/10.1093/ajcn/81.1.215S

Scalbert A, Manach C, Morand C, Remesy C, Jimenez L (2005b) Dietary polyphenols and the prevention of diseases. Crit Rev Food Sci Nutr 45(4):287–306. https://doi.org/10.1080/1040869059096

Scheltens P, Blennow K, Breteler MM, de Strooper B, Frisoni GB, Salloway S, Van der Flier WM (2016) Alzheimer's disease. Lancet 388(10043):505–517. https://doi.org/10.1016/S0140-6736(15)01124-1

Schramm DD, Karim M, Schrader HR, Holt RR, Cardetti M, Keen CL (2003) Honey with high levels of antioxidants can provide protection to healthy human subjects. J Agric Food Chem 51(6):1732–1735. https://doi.org/10.1021/jf025928k

Shelar A, Singh AV, Maharjan RS, Laux P, Luch A, Gemmati D, Tisato V, Singh SP, Santilli MF, Shelar A, Chaskar M, Patil R (2021) Sustainable agriculture through multidisciplinary seed nanopriming: prospects of opportunities and challenges. Cells 10(9). https://doi.org/10.3390/cells10092428

Sheng F, Hu B, Jin Q, Wang J, Wu C, Luo Z (2021) The analysis of phenolic compounds in walnut husk and pellicle by UPLC-Q-Orbitrap HRMS and HPLC. Molecules 26(10). https://doi.org/10.3390/molecules26103013

Siemann EH, Creasy LL (1992) Concentration of the phytoalexin resveratrol in wine. Am J Enol Vitic 43(1):49–52. https://doi.org/10.5344/ajev.1992.43.1.49

Sies H, Stahl W, Sevanian A (2005) Nutritional, dietary and postprandial oxidative stress. J Nutr 135(5):969–972. https://doi.org/10.1093/jn/135.5.969

Singla RK, Bishayee A (2025) Anticancer drug discovery based on natural products: from computational approaches to clinical studies. Cancers (Basel) 17(15). https://doi.org/10.3390/cancers17152507

Sinha K, Chaudhary G, Gupta YK (2002) Protective effect of resveratrol against oxidative stress in middle cerebral artery occlusion model of stroke in rats. Life Sci 71(6):655–665. https://doi.org/10.1016/s0024-3205(02)01691-0

Sivakumar D, Phan ADT, Slabbert RM, Sultanbawa Y, Remize F (2020) Phytochemical and nutritional quality changes during irrigation and postharvest processing of the underutilized vegetable african nightshade. Front Nutr 7:576532. https://doi.org/10.3389/fnut.2020.576532

Skrovankova S, Sumczynski D, Mlcek J, Jurikova T, Sochor J (2015) Bioactive compounds and antioxidant activity in different types of berries. Int J Mol Sci 16(10):24673–24706. https://doi.org/10.3390/ijms161024673

Smeds AI, Eklund PC, Sjoholm RE, Willfor SM, Nishibe S, Deyama T, Holmbom BR (2007) Quantification of a broad spectrum of lignans in cereals, oilseeds, and nuts. J Agric Food Chem 55(4):1337–1346. https://doi.org/10.1021/jf0629134

Song Z, Chen T, Wang S, Shen C, Ma Y, Li A, Chen Z, Li Y (2023) Large-scale preparation of five polar polyphenols including three isomers from Phyllanthus emblica Linn. by preparative high-speed counter-current chromatography. J Sep Sci 46(2):e2200748. https://doi.org/10.1002/jssc.202200748

Stalikas CD (2007) Extraction, separation, and detection methods for phenolic acids and flavonoids. J Sep Sci 30(18):3268–3295. https://doi.org/10.1002/jssc.200700261

Sun MF, Jiang CL, Kong YS, Luo JL, Yin P, Guo GY (2022) Recent advances in analytical methods for determination of polyphenols in tea: a comprehensive review. Foods 11 (10). https://doi.org/10.3390/foods11101425

Tajik N, Tajik M, Mack I, Enck P (2017) The potential effects of chlorogenic acid, the main phenolic components in coffee, on health: a comprehensive review of the literature. Eur J Nutr 56(7):2215–2244. https://doi.org/10.1007/s00394-017-1379-1

Tresserra-Rimbau A, Medina-Remon A, Perez-Jimenez J, Martinez-Gonzalez MA, Covas MI, Corella D, Salas-Salvado J, Gomez-Gracia E, Lapetra J, Aros F, Fiol M, Ros E, Serra-Majem L, Pinto X, Munoz MA, Saez GT, Ruiz-Gutierrez V, Warnberg J, Estruch R, Lamuela-Raventos RM (2013) Dietary intake and major food sources of polyphenols in a Spanish population at high cardiovascular risk: the PREDIMED study. Nutr Metab Cardiovasc Dis 23(10):953–959. https://doi.org/10.1016/j.numecd.2012.10.008

Tresserra-Rimbau A, Rimm EB, Medina-Remon A, Martinez-Gonzalez MA, de la Torre R, Corella D, Salas-Salvado J, Gomez-Gracia E, Lapetra J, Aros F, Fiol M, Ros E, Serra-Majem L, Pinto X, Saez GT, Basora J, Sorli JV, Martinez JA, Vinyoles E, Ruiz-Gutierrez V, Estruch R, Lamuela-Raventos RM, Investigators PS (2014) Inverse association between habitual polyphenol intake and incidence of cardiovascular events in the PREDIMED study. Nutr Metab Cardiovasc Dis 24(6):639–647. https://doi.org/10.1016/j.numecd.2013.12.014

Tsao R (2010) Chemistry and biochemistry of dietary polyphenols. Nutrients 2(12):1231–1246. https://doi.org/10.3390/nu2121231

Urquiaga I, Leighton F (2000) Plant polyphenol antioxidants and oxidative stress. Biol Res 33(2):55–64. https://doi.org/10.4067/s0716-97602000000200004

Vauzour D, Vafeiadou K, Rodriguez-Mateos A, Rendeiro C, Spencer JP (2008) The neuroprotective potential of flavonoids: a multiplicity of effects. Genes Nutr 3(3–4):115–126. https://doi.org/10.1007/s12263-008-0091-4

Vaya J, Mahmood S (2006) Flavonoid content in leaf extracts of the fig (Ficus carica L.), carob (Ceratonia siliqua L.) and pistachio (Pistacia lentiscus L.). Biofactors 28(3–4):169–175. https://doi.org/10.1002/biof.5520280303

Walle T (2011) Bioavailability of resveratrol. Ann N Y Acad Sci 1215:9–15. https://doi.org/10.1111/j.1749-6632.2010.05842.x

Wang X, Ouyang Y, Liu J, Zhu M, Zhao G, Bao W, Hu FB (2014) Fruit and vegetable consumption and mortality from all causes, cardiovascular disease, and cancer: systematic review and dose-response meta-analysis of prospective cohort studies. BMJ 349:g4490. https://doi.org/10.1136/bmj.g4490

Wang DD, Li Y, Bhupathiraju SN, Rosner BA, Sun Q, Giovannucci EL, Rimm EB, Manson JE, Willett WC, Stampfer MJ, Hu FB (2021) Fruit and vegetable intake and mortality: results from 2 prospective cohort studies of US men and women and a meta-analysis of 26 cohort studies. Circulation 143(17):1642–1654. https://doi.org/10.1161/CIRCULATIONAHA.120.048996

Wang L, Lan W, Chen D (2024) Blueberry (Vaccinium spp.) Anthocyanins and their functions, stability, bioavailability, and applications. Foods 13 (17). https://doi.org/10.3390/foods13172851

Westhuyzen J, Mills K, Healy H (2005) Predicting clinical outcomes in peritoneal dialysis patients using small solute modeling. Ann Clin Lab Sci 35(1):46–53

Wierzbicki AS (2008) A fishy business: omega-3 fatty acids and cardiovascular disease. Int J Clin Pract 62(8):1142–1146. https://doi.org/10.1111/j.1742-1241.2008.01781.x

Willett W, Rockstrom J, Loken B, Springmann M, Lang T, Vermeulen S, Garnett T, Tilman D, DeClerck F, Wood A, Jonell M, Clark M, Gordon LJ, Fanzo J, Hawkes C, Zurayk R, Rivera JA, De Vries W, Majele Sibanda L, Afshin A, Chaudhary A, Herrero M, Agustina R, Branca F, Lartey A, Fan S, Crona B, Fox E, Bignet V, Troell M, Lindahl T, Singh S, Cornell SE, Srinath Reddy K, Narain S, Nishtar S, Murray CJL (2019) Food in the anthropocene: the EAT-lancet Commission on healthy diets from sustainable food systems. Lancet 393(10170):447–492. https://doi.org/10.1016/S0140-6736(18)31788-4

Winkel-Shirley B (2002) Biosynthesis of flavonoids and effects of stress. Curr Opin Plant Biol 5(3):218–223. https://doi.org/10.1016/s1369-5266(02)00256-x

Yang D, Wang T, Long M, Li P (2020) Quercetin: its main pharmacological activity and potential application in clinical medicine. Oxid Med Cell Longev 2020:8825387. https://doi.org/10.1155/2020/8825387

Yates AA, Schlicker SA, Suitor CW (1998) Dietary Reference Intakes: the new basis for recommendations for calcium and related nutrients, B vitamins, and choline. J Am Diet Assoc 98(6):699–706. https://doi.org/10.1016/S0002-8223(98)00160-6

Zamora-Ros R, Knaze V, Rothwell JA, Hemon B, Moskal A, Overvad K, Tjonneland A, Kyro C, Fagherazzi G, Boutron-Ruault MC, Touillaud M, Katzke V, Kuhn T, Boeing H, Forster J, Trichopoulou A, Valanou E, Peppa E, Palli D, Agnoli C, Ricceri F, Tumino R, de Magistris MS, Peeters PH, Bueno-de-Mesquita HB, Engeset D, Skeie G, Hjartaker A, Menendez V, Agudo A, Molina-Montes E, Huerta JM, Barricarte A, Amiano P, Sonestedt E, Nilsson LM, Landberg R, Key TJ, Khaw KT, Wareham NJ, Lu Y, Slimani N, Romieu I, Riboli E, Scalbert A (2016) Dietary polyphenol intake in Europe: the European Prospective Investigation into Cancer and Nutrition (EPIC) study. Eur J Nutr 55(4):1359–1375. https://doi.org/10.1007/s00394-015-0950-x

Zhu L, Liu X, Tan J, Wang B (2013) Influence of harvest season on antioxidant activity and constituents of rabbiteye blueberry (Vaccinium ashei) leaves. J Agric Food Chem 61(47):11477–11483. https://doi.org/10.1021/jf4035892

# Chapter 2
# Food Sources and Action Mechanisms of Polyphenols

**Tapan Behl, Salvatore Parisi, and Sony George**

**Abstract** Polyphenols are a group of naturally evolved compounds which are found plenty in fruits, vegetables, and in many different medicinal plants. The remarkable presence of hydroxyl groups in the chemical structure can be easily considered as one of the main factors which contribute to the high reactivity and its biological significance. Should polyphenols-rich foods and derived products be easily available, the beneficial nature of phenolics would be demonstrated. Polyphenols are highly recommendable due to their ability to prevent oxidative damage, reduce inflammation, and their ability to alter pathways implicated in many chronic diseases. The mechanism of action includes direct antioxidant activity, modulation of cell signaling cascades, regulating gene expression, mitochondrial protection, apoptosis regulation, and microbiota interactions. This comprehensive application highlights the importance of polyphenols in both dietary and therapeutic contexts. Consequently, the usefulness of phenolics as therapeutic coadjuvant depends also on the availability of rich food sources. The consumption of selected food categories can give different beneficial effect on human health. Naturally, such a reflection is based on different action mechanisms. Consequently, a comparison between these mechanisms and their occurrence depending on one of another food source might be tried. However, the problem of health effects should be studied case by case. Each study concerning phenolics and health tends to be targeted within strict boundaries and taking into account selected targets, mechanisms of action, and (naturally) human subjects.

---

T. Behl (✉)
Amity Institute of Pharmacy, Amity University Mohali, Sahibzada Ajit Singh Nagar, India
e-mail: tapanbehl31@gmail.com

S. Parisi
Department of Nutrition and Food Processing Technology, Lourdes Matha Institute of Hotel Management and Catering Technology, Thiruvananthapuram, Kerala, India
e-mail: drparisi@inwind.it

S. George
Department of Food Technology, TKM Institute of Technology, Karuvelil, Kollam, Kerala, India
e-mail: sojen143@gmail.com

 
R. K. Singla et al. (eds.), *Polyphenols in Health*,
https://doi.org/10.1007/978-3-032-27324-6_2

**Keywords** Antioxidant activity · Apoptosis regulation · Cell signaling cascade · Food source · Microbiota · Mitochondrial protection · Polyphenol · Regulating gene expression

## Abbreviations

| | |
|---|---|
| Bcl-2 | B-cell lymphoma 2 |
| Bax | Bcl-2-associated X protein |
| COX-2 | Cyclooxygenase-2 |
| DNA | Deoxyribonucleic acid |
| ECG | Epicatechin gallate |
| EC | Epicatechin |
| EGCG | Epigallocatechin gallate |
| EGC | Epigallocatechin |
| SIRT1 | Sirtuin −1 |

## 2.1 Polyphenols as Antioxidant Agents. Natural and Commercial Availability

Polyphenols are a group of naturally evolved compounds which are found plenty in fruits, vegetables, and in many different medicinal plants. The remarkable presence of hydroxyl groups in the chemical structure (Chap. 6 and 12) can be easily considered as one of the main factors which contribute to the high reactivity and its biological significance (Rentzsch et al. 2009; Zhang et al. 2021). This statement can also explain the notable interest of pharmaceutical industries when speaking of good and beneficial effects against different human diseases and disorders, provided that the bioavailability of these compounds in the human diet is assured (Mechi et al. 2023; Neilson et al. 2017; Park 2014; Tihana et al. 2019). Should polyphenols-rich foods and derived products be easily available, the beneficial nature of phenolics would be demonstrated (Arshad et al. 2025; Domingues et al. 2024; Giglio et al. 2018; Hafiz 2025; Kardum et al. 2015; Kupczyk et al. 2025; Issaoui et al. 2020; Liva et al. 2025; Makarewicz et al. 2021; Martiniakova et al. 2025; Mechi et al. 2023; Mendes et al. 2025; Mileo and Miccadei 2016; Parisi 2019–2020; Rudrapal et al. 2022; Stoner and Mukhtar 1995; Tillman et al. 2025; Wang and Hu 2023).

On a general level, it can be assured that polyphenols are highly recommendable due to their ability to prevent oxidative damage, reduce inflammation, and their ability to alter pathways implicated in many chronic diseases. The mechanism of action includes direct antioxidant activity, modulation of cell signaling cascades, regulating gene expression, mitochondrial protection, apoptosis regulation, and microbiota interactions (Mishra et al. 2013; Nani et al. 2021; Rudrapal et al. 2022). This comprehensive application highlights the importance of polyphenols in both dietary and therapeutic contexts.

Consequently, the usefulness of phenolics as therapeutic coadjuvant is based on several factors. Two of these important pre-conditions are certainly the natural bioavailability, implying also the possibility of easy commercial availability in the modern industry of foods and beverages, and also the nature of antioxidant and other mechanisms which are often correlated with beneficial effects for human health. The first one problem is the localization of phenolics in foods.

## 2.2 Food Sources of Polyphenols in Major Foods

Polyphenols are a group of secondary metabolites characterized by the presence of different phenolic groups. The four major classes of polyphenols which dominate the food domain includes flavonoids (e.g., flavan-3-ols/catechins, anthocyanins, flavonols, flavones, flavanones, isoflavones), phenolic acids (hydroxycinnamic and hydroxybenzoic acids), stilbenes (notably resveratrol), and lignans. The substantial contributors to the total daily intake human include non-alcoholic beverages (tea, coffee, cocoa), fruits (berries, citrus, apples, and grapes), vegetables (onion, leafy brassicas), legumes, nuts, seeds, herbs, and wine as an essential contributor in Mediterranean Diet (Manach et al. 2004). The exact polyphenol profile depends on so many factors which include growing conditions, processing, and the different preparation methods (Rothwell et al. 2013).

### *2.2.1 Food Sources of Polyphenols. Beverages*

Tea (green, oolong, and black types) are a major source of polyphenols (Ganugapati et al. 2011; Karabulut 2025; Kyriacou et al. 2025). Tea leaves are rich in flavan-3-ols and green tea infusions can deliver a large amount of monomeric catechins such as epigallocatechin gallate (EGCG), epigallocatechin (EGC), epicatechin gallate (ECG), and epicatechin (EC). Black tea, which is the product of enzymatic oxidation, contains theaflavins and thearubigins. Polyphenols yield is significantly influenced by some variables, including the brew strength, leaf grade, and steep time (Bhagwat et al. 2014). Roasted coffee is a leading dietary source of chlorogenic acids (esters of caffeic and quinic acids), including 5-caffeoylquinic acid and dicaffeoylquinic acids. Coffee, reported to have potential antidiabetic effects (Hafiz 2025), is a leading source of chlorogenic acids including 5-caffeoylquinic acid and related dicaffeoylquinic acids. Roasting may lead to reduction of these acids, also generating melanoidins: these compounds help in the binding of phenolics and hence contribute to the antioxidant capacity (Gu et al. 2004). Cocoa powder and high-cocoa-content chocolates provide flavan-3-ols including catechin, EC, and oligomeric proanthocyanidins. Non-alkalized cocoa retains more native flavanols than alkalised powders

(Williamson 2017). Citrus fruits and juices supply flavanones such as hesperidin and narirutin, while wine contains a mixture of flavan-3-ols, anthocyanins, phenolic acids, and small amounts of stilbenes such as resveratrol. Related concentrations can vary depending on grape variety, climate, and winemaking) (Zamora-Ros et al. 2014).

### 2.2.2 *Food Sources of Polyphenols. Fruits*

Berries include blueberries, blackberries, raspberries, strawberries, and cranberries: all these foods are notable sources for anthocyanins, ellagitannins, and proanthocyanidins. Apples and pears contribute flavan-3-ols such as catechin and EC, hydroxycinnamic acids, and dihydrochalcones such as phloridzin. Grapes are rich in anthocyanins (mainly in skins), while white and green grapes contain flavan-3-ols and phenolic acids (Rothwell et al. 2013).

Citrus (such as oranges, mandarins, and lemon) are rich in flavanones, especially hesperidin, narirutin and naringin, while cherries, plums, and peaches (stone fruits) are good source of cyanidin-3-rutinoside and hydroxycinnamic acids. In addition, plums and peaches offer high amounts of phenolic acids and flavanols with high concentration near the skin (Mulvihill et al. 2016).

### 2.2.3 *Food Sources of Polyphenols. Vegetables*

Yellow and red shallots are the proven source of flavonols, especially quercetin and its glycosides (the main concentration is found outer layers). The majority of leafy vegetables (kale, broccoli, cabbage, and Brussels sprouts) are the maindietary source of quercetin, kaempferol, hydroxycinnamic acids, and hydrobenzoic acids (Manach et al. 2004). Similarly, eggplant skins are a rich source of anthocyanins (such as nausunin). Artichokes are rich in derivatives of chlorogenic acids, while spinach and asparagus contribute trace amounts of flavonoids and phenolic acids (Tsao 2010).

### 2.2.4 *Food Sources of Polyphenols.Legumes, Whole Grains, and Pseudocereals*

Different legumes (soy, lentils, chickpeas, and common beans) are a rich source of polyphenols (Hernández-Ruiz et al. 2025). Soy is rich in isoflavones includeng genistein, daidzein, and glycitein; fermentation may give a more abundant amount of bioavailable aglycones. Beans and lentils contain flavonols, phenolic acids, and proanthocyanidins, especially concentrated in pigmented seed coats.

Whole grains (wheat, oats, rice, barley, etc.) containremarkable amounts of phenolic acids—mainly ferulic acid—which are mainly bound to arabinoxylans in the bran. Oats are a notable source of avenanthramides. Millets are also a rich source of polyphenols. Sorghum varieties with dark pericarps contain remarkable amounts of condensed proanthocyanidins (class of tannins). Buckwheat is a rich source of rutin (quercetin-3-rutinoside) (Neveu et al.2010).

### *2.2.5 Food Sources of Polyphenols.Nuts, Seeds, and Plant Oils*

Major nuts, especially walnuts, supply a rich amount of flavan-3-ols, proanthocyanidins, phenolic acids, and ellagitannins. The most abundant amount of phenolics are concentrated near the skins (Williamson 2017). Flaxseed is renowned for lignans such as secoisolariciresinol diglucoside, which is turned into enterolignans because of gut microbiota action. Sesame contains lignans (sesamin, sesamolin, etc.) alongside minor phenolic acids. Olive oil, especially extra virgin olive oil, containS simple phenols (hydroxytyrosol, tyrosol, etc.), secoiridoids (including oleuropein and ligstroside aglycones), and lignans (Bucciantini et al. 2021; Liva et al. 2025; Mechi et al. 2023; Williamson 2017).

### *2.2.6 Food Sources of Polyphenols. Herbs and Spices*

Cloves, oregano, thyme, rosemary, sage, turmeric, and cinnamon are among the most polyphenols-rich foods and providing rosmarinic acids, caffeic acid, flavones, and curcuminoids (Tsao 2010). Other important essentials which contribute phenolic compounds include tomato (especially with reference to phenolic acids and flavonols). Pomegranate molasses and black garlic also make a better contribution of phenolic compounds through the daily intake of food.

### *2.2.7 Food Sources of Polyphenols. Fermented and Processed Foods*

Fermentation and similar microbial activity on food substrates can increase the bioavailability of flavonoids (e.g., isoflavone aglycones in fermented soy), and also forming different smaller phenolics due to tannin hydrolysis as in fermented teas and cocoa (Rothwell et al. 2013). Wine and ciderealso retain grape or apple phenolics (Zamora-Ros et al. 2014).

## 2.3 Polyphenols and Mechanisms of Action

Because of the correlation between the abundance of some peculiar phenolics category in certain foods instead of other ones, it may be inferred that the consumption of selected food categories can give different beneficial effect on human health. Naturally, such a reflection is based on different action mechanisms. Consequently, a comparison between these mechanisms and their occurrence depening on one of another food source might be tried. Before that this correlation is discussed, a brief explanation of these mechanisms has to be supplied.

### *2.3.1 Polyphenols and Mechanisms of Action. Antioxidant Power*

One of the most acknowledged mechanisms of polyphenols is their antioxidant activity which is mainly due to its ability to donate electrons and hydrogen atoms to neutralize free radicals such as reactive oxygen species (ROS) and reactive nitrogen species (RNS). This attitude helps when speaking of prevention of oxidative damages to lipids, proteins, and deoxyribonucleic acid (DNA) (Hussain et al. 2016; Rice-Evans et al. 1996). The rigid molecular structure, characterized by the presence of hydroxyl groups and the conjugated double bonds, helps them to stabilise the different free radicals.

Polyphenols can chelate pro-oxidant metal ions including iron and copper, and thus inhibiting Fenton-type reactions responsible for hydroxyl ions generation (Bulotta et al. 2013; Carluccio et al. 2007; Cicerale et al. 2012; Gavahian et al. 2019; Marcelino et al. 2019; Pietta 2000; Tihana et al. 2019). In addition to specified chemical interactions, they also influence antioxidant defence by triggering transcription factors such as erythroid 2-related factor 2. It regulates different antioxidant enzymes, including glutathione peroxidase,catalase, and superoxide dismutase which act against oxidative stress in the body (Pandey and Rizvi 2009).

### *2.3.2 Polyphenols and Mechanisms of Action. Modulation of Cell Signaling Pathways*

Polyphenols have significant influence over different cell signaling networks that regulates proliferation and inflammation in the body. They are well known modulators of mitogen-activated protein kinase pathways which thereby regulates or prevent cell growth, differentiation, and apoptosis (Kundu and Surh 2008).

The critical example for the activity of polyphenols involves the inhibition of nuclear factor kappa B by regulating the inflammation and immune responses. Phenolics reduce the expression of pro-inflammatory cytokines and different adhesion molecules, resulting in the mitigation of chronic inflammatory conditions (Baur and Sinclair 2006). Polyphenols, namely resveratrol, usually activate sirtuin −1 (SIRT1) which is connected to metabolic regulation and cellular stress resistance, thus helping in metabolic homeostasis and age-related diseases (Baur and Sinclair 2006).

### *2.3.3 Polyphenols and Mechanisms of Action. Anti-inflammatory Effects*

Chronic inflammation is an important phenomenon in the ambit of the pathogenesis of cardiovascular diseases, diabetes, tumours, and neurogenerative disorders. Phenolics reduce the expression of cyclooxygenase-2 (COX-2) and inducible nitric oxide synthase, enzymes that catalyse the production of pro-inflammatory mediators (Boots et al. 2008), so resulting in a reduction of prostaglandin and nitric oxide synthesis, and also alleviating inflammatory responses at the cellular level. Similarly, flavonoids such as quercetin help in downregulating cytokines including tumour necrosis factors, includingtumour necrosis factor-alpha and interleukin-6, thereby dampening inflammatory signaling (Singh et al. 2016).

### *2.3.4 Polyphenols and Mechanisms of Action. Mitochondrial Protection and Apoptosis Regulation*

Mitochondria, the said powerhouse of the cell, act as both generators and targets of oxidative stress. Involved actions include mitochondria protection by regulating the membrane potential, the reduction of excessive ROS production, and also the improvement of oxidative phosphorylation efficiency.

Apoptosis, a type of planned cell death, is carried out by caspases and controlled by members of the B-cell lymphoma 2 (Bcl-2) family.

Bcl-2 genetic family. This phenomenon is characterized by the morphological changes such as membrane blebbing, cell shrinkage, nuclear fragmentation, and chromatin condensation (Burz et al. 2009). Polyphenols exert a dual role in apoptosis regulation by preventing premature apoptosis in healthy cells by inhibiting oxidative damage and promoting programmed cell death in cancerous cells through upregulation of pro-apoptotic proteins like Bcl-2-associated X protein (Bax) and downregulation of anti-apoptotic proteins such as Bcl-2 (Scalbert et al. 2005). This targeted mechanism contributes to its anticancerous mechanism.

It has been observed that a variety of dietary substances, such as flavonoids, tea polyphenols, resveratrol, and curcumin, can cause apoptosis in a variety of tumour types (Kim et al. 2005). A flavone-3-ol-class flavonoid that is widely found in fruits and vegetables, quercetin (3,3′,4′,5,7-pentahydroxyflavone) has been shown to alter signal transduction pathways linked to apoptosis and cell division.

### 2.3.5 Polyphenols and Mechanisms of Action. Antimicrobial Actions and Gut Microbiota Modulation

Phenolics are well known for their antimicrobial properties: disruption of bacterial cell membranes; action with virulence factor expression; inhibition of quorum sensing; and finally, reduction of bacteria pathogenicity (Daglia 2012).

Therefore,the colonic microbiota is responsible for the extensive breakdown of the original polyphenolic structures into a series of absorbable low-molecular-weight phenolic metabolites. Consequently, they can act as responsible players for health effects derived from consuming foods rich in polyphenols. Phenolics are used to shape microbial composition by promoting beneficial bacterial such as *Lactobacillus* and *Bifidobacterium*spp, while inhibiting pathogenic species. This reciprocal interaction contributes to improved gut health, immune modulation, and metabolic regulation (Cardona et al. 2013).

The gut microbiota plays a key role in modulating the production, bioavailability and, thus, the biological activities of phenolic metabolites, particularly after the intake of food containing high-molecular-weight polyphenols. In addition, evidence is emerging on the activity of dietary polyphenols on the modulation of the colonic microbial population composition or activity.

## 2.4 Conclusions. Is it Possible to Correlate Foods with Health Effects?

Beyond their antioxidant effects, polyphenols have other mechanisms. Together, they affect apoptosis, protect mitochondria, control inflammation, alter gut microbiome, and modify important signaling pathways, all of which lead to better health outcomes. Their wide range of biological actions explains why they can help avoid chronic conditions like diabetes, cancer, heart disease, and neurological disorders. Further investigation is anticipated to uncover other molecular targets and expand their therapeutic uses, establishing polyphenols as crucial dietary and pharmacological constituents. On the other side, some preliminary correlation between original food sources and claimed health effects may be tried by means of the knowledge of reported action mechanisms as a linking ring. However, because of the role of external factors such as gut microbiota action and also of the multiplicity of interconnected and synergical mechanisms at the same time, such a correlation could be blurred enough and with indefinite boundaries. For this reason, the problem of health effects should be studied case by case, as described in the subsequent chapters of this book, taking into account that the consumption of vegetables, fruits, and derived products is generally favourable when speaking of human health. Many factors should be taken into account at the same time: for this reason, each study concerning phenolics and health tends to be targeted within strict boundaries and taking into account selected targets, mechanisms of action, and (naturally) human subjects.

## References

Arshad MT, Maqsood S, Ikram A, Abdullahi MA (2025) Synergistic effects of carrot (*Daucus carota* L.) root and papaya (*Carica papaya* L.) extracts in combating inflammation and enhancing immune functions—a new perspective. Food Sci Nutr 13(6):e70313. https://doi.org/10.1002/fsn3.70313

Baur JA, Sinclair DA (2006) Therapeutic potential of resveratrol: the in vivo evidence. Nat Rev Drug Discov 5(6):493–506. https://doi.org/10.1038/nrd2060

Bhagwat S, Haytowitz DB, Holden JM (2014) USDA database for the flavonoid content of selected foods, release 3.1. U.S. Department of Agriculture, Agricultural Research Service, Washington, D.C. https://www.ars.usda.gov/ARSUserFiles/80400535/Data/Flav/Flav_R03-1.pdf. Accessed 13 Oct 2025

Boots AW, Haenen GRMM, Bast A (2008) Health effects of quercetin: from antioxidant to nutraceutical. Eur J Pharmacol 585(2–3):325–337. https://doi.org/10.1016/j.ejphar.2008.03.008

Bucciantini M, Leri M, Nardiello P, Casamenti F, Stefani M (2021) Olive polyphenols: antioxidant and anti-inflammatory properties. Antioxidants 10(7):1044. https://doi.org/10.3390/antiox10071044

Bulotta S, Oliverio M, Russo D, Procopio A (2013) Biological activity of oleuropein and its derivatives. In: Ramawat KG, Mérillon JM (eds) Natural products. Springer, Berlin/Heidelberg, Germany, pp 3605–3638

Burz C, Berindan-Neagoe I, Balacescu O, Irimie A (2009) Apoptosis in cancer: key molecular signaling pathways and therapy targets. Acta Oncol 48(6):811–821. https://doi.org/10.1080/02841860902974175

Cardona F, Andrés-Lacueva C, Tulipani S, Tinahones FJ, Queipo-Ortuño MI (2013) Benefits of polyphenols on gut microbiota and implications in human health. J Nutr Biochem 24(8):1415–1422. https://doi.org/10.1016/j.jnutbio.2013.05.001

Carluccio MA, Massaro M, Scoditti E, de Caterina R (2007) Vasculoprotective potential of olive oil components. Mol Nutr Food Res 51(10):1225–1234. https://doi.org/10.1002/mnfr.200600305

Cicerale S, Lucas LJ, Keast RS (2012) Antimicrobial, antioxidant and anti-inflammatory phenolic activities in extra virgin olive oil. Curr Opin Biotechnol 23(2):129–135. https://doi.org/10.1016/j.copbio.2011.09.006

Daglia M (2012) Polyphenols as antimicrobial agents. Curr Opin Biotechnol 23(2):174–181. https://doi.org/10.1016/j.copbio.2011.08.007

Domingues YO, Lemes GA, de Oliveira FL, de Souza TR, Silva B, Bento JAC, Morzelle MC (2024) In vitro simulated gastrointestinal digestion and bioaccessibility of phenolic compounds and antioxidants of soursop (*Annona muricata* L.) peel and pulp. Lwt 208:116694. https://doi.org/10.1016/j.lwt.2024.116694

Ganugapati J, Mukkavalli S, Sahithi A (2011) Docking studies of green tea flavonoids as insulin mimetics. Int J Comp App 30(4):48–52

Gavahian M, Khaneghah AM, Lorenzo JM, Munekata PES, Garcia-Mantrana I, Collado MC, Meléndez-Martínez AJ, Barba FJ (2019) Health benefits of olive oil and its components: impacts on gut microbiota antioxidant activities, and prevention of noncommunicable diseases. Trends Food Sci Technol 88:220–227. https://doi.org/10.1016/j.tifs.2019.03.008

Giglio RV, Patti AM, Cicero AF, Lippi G, Rizzo M, Toth PP, Banach M (2018) Polyphenols: potential use in the prevention and treatment of cardiovascular diseases. Curr Pharm Des 24(2):239–258. https://doi.org/10.2174/1381612824666180130112652

Gu L, Kelm MA, Hammerstone JF, Beecher G, Holden J, Haytowitz D, Prior RL (2004) Concentrations of proanthocyanidins in common foods and estimations of normal consumption. J Nutr 134(3):613–617. https://doi.org/10.1093/jn/134.3.613

Hafiz MS (2025) Anti-diabetic potentials of coffee polyphenols: a narrative review. Curr Nutr Food Sci 21(2):202–211. https://doi.org/10.2174/0115734013292395240219091340

Hernández-Ruiz RG, Olivares-Ochoa XC, Salinas-Varela Y, Guajardo-Espinoza D, Roldán-Flores LG, Rivera-Leon EA, López-Quintero A (2025) Phenolic compounds and anthocyanins in legumes and their impact on inflammation, oxidative stress, and metabolism: comprehensive review. Molecules 30(1):174. https://doi.org/10.3390/molecules30010174

Hussain T, Tan B, Yin Y, Blachier F, Tossou MC, Rahu N (2016) Oxidative stress and inflammation: what polyphenols can do for us? Oxidative Med Cell Longev 2016(1):7432797. https://doi.org/10.1155/2016/7432797

Issaoui M, Delgado AM, Iommi C, Chammem N (2020) Polyphenols and the mediterranean diet. Springer International Publishing, Cham

Karabulut G (2025) Tea polyphenols: their role in human health and drug interactions. Dissertation, University of Padova, Italy

Kardum N, Milovanović B, Šavikin K, Zdunić G, Mutavdžin S, Gligorijević T, Spasić S (2015) Beneficial effects of polyphenol-rich chokeberry juice consumption on blood pressure level and lipid status in hypertensive subjects. J Med Food 18(11):1231–1238. https://doi.org/10.1089/jmf.2014.0171

Kim WK, Bang MH, Kim ES, Kang NE, Jung KC, Cho HJ, Park JH (2005) Quercetin decreases the expression of ErbB2 and ErbB3 proteins in HT-29 human colon cancer cells. J Nutr Biochem 16(3):155–162. https://doi.org/10.1016/j.jnutbio.2004.10.010

Kundu JK, Surh YJ (2008) Cancer chemopreventive and therapeutic potential of resveratrol. Cancer Lett 269(2):243–261. https://doi.org/10.1016/j.canlet.2008.03.057

Kupczyk D, Bilski R, Szeleszczuk Ł, Mądra-Gackowska K, Studzińska R (2025) The role of diet in modulating inflammation and oxidative stress in rheumatoid arthritis, ankylosing spondylitis, and psoriatic arthritis. Nutrients 17(9):1603. https://doi.org/10.3390/nu17091603

Kyriacou NM, Gross AS, McLachlan AJ (2025) Green tea Catechins as perpetrators of drug pharmacokinetic interactions. Clin Pharmacol Ther 118(1):45–61. https://doi.org/10.1002/cpt.3547

Liva K, Panagiotopoulos AA, Foscolou A, Amerikanou C, Vitali A, Zioulis S, Argyri K, Panoutsopoulos GI, Kaliora AC, Gioxari A (2025) High polyphenol extra virgin olive oil and metabolically unhealthy obesity: a scoping review of preclinical data and clinical trials. Clin Pract 15(3):54. https://doi.org/10.3390/clinpract15030054

Makarewicz M, Drożdż I, Tarko T, Duda-Chodak A (2021) The interactions between polyphenols and microorganisms, especially gut microbiota. Antioxidants 10(2):188. https://doi.org/10.3390/antiox10020188

Manach C, Scalbert A, Morand C, Rémésy C, Jiménez L (2004) Polyphenols: food sources and bioavailability. Am J Clin Nutr 79(5):727–747. https://doi.org/10.1093/ajcn/79.5.727

Marcelino G, Hiane PA, Freitas KC, Santana LF, Pott A, Donadon JR, RCA G (2019) Effects of olive oil and its minor components on cardiovascular diseases, inflammation, and gut microbiota. Nutrients 11(8):1826. https://doi.org/10.3390/nu11081826

Martiniakova M, Sarocka A, Penzes N, Biro R, Kovacova V, Mondockova V, Sevcikova A, Ciernikova S, Omelka R (2025) Protective role of dietary polyphenols in the management and treatment of type 2 diabetes mellitus. Nutrients 17(2):275. https://doi.org/10.3390/nu17020275

Mechi D, Baccouri B, Martín-Vertedor D, Abaza L (2023) Bioavailability of phenolic compounds in Californian-style table olives with Tunisian aqueous olive leaf extracts. Molecules 28(2):707. https://doi.org/10.3390/molecules28020707

Mendes I, Ribeiro MGC, de Souza LF, Rosa CDOB, Hermsdorff HHM, Bressan J (2025) Effect of polyphenol supplementation on adiposity: a systematic review of randomized clinical trials. Curr Nutr Rep 14(1):36. https://doi.org/10.1007/s13668-025-00626-3

Mileo AM, Miccadei S (2016) Polyphenols as modulator of oxidative stress in cancer disease: new therapeutic strategies. Oxidative Med Cell Longev 2016(1):6475624. https://doi.org/10.1155/2016/6475624

Mishra A, Sharma AK, Kumar S, Saxena AK, Pandey AK (2013) *Bauhinia Variegata* leaf extracts exhibit considerable antibacterial, antioxidant, and anticancer activities. Biomed Res Int 2013(1):915436. https://doi.org/10.1155/2013/915436

Mulvihill E, Burke A, Huff M (2016) Citrus flavonoids as regulators of lipoprotein metabolism and atherosclerosis. Ann Rev Nutr 36:275–299. https://doi.org/10.1146/annurev-nutr-071715-050718

Nani A, Murtaza B, Sayed Khan A, Khan NA, Hichami A (2021) Antioxidant and anti-inflammatory potential of polyphenols contained in mediterranean diet in obesity: molecular mechanisms. Molecules 26(4):985. https://doi.org/10.3390/molecules26040985

Neilson AP, Goodrich KM, Ferruzzi MG (2017) Bioavailability and metabolism of bioactive compounds from foods. In: Coulston AM, Boushey CJ, Ferruzzi MG, Delahanty LM (eds) Nutrition in the prevention and treatment of disease. Academic Press, Cambridge, MA, pp 301–319. https://doi.org/10.1016/B978-0-12-802928-2.00015-1

Neveu V, Pérez-Jiménez J, Vos F, Crespy V, du Chaffaut L, Mennen L, Knox C, Eisner R, Cruz J, Wishart D, Scalbert A (2010) Phenol-explorer: an online comprehensive database on polyphenol contents in foods. Database 2010:bap024. doi:https://doi.org/10.1093/database/bap02438

Pandey KB, Rizvi SI (2009) Plant polyphenols as dietary antioxidants in human health and disease. Oxidative Med Cell Longev 2(5):270–278. https://doi.org/10.4161/oxim.2.5.9498

Parisi S (2019) Analysis of major phenolic compounds in foods and their health effects. J AOAC Int 102(5):1354–1355. https://doi.org/10.5740/jaoacint.19-0127

Parisi S (2020) Characterization of major phenolic compounds in selected foods by the technological and health promotion viewpoints. J AOAC Int 103(4):904–905. https://doi.org/10.1093/jaoacint/qsaa011

Park J (2014) Flaxseed secoisolariciresinol diglucoside and visceral obesity, a closer look at its chemical properties, absorption, metabolism, bioavailability, and effects on visceral fat, lipid profile, systemic inflammation, and hypertension. In: Watson RR (ed) Nutrition in the prevention and treatment of abdominal obesity book, 1st edn. Academic Press, San Diego, CA, USA, pp 317–328

Pietta PG (2000) Flavonoids as antioxidants. J Nat Prod 63(7):1035–1042. https://doi.org/10.1021/np9904509

Rentzsch M, Wilkens A, Winterhalter P (2009) Non-flavonoid phenolic compounds. In: Moreno Arribas MV, Polo MC (eds) Wine chemistry and biochemistry. Springer, New York, NY, USA, pp 509–527

Rice-Evans CA, Miller NJ, Paganga G (1996) Structure-antioxidant activity relationships of flavonoids and phenolic acids. Free Radic Biol Med 20(7):933–956. https://doi.org/10.1016/0891-5849(95)02227-9

Rothwell JA, Pérez-Jiménez J, Neveu V, Medina-Remón A, M'hiri N, García-Lobato P, Manach C, Knox C, Eisner R, Wishart DS, Scalbert A (2013) Phenol-explorer 3.0: a major update of the phenol-explorer database to incorporate data on the effects of processing on polyphenol content in foods. Database 2013:bat070.doi:https://doi.org/10.1093/database/bat070

Rudrapal M, Khairnar SJ, Khan J, Dukhyil AB, Ansari MA, Alomary MN, Alshabrmi FM, Palai S, Deb PK, Devi R (2022) Dietary polyphenols and their role in oxidative stress-induced human diseases: insights into protective effects, antioxidant potentials and mechanism (s) of action. Front Pharmacol 13:806470. https://doi.org/10.3389/fphar.2022.806470

Scalbert A, Manach C, Morand C, Rémésy C, Jiménez L (2005) Dietary polyphenols and the prevention of diseases. Crit Rev Food Sci Nutr 45(4):287–306. https://doi.org/10.1080/1040869059096

Singh G, Passsari AK, Leo VV, Mishra VK, Subbarayan S, Singh BP, Kumar B, Kumar S, Gupta VK, Lalhlenmawia H, Nachimuthu SK (2016) Evaluation of phenolic content variability along with antioxidant, antimicrobial, and cytotoxic potential of selected traditional medicinal plants from India. Front Plant Sci 7:407. https://doi.org/10.3389/fpls.2016.00407

Stoner GD, Mukhtar H (1995) Polyphenols as cancer chemopreventive agents. J Cell Biochem 59(S22):169–180. https://doi.org/10.1002/jcb.240590822

Tihana Ž, Abdelkebir R, Alcantara C, Carmen M, García-Pérez JV, Meléndez-Martínez AJ, Re A, Lorenzo JM, Barba FJ (2019) From extraction of valuable compounds to health promoting

benefits of olive leaves through bioaccessibility, bioavailability and impact on gut microbiota. Trends Food Sci Technol 83:63–77. https://doi.org/10.1016/j.tifs.2018.11.005
Tillman L, Rieres JM, Ahjem E, Bishop-Guest F, McGrath M, Hatrick H, Pranjol MZI (2025) Thinking outside the therapeutic box: the potential of polyphenols in preventing chemotherapy-induced endothelial dysfunction. Cells 14(8):566. https://doi.org/10.3390/cells14080566
Tsao R (2010) Chemistry and biochemistry of dietary polyphenols. Nutrients 2(12):1231–1246. https://doi.org/10.3390/nu2121231
Wang K, Hu S (2023) The synergistic effects of polyphenols and intestinal microbiota on osteoporosis. Front Immunol 14:1285621. https://doi.org/10.3389/fimmu.2023.1285621
Williamson G (2017) The role of polyphenols in modern nutrition. Nutr Bull 42(3):226–235. https://doi.org/10.1111/nbu.12278
Zamora-Ros, Knaze V, Rothwell JA, Hémon B, Moskal A, Overvad K, Tjønneland A, Kyrø C, Fagherazzi G, Boutron-Ruault MC, Touillaud M, Katzke V, Kühn T, Boeing H, Förster J, Trichopoulou A, Valanou E, Peppa E, Palli D, Agnoli C, Ricceri F, Tumino R, Santucci de Magistris M, Peeters PHM, Bueno-de-Mesquita HB, Engeset D, Skeie G, Hjartåker A, Menendez V, Agudo A, Molina-Montes E, Huerta JM, Barricarte A, Amiano P, Sonestedt E, Nilsson LM, Landberg R, Key TJ, Khaw KT, Wareham NJ, Lu Y, Slimani N, Romieu I, Riboli E, Scalbert A (2014) Dietary polyphenol intake in Europe: the European prospective investigation into cancer and nutrition (EPIC) study. Eur J Nutr 55(4):1359–1375.doi:https://doi.org/10.1007/s00394-015-0950-x
Zhang S, Xu M, Zhang W, Liu C, Chen S (2021) Natural polyphenols in metabolic syndrome: protective mechanisms and clinical applications. Int J Mol Sci 22(11):6110. https://doi.org/10.3390/ijms22116110

# Chapter 3
# Polyphenols in Cancer Prevention: Mechanisms of Action Against Carcinogenesis and Epidemiological Evidence of Risk Reduction

**Vivek P. Chavda, Joyeeta T. Khan, Modi Krishna, Kotadiya Dhruvil, Panara Gunja, and Amit K. Tiwari**

**Abstract** This chapter provides an extensive review of polyphenols and their preventive role against cancer, highlighting both mechanistic and epidemiological perspectives. It starts with an overview of polyphenol types, their natural dietary origins, and the worldwide burden of cancer, emphasizing the importance of effective prevention strategies. The discussion covers polyphenol structures, metabolism, and bioavailability, including the critical influence of gut microbiota on their anticancer activity. Key cancer inhibitory mechanisms are explored, such as antioxidant and anti-inflammatory effects, regulation of cellular detoxification and apoptosis, epigenetic modifications, and suppression of tumor angiogenesis and metastasis. The chapter integrates evidence from laboratory studies, population-based research, and clinical trials examining polyphenol impact on various cancer types. Furthermore, it addresses synergistic and antagonistic interactions with established cancer treatments, challenges in clinical application, and advances in delivery technologies. The goal is to inform researchers and healthcare professionals about optimizing polyphenol use for cancer prevention and therapeutic advancements.

**Keywords** Anticancer mechanism · Antimetastatic effect · Antioxidant property · Cancer cell inhibition · Epigenetic modulation · Polyphenol · Pro-apoptotic activity

V. P. Chavda (✉) · M. Krishna · K. Dhruvil · P. Gunja
Department of Pharmaceutics, L. M. College of Pharmacy, Ahmedabad, Gujarat, India
e-mail: vivek.chavda@lmcp.ac.in

J. T. Khan · A. K. Tiwari
Department of Pharmaceutical Sciences, University of Arkansas for Medical Science, Little Rock, AR, USA
e-mail: JTKhan@uams.edu

A. K. Tiwari
e-mail: ATiwari@uams.edu

 
R. K. Singla et al. (eds.), *Polyphenols in Health*,
https://doi.org/10.1007/978-3-032-27324-6_3

## Abbreviations

| | |
|---|---|
| AMP | AMP-activated protein kinase |
| ARO | K Apoptosis research observer |
| CSC | Cancer stem cell |
| Cdk1 | Cyclin-dependent kinase 1 |
| COX | Cyclooxygenase |
| COX-2 | Cyclooxygenase-2 |
| DNA | Deoxyribonucleic acid |
| DNMT | DNA methyltransferase |
| EGFR | Endothelial growth factor receptor |
| EGCG | Epigallocatechin gallate |
| EPIC | European Prospective Investigation into Cancer and Nutrition |
| EU | European Union |
| ERK | Extracellular signal-regulated kinase |
| G1/S | Gap1/synthesis |
| G2/M | Gap 2/mitosis |
| $IC_{50}$ | Half-maximal inhibitory concentration |
| HER-2/neu | Human epidermal growth factor receptor 2 |
| CWR22R | Human primary prostate cancer xenograft |
| IL | Interleukin |
| LoVo | Large intestine adenocarcinoma |
| mTOR | Mammalian/mechanistic target of rapamycin |
| MMP | Matrix metalloproteinase |
| μRNA | Microribonucleic acid |
| MAPK | Mitogen-activated protein kinase |
| MDR | Multidrug resistance |
| NMU | Neuromedin U |
| NOS | Nitric oxide synthase |
| NF-κB | Nuclear factor kappa-light-chain-enhancer of activated B cells |
| PI3K | Phosphatidylinositol 3-kinase |
| PARP | Poly(ADP-ribose) polymerase |
| PSA | Prostate-specific antigen |
| Akt | Protein kinase B |
| Ras | Ras sarcoma GTPase |
| RNS | Reactive nitrogen species |
| ROS | Reactive oxygen species |
| TP21 | Tumor protein p21 |
| TP27 | Tumor protein p27 |
| TP53 | Tumor protein p53 |
| UV | Ultraviolet |
| VEGF | Vascular endothelial growth factor |
| Wnt | Wingless-type integration site |

## 3.1 Introduction to Polyphenols

Polyphenols are part of vegetative nature and contains multiple phenol groups and exerts anticancer and antioxidant properties (Fig. 3.1). They help in the prevention of cancer through regulating the cell signaling, triggering apoptosis, and increasing the efficiency of chemotherapy.

There are a number of major classifications of polyphenols. Examples are flavones (e.g., apigenin), flavonols (e.g., quercetin), flavanones, flavanols (e.g., catechins), isoflavones (e.g., genistein), anthocyanidins and chalcones. Phenolic acids can be classified as benzoic acid derivatives (e.g., gallic acid), and cinnamic acid derivatives (e.g., caffeic and chlorogenic acids). Such compounds as stilbenes (resveratrol) are characteristically present in grapes and red wine. Seeds such as flaxseed contain lignans that are antioxidant and related to hormonal activity. Tannins (e.g., hydrolyzable and condensed, e.g., proanthocyanidins) which have protein-binding and anticancer activities (Dei Cas and Ghidoni 2018).

There are some dietary sources which are rich in polyphenol content including fruits, vegetables, tea, coffee, cocoa, nuts, whole grains, legumes, herbs, and spices.

Fruits such as berries, apples, grapes, cherries, and pomegranate contains flavonoids and phenolic Acids. Flavonols and some polyphenols are present in vegetables like onions, spinach, broccoli, kale, and artichokes. Green tea and Black tea contain very high amount of catechins and theaflavins which are strong antioxidants and anticarcinogen agents. Coffee contains a good amount of phenolic acid

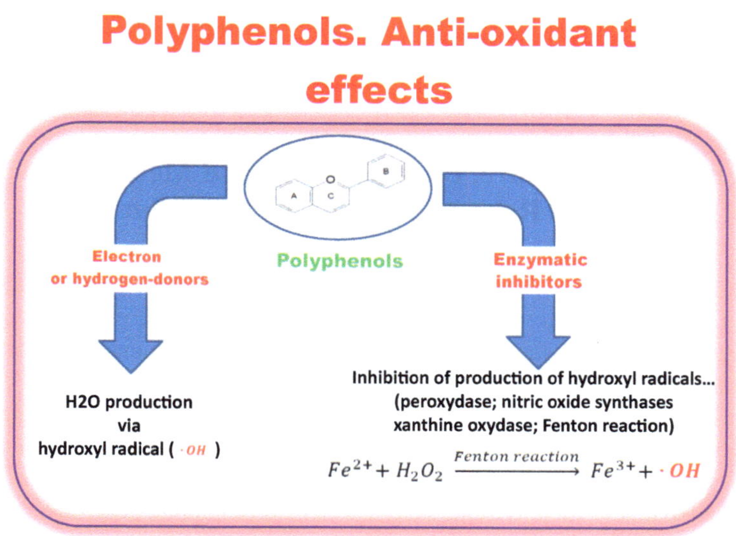


**Fig. 3.1** Polyphenols are reported to have interesting anticancer actions, including antioxidants properties by neutralizing harmful free radicals (Maleki Dana et al. 2022). This picture has been created by the authors

and chlorogenic acid. Cocoa is rich in flavanols, specially epicatechin which has anti-inflammatory and vascular health-promoting values. Walnuts, almonds, and flaxseeds contain lignans and tannins. Walnuts and hazelnuts contain a big range of polyphenols. Whole grains like beans, lentils, soya products, oats, and wheat contain phenolic acids and isoflavones. Concentrated source of polyphenols in herb and spices are cloves, cinnamon, turmeric, oregano, thyme, etc. These sources exert anti-inflammatory as well as anticancer activity (Zelicha et al. 2022).

Cancer is one of the biggest global health problems, and the number of its cases is continued to increase with aging, inappropriate lifestyle, and the environment. The areas, specifically low- and middle-income countries, experience high death rates due to late diagnosis and improper treatment.

Careful planning-such as healthy diet, avoiding tobacco and alcohol, vaccines and screenings is important. There are ability to prevent cancer, prevent money spent in treatment, and generate better outcomes, preventing a cause of global concern (Luo et al. 2025).

In the past, it has been researched that populations especially mediterranean and Asian people with high intake of fruits, vegetables, tea, whole grains etc. shows low rate of cancer cause. Epidemiological studies prove the fact that high consumption of polyphenols is associated with low risks of cancer-like breast, colon and prostate. These data show the preventive nature of polyphenols in the prevention of cancer (Hazafa et al. 2020).

This chapter points out the role of dietary polyphenols in the prevention of cancer including the mechanisms, sources, and types of polyphenols. It also discusses the state of cancer in the world and the need to prevent it. The chapter trends historical, epidemiological, and scientific evidence with the aim of ensuring that dietary interventions that rely on polyphenols are helpful toward cancer prevention.

## 3.2 Chemical Structure, Bioavailability, and Metabolism of Polyphenols Relevant to Cancer Prevention

Polyphenols are compounds derived from plants. They contain at least two aromatic hydroxyl groups in their chemical structure. They are differing in degrees of complexity according to the number of phenol molecules, presence of substitution of different groups as well as linkages.

There are some major classes of polyphenols. Flavonoids (contain both flavanols and flavones as well as isoflavones), flavanones, anthocyanins, and flavanols, but common ones include quercetin and catechins. Phenolic acid is a combination of hydroxybenzoic acid such as gallic acid and hydroxycinnamic acid such as caffeic acid. Resveratrol structures are simple 1,2-diarylethenes and have a simple bioactivity. Mainly present in flaxseed and grains and mostly in the form of a phenylpropanoid unit, lignans are the basis of a second category of these compounds. Large polyphenols are tannins, hydrolysable, e.g., ellagitannins in size or condensed, e.g., proanthocyanidins commonly displaying polymeric structures.

These chemical features are responsible for their biological activity such as antioxidant, anti-inflammatory and anticancer effect (Singla et al. 2019).

Polyphenols tend to have poorly absorbable complex structures in the small intestine. Mostly, they are digested in the colon by intestinal microbiota to get smaller, bioactive molecules. These processes breakdown them enhancing their absorption and biological activity. These produced metabolites may have antioxidant, anti-inflammatory, and anticancer potential hence the intestinal microbiota are the key to exploring the full potential of dietary polyphenols to benefit human health (Luca et al. 2020).

Bioavailability of polyphenols is dependent upon various factors. Polyphenol liberation and absorption can be influenced by the food matrix that contains fats, fibers, and proteins. Once they are absorbed they are metabolized in the intestine and the liver, here alteration of activity may take place. Conjugation processes such as glucuronidation, sulfation, and methylation change their stability and biological activity. Another important factor is represented by gut microbiota which converts polyphenols into more bioavailable active forms. Furthermore, the efficiency of polyphenols absorption and their usage in the body depends on their chemical structure as well. All this factors contribute to the extent that polyphenols can play their benefits that are related to health (Wan et al. 2021).

After absorption, polyphenols circulate via the bloodstream to the different tissues. They are distributed to an extent which relies on the structure, metabolism, and plasma protein binding. Research shows that polyphenols and its metabolites have abilities to accumulate in organs including liver, colon, prostate, and brain, which are commonly associated with cancer growth. The targeting of anticancer properties is supported by this selective bioaccumulation allowing local antioxidant, anti-inflammatory, and proapoptotic activity in those tissues (Di Lorenzo et al. 2021).

It is challenging to link between in vitro observations to in vivo effects due to the high purity of dose of polyphenols which are commonly administered in lab studies generally exceeds the potential of the body to assimilate in the diet. Polyphenols are digested, metabolized, and transformed in the body, which transforms their activity. Their true effect is also dependent on factors such as bioavailability, tissue specificity and the interaction with the gut microbiota so comparing results directly between the lab and a human is complex (Roszkowski 2023).

## 3.3 Molecular Mechanism of Polyphenols in Carcinogenesis Inhibition

Polyphenols, dietary sources that are acquired through the diet, play a significant role in the mechanism of carcinogenesis, which is a multistep process when some normal cells become cancerous cells due to a violation of some key molecular pathways at different stages of tumor formation. The carcinogenesis is a multistep phenomenon which incorporates initiation, promotion and progression. During the initiation stage, normal cell deoxyribonucleic acid (DNA) is exposed to genetic mutations that are usually through oxidative damage or exposure to carcinogens. The next stage is the promotion step, where by these initiated cells are grown selectively and cloned further thus increasing the chances of other mutations. During the progression stage, genetic and epigenetic alternations builds up, leading to malignant transformations, making it easier for cancer cells to invade, angiogenesis, and metastasis (Ahmed et al. 2025). Plant compounds—polyphenols which are abundant in the diet of human beings—have been shown to intervene at every level during carcinogenesis by regulating cellular and molecular pathway (D'Archivio et al. 2008).

Modulating oxidative stress is one of the very important processes in which polyphenols mediate their anticarcinogenic effects, and oxidative stress is central to the process of DNA damage and repair. Polyphenols negate the reactive oxygen species (ROS) as well as reactive nitrogen species (RNS) and help lower DNA strands distortions and oxidative base alterations that cause carcinogenesis initiation. Certain polyphenols such as catechins, quercetin, and resveratrol are majorly protective of oxidative stress on DNA, lipids, and proteins (Angelé-Martínez et al. 2022). Polyphenols may also produce pro-oxidants effects in microenvironments especially where transition metals are present, and locally increase ROS concentration in cancer cells to a point where their already impaired antioxidant defense is no longer able to cope up. Such a rise in oxidative stress in tumor cells may help cancer cell selective death without touching normal cells (Dzah et al. 2024).

In addition to the effect of reducing the concentration of oxidative stress and the repair of the DNA, polyphenols also have the chemoprotective potential through the regulation of major pathways of detoxification to metabolize potential carcinogens (Fig. 3.2). Xenobiotics metabolism enzymes are considered to have two large steps: Phase 1 concerning cytochrome P450 which is the barrier to carcinogens by activation and detoxification. This activity is not controlled and may lead to the formation of mutagenic metabolites. Phase 2 concerns glutathione S-transferase which converts actual metabolic compounds that are used in the detoxification and excretory mechanism (Hodges and Minich 2015). Polyphenols suppress some of Phase 1-enzymes, limiting DNA damaging metabolites, and also stimulate Phase 2 enzymatic activity, thus detoxification. Such activities mitigate the risks of mutagenesis and development of cancer (D'Archivio et al. 2008).

Besides detoxification, the polyphenols also affect vital processes controlling cell survival and division, such as control of the cell cycle and programmed cell death. Polyphenols act on cell proliferation through altering pivotal players involved in cell cycle -cyclins, cyclin-dependent kinases, and other cell cycle regulatory proteins like tumor protein p21 (TP21) and tumor protein p27 (TP27) that lead to cell cycle arrest at gap1/synthesis (G1/S) or gap2/mitosis (G2/M) phases of cell cycle. This action stops the spread of genetically damaged or transformed cells (Briguglio et al. 2020). They also trigger apoptosis programmed cell death by: activation of pro-apoptotic proteins (Bax, Bak), reducing anti-apoptotic proteins (Bcl-2, Bcl-xL), and activation of caspases, with mitochondrial membrane destabilization (Chimento et al. 2023).

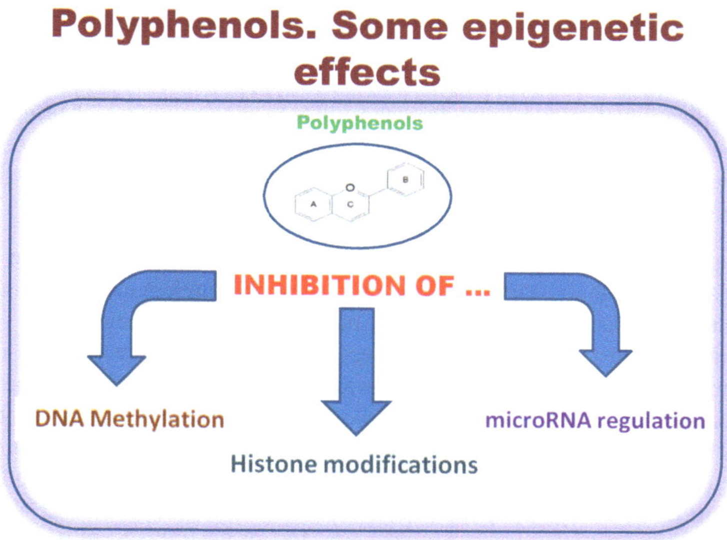


**Fig. 3.2** Polyphenols are reported to have interesting anticancer actions, including the modification of epigenetic mechanisms, inhibiting the spread of cancer cells (antimetastatic action) and promoting programmed cell death (pro-apoptotic action) (Maleki Dana et al. 2022). This picture has been created by the authors

Both inhibition of cell growth and promotion of apoptosis are essential to the removal of precancerous and cancerous cells.

In addition to their ability to control the fate of cells, polyphenols play an important role in inhibiting inflammation, modulating highly influential signaling cascades that direct tumor evolution and development. Tumorigenesis is known to be a contributor of inflammation. Polyphenols suppress dormant inflammation in the following ways: prevention of the activation of nuclear factor kappa-light-chain enhancers of activated B cells (NF-κB) and DNA-binding; down-regulation of transcription of pro-inflammatory genes and cytokines (e.g., TNF-alpha, IL-6), inhibiting the expression of cyclooxygenase-2 (COX-2) and therefore the synthesis of prostaglandin E2 associated with the development of tumors (Karunaweera et al. 2015), causing a decrease in mitogen-activated protein kinase (MAPK), and PI3/Akt cascade plus the down regulation of inflammatory events (Yahfoufi et al. 2018). Such events are used to sustain an antitumorigenic microenvironment.

The benefits of polyphenols extend in addition to the anti-inflammatory effect as they also play a significant role at the epigenetic level, which further demonstrates their role to prevent cancer occurrence. Polyphenols modify epigenetic marks including:

(a) DNA methylation, by regulation of DNMT activity, reversing hypermethylation of tumor suppressor genes, and restoring their expression (Adelipour et al. 2022)

(b) Histone modifications, by inhibition of histone acetyltransferases or histone deacetylases, which impact chromatin structure and gene transcription (Borsoi et al. 2023)
(c) Modification of microribonucleic acid (μRNA) expression, which affects oncogene and tumor suppressor gene regulation (Borsoi et al. 2023). These epigenetic alterations assist in reactivating tumor suppressor pathways and silencing oncogenes.

Along with the influencing gene activity, polyphenols also help stop the growth and spread of tumor by targeting key processes like angiogenesis and invasion. Polyphenol inhibits tumor growth and metastasis by downregulating vascular endothelial growth factor (VEGF) and receptors to restrict new vessel formation critical to tumor survival (Caban and Lewandowska 2021), blocking some matrix metalloproteinases and extracellular matrix degrading enzymes that promote invasion and metastasis, modulating signaling pathways such as phosphatidylinositol 3-kinase (PI3K), protein kinase B (AKT) and extracellular signal-regulated kinase (ERK), further downregulating metastatic capability (Subbaraj et al. 2021).

In addition to inhibiting tumor growth, polyphenols have also been shown to bind directly to the main cancer controlling genes and pathways. They act on key cancer pathways by:

(1) Stabilization and activation of tumor protein p53 (TP53), inducing DNA repair, cell cycle arrest, and apoptosis, inhibiting ras sarcoma GTPase (Ras)/MAPK and wingless-type integration site (Wnt)/β-catenin pathways, and suppressing cell proliferation and tumorigenic signaling (Cháirez-Ramírez et al. 2021)
(2) Modulation of protein kinases and downstream effectors associated with oncogenic transformation (Kang et al. 2011).

In combination with these genetic influences, polyphenols also address more fundamental causes of cancer in a new way by addressing cancer stem cells and remodeling the environment around the tumor. Cancer stem cells (CSC) promote recurrence and treatment resistance. Polyphenols can:

(a) Inhibit CSC (Notch, Hedgehog, and Wnt) pathway, inhibiting self-renewal and survival (Ahmed et al. 2025) and
(b) Promote CSC differentiation or apoptosis, lowering the CSC pool.

They reprogram the tumor microenvironment, changing immune cells and cytokine profiles to deter tumor growth and metastasis (Yahfoufi et al. 2018).

Thus, polyphenols fight cancer in numerous ways—by alleviating inflammation, safeguarding our genes, regulating cell growth, and even addressing cancer stem cells. On multiple fronts, they are able to act; thus, they are potent natural agents for the prevention of cancer and aid in treatment. They just need additional clinical trials to realize their full potential in actual cancer care.

## 3.4 Polyphenol Classes and Their Specific Anticancer Properties

### *3.4.1 Flavonoids: The Predominant Polyphenolic Class*

Flavonoids are known as the biggest and most thoroughly studied type of polyphenolic compounds, being over 4,000 in number as well as structurally varied molecules with noteworthy anticancer properties (Kumar and Pandey 2013). Besides, flavonoids have the fundamental structure of having two phenyl rings linked to a three-carbon chain forming a saturated (3,4)-dihydro-(2H)-1-benzopyran (chromane) ring system that contributes to its biological activity (Pereira et al. 2025).

Quercetin, an abundant flavanol present in apples, berries and onions, exhibits potent anticancer effects via its ability to interact with multiple cellular pathways simultaneously. Previous studies have reported that quercetin can induce apoptosis in cancer cells via activation of TP53-mediated pathway while suppressing NF-κB signaling at the same time, thereby decreasing the inflammatory responses which enhances tumorigenesis (Aghababaei and Hadidi 2023). Additionally, quercetin also elicit anti-angiogenic effects via down-regulation of VEGF expression and matrix metalloproteinase (MMP) activity, affecting the tumor metastasis and limiting tumor vascularization (Esteghlal et al. 2021; Reyes-Farias and Carrasco-Pozo 2019).

Epigallocatechin gallate (EGCG), the main catechin found in green tea, has been marked as one of the most effective chemo preventive agents. EGCG exhibits its anticancer effects via direct binding to 67-kDa laminin receptor on the cancer cell surfaces followed by altered membrane dynamics and increased sensitivity to apoptosis (Kciuk et al. 2023; Singh et al. 2011). Furthermore, EGCG modulates the epigenetic pathways by blocking DNMT and histone deacetylases that results in the reactivation of frequently silenced tumor suppressor genes. Also, the compound demonstrates can inhibit telomerase activity, particularly in highly proliferative cancer cells and attenuate their replicative potential (Meeran et al. 2010; Talib et al. 2024).

Genistein, the predominant isoflavone present in soybeans, elicits anticancer properties in a unique way by its simultaneous role as both weak estrogen receptor agonist and antagonist, although attributed to endogenous and tissue estrogen content (Atanasov et al. 2021; Molteni et al. 1995). This selective estrogen receptor modulator can be effective against aggressive hormone-dependent cancers, particularly breast and prostate malignancies. Genistein is also known for inhibiting protein tyrosine kinases such as epidermal growth factor receptor (EGFR) and human epidermal growth factor receptor 2 (HER-2/neu) (Pavese et al. 2010; Sakla et al. 2007). Additionally, genistein induces G2/M arrest via TP21 up-regulation and cyclin B1 down-regulation (Choi et al. 1998).

Anthocyanins, responsible for the reddish colors of cherries, pomegranates, berries and purple vegetables, is high in antioxidant content that affect the cellular redox homeostasis. These compounds can scavenge reactive nitrogen species (RNS) and reactive oxygen species (ROS) and prevent oxidative DNA damage that contributes to carcinogenesis (Dong et al. 2022). Anthocyanins also exert anti-inflammatory via inhibition of COX-2 and inducible nitric oxide synthase expression and production of inflammatory cytokines (Ngamsamer et al. 2022).

### *3.4.2 Phenolic Acids: Simple Yet Effective Anticancer Agents*

Caffeic acid and ferulic acid are the most abundant hydroxycinnamic acids in the human diet, particularly found in whole grains, fruits and coffee. These compounds can modulate phase II-detoxification enzymes such as glutathione S-transferases and quinone reductase, and can also help in cellular neutralization of carcinogenic metabolites (Baroni et al. 2021; Sova and Saso 2020). Caffeic acid, in particular, blocks lipoxygenase activity and decreases arachidonic acid metabolism and production of pro-inflammatory eicosanoids (Armutcu et al. 2015). On the other hand, ferulic acid has photoprotective effects which is shown via absorption of ultraviolet radiation and prevention of ultraviolet (UV)-induced DNA damage or UV-dependent cancers such as melanoma (Jesus et al. 2023).

### *3.4.3 Stilbenes: Resveratrol and Beyond*

Resveratrol, abundantly present in grape skins and red wine, has received popularity recently for its pleiotropic anticancer properties. On the other hand, stilbene has resveratrol that can activate sirtuin 1, a nicotinamide adenine dinucleotide-dependent deacetylase that regulates cellular stress responses and promote DNA repair mechanisms (Farhan and Rizvi 2023; Wiciński et al. 2023). Resveratrol can also modulate autophagy pathways which will increase the degradation of damaged cellular components while mitochondria-dependent apoptosis is being induced (Fan et al. 2018). The compound can also suppress epithelial–mesenchymal transition and decrease the population of cancer stem cells (Guo et al. 2021).

### *3.4.4 Lignans and Tannins: Emerging Anticancer Compounds*

Enterolactone and enterodiol, the gut microbiota produced mammalian lignans from lignan precursors, expresses weak estrogenic activity selectively relevant against protection to hormone-dependent cancers (Carreau et al. 2008). These compounds also exert telomerase activity and modulate steroid hormone metabolism followed by attenuation of risk in cancer (Ilbeigi et al. 2017).

Tannins, such as hydrolysable and its condensed forms, elicit activity via their strong protein binding capacity and modulation of cellular signaling pathways. Ellagitannins, present in beets, pomegranates and berries, metabolized to ellagic acid, demonstrates anticancer activity by activating TP53 and TP21, also inducing DNA damage and cell cycle arrest (Ceci et al. 2018).

### *3.4.5 Comparative Potency and Synergistic Interactions*

Combined EGCG and resveratrol act together to enhance mitochondrial function and promote metabolic balance, creating a cellular environment less favorable for cancer growth in clinical trial (Most et al. 2016). The combination also showed a strong synergy against head and neck cancer by inducing apoptosis and suppressing AKT-mammalian/mechanistic target of rapamycin (mTOR) signaling in vivo and in vitro (Ruhul Amin et al. 2021). When speaking of ovarian cancer and human epithelial carcinoma cell line of the ovary (OVCAR-5) cells in particular, quercetin (causing G1/S arrest) and genistein (suppressing G2/M) showed synergistic effect in inhibition of cell growth and colony formation. The enhanced effect was attributed to their complementary actions on enzymes in the inositol 1,4,5-trisphosphate-signaling pathway, especially when given in sequence (Shen and Weber 1997). In another study with triple negative breast cancer cells, quercetin and curcumin combination treatment together were reported to have synergic actions, including inhibition of cell survival and invasion via alteration in histone acetylation at its promoter (Tang et al. 2010). Therefore, the rational design of polyphenol combinations can be proposed as a therapeutic approach for more effective and less toxic anticancer treatments (Niedzwiecki et al. 2016).

## 3.5 Preclinical Studies. In Vitro and In Vivo Evidence of Polyphenol Activity

### *3.5.1 Cell Culture Studies: Mechanistic Insights into Anticancer Activity*

In depth in vitro investigation of cancer cell lines derived from humans have provided significant insights into the mechanistic pathways attributed to polyphenol's anticancer activities such as cell cycle arrest, antiproliferative, antimigratory effect, DNA damage, apoptosis induction, and inhibition of survival signaling cascades (Sharma et al. 2022).

In SK-Br3 and MDA-MB-453 breast cancer cells, quercetin at 10 μM selectively inhibited proliferation by inducing G1/S and G2/M cell cycle arrest, while not being cytotoxic to normal breast epithelial cells MCF-10A. It also induced DNA damage, inhibited cyclin-dependent kinase 1 (Cdk1) and cyclin B1 in immunoblotting, and mRNA level of the cyclin B1 gene in reverse transcription polymerase chain reaction. Similar effects were reported with reference to selected breast cancer cells when speaking of EGCG (Liang et al. 1999). In another study, EGCG, resveratrol, genistein, and caspacin at 50 μM were investigated, and EGCG was found to be most effective in cytotoxicity. It also inhibited EGFR and its downstream signaling pathways, cyclin B1, Cdk1, increased G1/S arrest, and enhanced apoptosis-related proteins, in apoptosis research observer (ARO) cells at 50–200 μM (Lim and Cha 2011; Tsai et al. 2023).

On the other side, *hibiscus* anthocyanin-rich extracts at 1–3 mg/mL have enhanced dose- and time-dependent apoptosis in large intestine adenocarcinoma (LoVo) colorectal cancer cells by upregulating AMP-activated protein kinase (AMPK) signaling, triggering Fas-mediated caspase-8/tBid pathways, cytochrome c release, and caspase-3 cleavage (Tsai et al. 2023). Similarly, in HT-29 human colon cancer cells, resveratrol at 75–150 μM decreased anchorage-dependent and -independent growth and triggered apoptosis via caspase-8 and caspase-3 activation alongside poly(ADP-ribose) polymerase (PARP) cleavage (Miki et al. 2012). In HCT-15 colon cancer cells, exposure to caffeic acid—with half maximal inhibitory concentration ($IC_{50}$) ≈ 800 μM—also led to a marked, time-dependent cytostatic effect and colony formation decline, along with visible apoptotic features such as ROS generation, membrane blebbing and cell shrinkage (Jaganathan 2012). At 40–60 μM, ellagic acid can also suppress proliferation and induce apoptosis in HCT-15 colon cancer cells through ROS generation, G2/M arrest, PI3K/Akt inhibition, and pro-apoptotic signaling activation (Umesalma et al. 2015).

### 3.5.2 *Animal Models: Translational Evidence of Chemopreventive Efficacy*

Due to consistent promising effects, animal studies were incorporated for notable anticancer effects ascribed to polyphenols. In a human primary prostate cancer xenograft (CWR22R) prostate cancer -xenograft nude mouse model, regular 50 mg/kg intraperitoneal administration of EGCG or its derivative (EGCG palmitate) for 20 days significantly inhibited tumor growth and angiogenesis while inducing apoptosis (Lee et al. 2008). Quercetin at 10–120 μM in 24–48 h decreased proliferation and induced apoptosis in multiple cancer cell lines. The apoptosis was confirmed with Annexin V/PI-apoptosis assay, and in vivo treatment decreased CT-26[1]- and MCF-7[2]-induced tumors' volumes via apoptotic signaling and inhibition of tumor growth pathways at 50–200 mg/kg dose (Hashemzaei et al. 2017). On the other

---

[1] This acronym is for a *N*-nitroso-*N*-methylurethane-induced, undifferentiated colon carcinoma cell line (Brattain et al. 1980).

[2] This acronym is for an estrogen-sensitive human breast cancer cell line (Kawasaki and Sendo 2021).

hand, local peritumoral injection of genistein (0.5, 1, and 1.5 mg/kg) on a nude mouse xenograft model using selected human gastric cancer cells, dose-dependently reduced tumor growth (10.8–39.6%), enhanced apoptosis (28.9–37.7%), suppressed Bcl-2, and enhanced Bax at both protein and mRNA levels (Zhou et al. 2008). Similarly, resveratrol at 10–100 mg/kg in Sprague–Dawley rats was given 7 days before initiation to 120 days post- neuromedin U (NMU): it notably inhibited mammary tumor development over 42 days. In another study, a water solution of resveratrol ($10^{-4}$% in drinking water) with daily supplementation approximately of 4 μg/mouse has markedly decreased spontaneous mammary tumor formation in female HER2/neu mice in a treatment duration of 20 weeks (Carter et al. 2014). As well as caffeic acid 3,4-dihydroxyphenethyl ester-supplementation (3.5 mg/kg intraperitoneal dose in 7th, 12th, and 24th h dose for 8 days) in H22 hepatoma and S180 sarcoma mice showed significant inhibition of tumor growth and ascites through rapid distribution and hydrolysis to caffeic acid and hydroxytyrosol (Guo et al. 2013). Therefore, polyphenolic compounds are prominent with their dose- and time-dependent effects in both in vitro and in vivo models. The findings of these studies are summarized in Table 3.1.

### *3.5.3 Dose-Dependent Responses and Bioefficacy Considerations*

Preclinical studies have shown that linear dose-dependent relationship with polyphenols is a myth. In detail, such a relationship is often complex and nonlinear as many polyphenols follow a biphasic pattern, where medium doses offer the most benefit, while higher doses may be less effective or even produce unintended negative effects (Renaud and Martinoli 2019). This hormetic response pattern has been particularly well-documented for resveratrol, where low doses show superior chemopreventive activity compared to higher doses in rodent models (Juhasz et al. 2010).

Animal studies have disclosed that effective supplementation of polyphenols for human subjects is usually challenging because—despite strong in-the-lab activity, most of these compounds are poorly absorbed when taken orally (Table 3.1). This fact is mainly due to rapid metabolism, quick conversion into inactive forms, and limited distribution to tissues (Aatif 2023). For example, quercetin aglycone is the most powerful agent: significant amounts of accumulated quercetin were reported in pancreas, liver, lung, and tumor tissues in one study. Although quercetin glucosides are known for its remarkable absorption in humans, rodent studies show low oral bioavailability of the aglycone, with most circulating as glucuronide or sulfate conjugates (Aghababaei and Hadidi 2023; Zhang et al. 2010).

**Table 3.1** Summary of anticancer effects of selected polyphenols in in vitro and in vivo models (Aatif 2023; Aghababaei and Hadidi 2023; Carter et al. 2014; Hashemzaei et al. 2017; Lee et al. 2008; Zhang et al. 2010; Zhou et al. 2008)

| Polyphenol name | Dose | Model | Duration of treatment | Anticancer effects | Study type |
|---|---|---|---|---|---|
| Quercetin | 10 μM | SK-Br3, MDA-MB-453 breast cancer cells | 24–48 h | G1 & G2/M arrest, selective inhibition of proliferation, DNA damage, Cdk1 & cyclin B1 inhibition | In vitro |
| EGCG | 30 μM | MCF-7 breast cancer cells | – | G1 arrest, hypophosphorylated Rb, downregulated Cdk2/Cdk4, upregulated TP21, TP27, and TP53 | In vitro |
| EGCG | 50–200 μM | ARO cells | – | Cytotoxicity, inhibited EGFR & downstream pathways, Cyclin B1/ Cdk1 inhibition, G1/S arrest, enhanced apoptosis | In vitro |
| Resveratrol | 75–150 μM | HT-29 colon cancer cells | – | Reduced anchorage-dependent/ -independent growth, caspase-8/-3 activation, PARP cleavage | In vitro |
| Hibiscus anthocyanin-rich extracts | 1–3 mg/mL | LoVo colorectal cancer cells | Dose- and time-dependent | AMPK activation, Fas-mediated apoptosis (caspase-8/ tBid), cytochrome c release, caspase-3 cleavage | In vitro |
| Caffeic acid | $IC_{50} \approx$ 800 μM | HCT-15 colon cancer cells | – | Time-dependent cytostatic effect, decreased colony formation, ROS generation, membrane blebbing, cell shrinkage | In vitro |

(continued)

**Table 3.1** (continued)

| Polyphenol name | Dose | Model | Duration of treatment | Anticancer effects | Study type |
|---|---|---|---|---|---|
| Ellagic acid | 40–60 μM | HCT-15 colon cancer cells | – | ROS generation, G2/M arrest, PI3K/Akt inhibition, pro-apoptotic signaling | In vitro |
| EGCG/EGCG-P | 50 mg/kg | CWR22R prostate cancer xenograft nude mice | 20 days | Tumor growth inhibition, reduced angiogenesis, increased apoptosis | In vivo |

### 3.5.4 *Toxicological Profiles and Safety Margins*

Comprehensive preclinical studies have revealed that dietary polyphenols are well tolerated at general chemopreventive doses. For example, toxicology data for resveratrol in rats established a safe benchmark dose lower bound of 344 mg/kg body weight/day, and acute toxicity studies indicate that many polyphenols have notable median lethal dose values (1–5 g/kg), offering a wide safety margin relative to effective doses (Quesada-Vázquez et al. 2024).

Chronic toxicity studies revealed that high doses of green tea extracts may pose some risk with concern to liver toxicity. In animal models, daily intake exceeding 500 mg/kg for extended periods has been associated with severe hepatotoxicity likely caused by the pro-oxidant effects of catechins at elevated concentrations. However, in humans, typical consumption of green tea infusions (90–300 mg EGCG/day) is generally safe, though doses at or above 800 mg/day from supplements have been related to increased liver enzyme levels in some individuals, suggesting caution with high supplemental intake.

Genotoxicity studies such as the Ames test, micronucleus assay, and chromosome aberration tests, reveal that dietary polyphenols are usually considered nonmutagenic and often protective agents against DNA damage caused by known carcinogens, supporting their safety and potential antimutagenic effects (Al-Naqeb et al. 2024; Monazzah and Lachenmeier 2025).

### 3.5.5 *Key Preclinical Findings Supporting Clinical Translation*

Overall, preclinical studies provide a strong support for the investigation of polyphenols in cancer prevention. Key insights include (Guo et al. 2024; Singaravelan and Tollefsbol 2025):

(1) Clear, dose-dependent anticancer effects across different cancer types and animal models
(2) Identification of specific molecular targets and pathways through which these compounds act
(3) Generally good safety at effective doses, and
(4) Evidence of synergistic effects, highlighting the potential of combining polyphenols for enhanced outcomes.

Preclinical studies have also revealed some key challenges, such as low bioavailability, differences in individual metabolization, and the need for long-term use to see protective effects. These insights have guided the design of clinical trials and the creation of new delivery methods to improve their absorption and effectiveness (Aatif 2023; Quesada-Vázquez et al. 2024).

## 3.6 Epidemiological Evidence Linking Polyphenol Intake to Cancer Reduction

Epidemiological studies have thoroughly investigated the correlation between the intake of dietary polyphenol and the risk of cancer, utilizing a variety of study designs including cohort studies, case control and cross-sectional methodologies to elucidate possible casual relationships across different population (Zamora-Ros et al. 2018). Cohort studies, such as the European Prospective Investigation into Cancer and Nutrition (EPIC), monitor large groups for many years, providing robust data but potentially subject to change in dietary patterns and confounding over time (Fike et al. 2022; Zamora-Ros et al. 2018). On the other hand, case–control studies compare dietary polyphenol intake of individuals diagnosed with cancer to matched controls, often offering preliminary associations but carrying risk of recall bias and difficulty establishing temporal relationship (Grosso et al. 2017). Cross-sectional studies, while less prevalent in this area, provide valuable insights into polyphenol intake and cancer occurrence within specific populations, yet they are inherently restricted in their ability to determine causality (Xu et al. 2021).

Different populations and geographic origins significantly enhances existing epidemiological research, large-scale cohort and case–control studies, with significant contribution from European, Asia, North American, and Mediterranean cohorts (Leong et al. 2015). For example, investigation within EPIC cohort reveal difference in polyphenol consumption across European Countries, influenced by dietary habit such as coffee and fruit consumption (Fike et al. 2022; Zamora-Ros et al. 2018). Mediterranean population are particularly informative, with their high habitual polyphenol intake and unique genetic background, which may influence the observed relationship for cancer risk (Gardeazabal et al. 2019; Xu et al. 2021).

Regarding breast cancer, recent meta-analyses suggest no significant association for total polyphenol intake, though postmenopausal women in highest intake tertile occasionally show reduced risk (Gardeazabal et al. 2019; Grosso et al. 2017).

The evidence concerning colorectal cancer is similarly complex. Some large-scale cohort analyses and meta-analyses suggest possible decrease in risk with high intake of specific polyphenol subclasses (particularly phenolic acid and hydroxybenzoic acids), and specifically in men, although overall finding frequently remain nonsignificant in adjusted models (Fike et al. 2022; Zamora-Ros et al. 2018). For prostate cancer, both prospective and case–control studies consistently report null findings (Table 3.2), with meta-analyses indicating no clear benefits from higher polyphenol intake and highlighting the need of further research on potential risks associated with some subclasses (Almanza-Aguilera et al. 2023; Huang et al. 2024). The risk for lung cancer may be positively influenced by high consumption of flavonoids, isoflavones, and specific polyphenol subclass, with pooled odds ratios indicating lower incidence among those with high intakes groups (Grosso et al. 2017; Rostampour et al. 2025). For gastric and esophageal cancers, systematic studies report that moderate risk of reduction may be associated with high dietary polyphenol intake, particularly in women, though variability in study settings and populations necessitates cautious interpretation (Fagundes et al. 2022; Liang et al. 2025). For liver cancer and differentiated thyroid cancer, research is limited enough, generally revealing null or inconsistent associations, highlighting the need for larger prospective datasets (Zamora-Ros et al. 2020).

Dietary questionnaires and food frequency survey widely use tools which are prone to measurement errors, inter individual variability in absorption, and recall bias, and highlighting the value of biomarker-based methodologies and the necessity for through adjustments for dietary confounders such as fiber, alcohol, and red meat, and energy intake (Rothwell et al. 2017; Wang et al. 2015; Xu et al. 2021). Confounding

**Table 3.2** Polyphenols and related therapeutic action (Almanza-Aguilera et al. 2023; Fagundes et al. 2022; Fike et al. 2022; Gardeazabal et al. 2019; Grosso et al. 2017; Huang et al. 2024; Liang et al. 2025; Rostampour et al. 2025; Zamora-Ros et al. 2018, 2020)

| Cancer type | Polyphenol subclass | Observed effect | Key study limitations/ notes |
|---|---|---|---|
| Breast cancer | Flavonoids (e.g., quercetin) | Reduced risk associated with higher intake | Some inconsistent findings; more research needed |
| Colorectal cancer | Phenolic acids | Lower incidence linked to polyphenol-rich diets | Variability in study designs |
| Lung cancer | Flavonoids and others | Modest protective effect | Confounding factors such as smoking |
| Prostate cancer | Isoflavones | Possible risk reduction | Limited number of large-scale studies |
| Gastric cancer | Various polyphenols | Potential protective effects reported | Mixed results; population-specific differences |

factors, such as smoking physical activity, body mass index, and hormonal influence in women, can significantly hide genuine associations if not properly managed (Gardeazabal et al. 2019; Zamora-Ros et al. 2018).

Dose-response relationship is a critical aspect of nutrient cancer epidemiology, although some meta-analyses indicate a trend toward reduction in cancer risk with higher polyphenol consumption. Threshold effects and potential U-shaped curves are frequently observed; absolute intake for protective effect have yet to be standardized across various cancer types (Fagundes et al. 2022; Grosso et al. 2017; Zamora-Ros et al. 2018). Genetic variation influencing polyphenol metabolism, along with lifestyle factor such as alcohol consumption and adherence to traditional dietary practice, may further influence cancer preventive efficacy of polyphenols both within and among different populations (Zamora-Ros et al. 2018; Wang et al. 2015).

Epidemiological studies report that high dietary polyphenol intake -specifically flavonoids and phenolic acids—may reduce risk of some cancer such as lung, gastric, and colorectal cancers. However, evidence is limited for other types of cancer such as breast and prostate cancer. More research required to clarify this linkage, while considering factors like study design, population differences, confounding variables, and genetic interaction.

## 3.7 Clinical Trial and Intervention Studies

Recently, there has been notable increase in clinical trials focusing on anticancer effects of polyphenol supplements and foods rich in polyphenols, green tea extracts, resveratrol, soy isoflavone, and curcumin (Bishayee 2009; Farhan 2023). Polyphenols widely found in fruits, vegetables, tea, and wine, have anticancer properties through their antioxidant, anti-inflammatory, and signaling modulatory effects, found in several preclinical and human study (Cháirez-Ramírez et al. 2021; Huang et al. 2024). However, challenge of translating these promising results from cellular and animal studies to human application persists, primarily due to variations in metabolism and bioavailability, with complex interactions between diet and genetics (D'Archivio et al. 2010; Farhan 2023).

Clinical trial evidence for polyphenols is increasing. Some examples such as green tea extracts, particularly its main constituent (EGCG), have been tested for their efficacy and safety in preventing or slowing the progression of various cancers, including prostate, breast, and bladder cancer. Some trials suggest advantages in biomarker reduction, including lower prostate-specific antigen (PSA) levels or reduced DNA damage, while other studies reported that there is no significant clinical benefit over placebo. Results are still insufficient, with meta-analyses pending; strong and uniform data are currently needed to validate any preventive or therapeutic effects (Cheng et al. 2019; Wang et al. 2015).

Resveratrol, a polyphenolic present in grapes, berries, and peanuts, drawn attention due to its capacity to affect multiple molecular targets involve in carcinogenesis. Although early phase clinical trial ensures its safety and effective bioaccumulation in serum, available data about its impact on cancer are limited, primarily focusing on alteration in molecular biomarkers rather than clear clinical endpoints. For example, resveratrol supplementation has been linked to change in the epigenetic status of genes of cancer-related genes, but direct impact on cancer incidence or long-term outcome yet to be established in larger trials (Berman et al. 2017; Bishayee 2009).

Soy isoflavones, such as genistein and daidzein, have also been investigated by means of population-based studies and intervention trials, particularly when speaking of breast and prostate cancers (Khan et al. 2012; Wu et al. 2013). Some intervention in women at high risk of breast cancer showed no significant changes in markers of breast cell proliferation or cytologic atypia overall, though increase in proliferation was observed in postmenopausal women, raising safety concerns in this demographic ambit (Khan et al. 2012). On the opposite hand, observational studies and some trials suggested that isoflavones might modestly lower incidence rate or biomarkers in specific populations, but results remain inconsistent, highlighting need of precision in both subject selection and end-point measurements (Huang et al. 2024; Wu et al. 2013).

Curcumin, the major polyphenol found in turmeric, is currently used for testing studies in human trials for various cancer types: apparently, curcumin can inhibit critical cell signaling pathways involved in cancer progression, such as Wnt/β-catenin and Hedgehog, across multiple malignancies. Many trials indicate that curcumin is safe and may increase surrogate endpoints like reduced inflammatory markers or tumor cell proliferation. On the other side, there are missing information concerning bioavailability and variegated evidences in favor of cancer reduction and survival-related benefits (Ameer et al. 2024; Zoi et al. 2021).

Some major limitation in these trials includes polyphenol bioavailability, which is generally low due to poor gastrointestinal absorption and extensive metabolism (D'Archivio et al. 2010; Ozkan et al. 2024). The actual biological effects depend on levels circulating in bloodstream and their distribution in tissues, which is influenced by food matrix types, individual metabolism, gut microbiota, and formulation (Table 3.3). Many polyphenols, such as anthocyanins and flavonols, are poorly absorbed, and their beneficial effects may be lost if only minimal amounts reach target tissues. In addition, variability in trial design, small sample sizes, short intervention periods, inadequate control groups, and absence of reliable biomarkers are able to limit the interpretability and reproducibility of results (D'Archivio et al. 2010; Wu et al. 2021).

Long-term compliance is a recurring challenge, as polyphenol-rich foods or supplements often requires sustained intake for measurable effects, but adherence may diminish due to factors such as taste fatigue or gastrointestinal discomfort (Khan et al. 2012). Furthermore, while polyphenol-rich foods are generally regarded as safe options for the majority of peoples, certain high-dose supplements may cause adverse effects such as gastrointestinal symptoms, nutrient malabsorption, and (potentially) an increased risk of specific health issues (imbalance in absorption of iron, thiamine, or folate). Research derived from animal studies suggest that excessive intakes of polyphenol supplements may result in organ toxicity or hormonal changes, although

**Table 3.3** Polyphenols for cancer targeting (D'Archivio et al. 2010; Khan et al. 2012; Ozkan et al. 2024; Petre 2019; Wu et al. 2021)

| Polyphenol compound | Source | Cancer targeted | Key trial outcomes | Challenges/notes |
|---|---|---|---|---|
| EGCG (Green tea extract) | Green tea | Prostate, breast, bladder | Some trials showed reduced PSA levels, DNA damage; others showed no significant effect | Inconclusive results; further meta-analysis needed |
| Resveratrol | Grapes, red wine | General anticancer focus | Positive outcomes in rodent studies; limited but growing clinical trial data | Poor bioavailability in humans |
| Soy isoflavones | Soy products | Breast cancer (risk reduction) | Randomized Phase II trial showed some promise in risk markers | Not yet definitive; more longitudinal studies needed |
| Curcumin | Turmeric | Various cancers | Known antioxidant and anti-inflammatory roles; referenced in literature | Not directly evaluated in detail in this excerpt |

such event in humans are not frequent and typically arise at doses significantly higher than those found in dietary sources (Petre 2019).

Anticancer property is reported for polyphenols on the basis of mechanistic and biomarker-related data. Transfer from laboratories to clinic environments is still uncertain and require better trial design, dose standardization, and long-term outcome tracking. Human trials often face difficulty with low systemic absorption and variation among participants.

Polyphenol-rich foods and supplements, such as green tea extracts, resveratrol, soy isoflavones, and curcumin, are gaining attention for cancer prevention because of their antioxidant and anti-inflammatory properties. However, current clinical evidence is inconsistent due to poor bioavailability, individual differences, and limitation in study design. In addition, many trials do not show records of long-term data, standardized dosages, and clear clinical outcomes. In other terms, controlled supplementation of polyphenols may support overall health without reliable effects as collateral therapies for cancer. More advanced and properly designed trials needed to confirm their true benefits and safety.

## 3.8 Interactions Between Polyphenols and Conventional Cancer Therapies

Over the past few years, researchers have been in pursuit of finding natural compounds which could function as chemo-sensitizers by targeting multiple signaling pathways in cancerous cells, and often boost the effectiveness of conventional polyphenol treatment with a diverse group of natural molecules working in the same manner (Hon and Naidu 2024). Polyphenols have a direct effect on the efficacy of anticancer drugs through various mechanisms. For instance, curcumin given along with Gefitinib for lung cancer treatment which leads to inhibition of Sp1 transcription factor/EGFR activity to induce autophagy-mediated apoptosis which results in reduction of tumor volume and increased sensitivity of Gefitinib in nonsmall cell lung cancer patients with mutated EGFR (Chen et al. 2019; Farhan 2023). Similarly, resveratrol shows synergistic effect with doxorubicin which leads to apoptosis and chemo-sensitization of colorectal cancerous cells by suppressing the level and activity of multidrug resistance (MDR) proteins (breast cancer resistance protein, MDR-associated protein-1, and P-glycoprotein) and enzymes which metabolizes the drug (Rai et al. 2016). Furthermore, epigallocatechin gallate lightens inflammation by modulating some signaling pathways like and NF-KB, and neutralizes ROS antioxidative effects, key elements involved in tumor development and chemotherapy-induced toxicity (Alam et al. 2024; Hon and Naidu 2024; Agarwal et al. 2008).

Polyphenols have shown potential protective effects against toxic side effects associated with chemotherapy and radiotherapy. During radiotherapy, they safeguard vital organs from radiation-induced stress by neutralizing harmful free radicals, promoting DNA healing, and controlling inflammatory responses. These benefits not only decrease toxicity but also make therapy bearable for patients which enables doctor to increase the dosing of radio- and chemo-therapy (Prades-Sagarra et al. 2023).

Polyphenols can have the potential to alter or reduce the absorption of cancer drugs in the body. For instance, quercetin—which block certain enzymes necessary for processing chemotherapeutic agents—ultimately reduces the drug's potency and leads to poorer outcomes (Duda-Chodak and Tarko 2023). The combination therapy with polyphenols is effective in boosting drug activity, reducing toxic effects, and sensitizing the resistant tumor. Personalized medicines allow for different options on the basis of individual genetic makeup. For instance, curcumin with 5-fluorouracil can act synergically synergistic in colorectal cancer, yet shows high variability making it necessary to use carefully (Vladu et al. 2022).

While polyphenols serve to have great capability of improving the effectiveness and—on the same side—minimizing toxic effects of conventional chemotherapies, subsequent clinical investigations would be required to comprehend their interaction, optimize their integration, and ensures both safety as well as therapeutic efficacy in cancer patients.

## 3.9 Challenges in Translating Polyphenol Research into Cancer Prevention Practice

The daily dietary pattern consists of complex combinations of polyphenols, which makes it challenging to detect compounds or to figure out effective doses matching regular food consumption. Polyphenols undergo considerable modifications because of cooking, aging, fermentation, and storage (Condezo-Hoyos et al. 2021). For instance, when black tea is fermented, it develops unique compounds like thearubigins and theaflavin which is depicted by its rich color and flavor. Similarly, content of polyphenols varies during aging time (Cheynier 2005). Most part of polyphenols are bound with in food matrices, and their absorption depends on how they react with other nutrients, and how our body breaks them down, along with the help of gut microbes (Bertelli et al. 2021). Factors like genes, age, gender, ethnicity, overall health, and especially the balance of bacteria in our gut define how well they can be absorbed, with some variability from person to person. Our bodies absorb only small number of polyphenols during digestion, and the majority get transformed by gut microbes. The wide range of variations complicates the process of transforming scientific research into straightforward and universal recommendations for cancer prevention (Favari et al. 2024).

Monitoring the uptake of polyphenols with precision and its influence on human biology remains a complex issue in nutrition research. Having clear, consistent, and reliable biomarkers to track how much polyphenols we consume and how these influence health especially when it comes to things like preventing cancer. In absence of standardized biomarkers, researchers have to rely on food diaries or questionnaires, often very tedious and even inaccurate, because people may forget to report what they eat correctly (Spencer et al. 2008). Total urinary polyphenol excretion is one of the most reliable biomarkers: this parameter is measured using Folin–Ciocalteu method in urine samples, reflecting consumption from all sources (Guo et al. 2016). The current challenge is the need of well-validated biomarkers which can give path to research and enables accurate investigation (Picó et al. 2019). Unavailability of universally validated biomarkers is not only a drawback to study of polyphenols but also influences monitoring and management of responsible use of supplements and functional foods. According to the Regulation (EC) No 1924/2006, there should be proper scientific backing for any health claims about foods, including those with polyphenols. Therefore, at this point, only olive oil polyphenols—especially hydroxytyrosol—have approved health statements, where other claims concerning tea or apple polyphenols were rejected because of lack or incomprehensibility of evidence (Patanè et al. 2019). In the United States, regulatory controls over polyphenol-containing supplements appear to allow broad and sometimes unverified claims despite concerns about liver damage when taken in high amounts (Cory et al. 2018).

During a 2022 Romanian survey, 47.6% of participants were unaware with the term polyphenol while 45.6% did not even know even common food as polyphenol sources (Haş et al. 2023). Challenges like difference in bioavailability, its metabolism, interaction with gut flora, pressing need to enhance public education, standardization of research methodology, and easier regulatory structure so forth to translate science of polyphenols into meaningful health messages.

## 3.10 Future Directions and Emerging Technologies

The field of cancer prevention is progressively moving toward personalized strategies, where polyphenol-based interventions stand to gain substantially from advancements in nutrigenomics and metabolomics (Nicastro et al. 2012). Nutrigenomics aids in uncovering genetic variations that affect individual responses to polyphenols, allowing more tailored dietary guidance and supplementation regimens. At the same time, metabolomics provides a detailed analysis of polyphenol metabolites and their bioactive forms within the body, shedding light on factors influencing bioavailability and biological activity on a personalized level. Utilizing these omics tools promises to improve the precision and effectiveness of polyphenol-driven cancer prevention (Zeisel 2007).

The gut microbiome's pivotal role in transforming dietary polyphenols into more bioactive or bioavailable metabolites opens another avenue for enhancing their anticancer properties (Sejbuk et al. 2024). Future investigations aiming to modulate the gut microbial community or its functions could potentiate polyphenol efficacy through host-microbiome synergy, possibly by combining polyphenol intake with targeted prebiotic or probiotic therapies designed to optimize metabolic conversion.

Challenges related to polyphenol stability and absorption in clinical settings are still significant. Innovative delivery technologies—such as nanoparticles, liposomes, and other nanocarriers—offer solutions by protecting polyphenols from degradation, improving their solubility, enabling targeted delivery to tumor tissues, and facilitating controlled release. These approaches help to achieve higher therapeutic concentrations locally while reducing systemic exposure (Kim et al. 2023).

The use of systems biology methods offers a comprehensive perspective for unraveling the complex and multifaceted interactions of polyphenols within cancer-related molecular pathways. The integration of multiomics datasets, computational network modeling, and pharmacological network analyses will enhance mechanistic understanding and support the rational design of polyphenol-based therapies targeting simultaneously multiple carcinogenic processes (Heo et al. 2021).

Lastly, incorporating polyphenols into integrative cancer prevention strategies—including lifestyle adjustments, conventional chemoprevention, immune modulation, and microbiome-directed approaches—holds promise for synergistic effects. Future work must emphasize interdisciplinary collaboration and rigorously designed clinical trials to validate these combinational paradigms and translate them into individualized, effective cancer prevention approaches.

These perspectives highlight a promising future where polyphenols, empowered by cutting-edge technologies and systems integration, might serve as vital components in precise, effective cancer prevention and management.

## 3.11 Practical Recommendations for Clinicians, Nutritionists, and Consumers

There is a strong and growing agreement that plenty of polyphenol-rich natural foods in everyday meals play a key role in lowering the chances of developing cancer. Clinicians and nutritionists should advise people to consume more fruits, vegetables, whole grains, nuts, and legumes as main source of various polyphenols. As per the latest studies, diets rich in these food substance are greatly linked to low incidences of diverse type of cancer, such as the colorectal, breast, liver, and digestive tract cancers (Kishino et al. 2025). For example, diets focused on fiber-rich vegetable foods with abundance of natural antioxidants and protective compounds have been linked to roughly 7–12% drop in both cancer risk and death rates (Nandakumar et al. 2025). Food like berries, green tea, and broccoli containing high polyphenol amounts, adds their benefit in complex nutrient matrices, where isolated supplements fail to do so.

The best strategy for cancer prevention is through the combination of dietary prescriptions with other lifestyle changes. Regular physical activity is crucial in the prevention of breast, colorectal, lung, and endometrial cancer because it balances hormone levels, moderates inflammation, and favors healthy weight. The most significant behavioral modifications in cancer prevention are smoking cessation and further reduction in consumption of alcohol (such a reduction strengthens its protective effect), as global recommendations have confirmed (Motevalli and Stanford 2025). Stress management, healthy body weight, and sufficient sleep also affect the risk of cancer.

When it comes to supplements, a careful approach is essential. Polyphenol capsules or powders should not be seen as a replacement for eating whole foods. The real benefits of these compounds show up most clearly when they are part of a full, natural diet alongside fibers, vitamins, and other nutrients that work together. The consumption of high-dose extracts like green tea or resveratrol pills is not without risk either; in some cases, they've been linked to liver problems. There have also been ongoing concerns about contamination and adulteration in supplement products. By 2025, analysis of weight-loss supplements found an estimated 30% of deliberately adulterated products with unspecified pharmaceuticals including sibutramine, which is majorly dangerous and shows huge gaps in quality control regulation (Phan et al. 2025). Because of these issues, applications of supplements could be justified only in scenarios when clinically necessary and with the assistance of a professional.

Notably, the advantages of cancer prevention are the whole diet-based strategies and not individual compounds. Such models like Mediterranean diet which includes diverse types of vegetable foods show better protective effects due to the additive action of nutrients and positive lifestyle (Wang et al. 2025). Promoting food diversification leads to the consumption of various polyphenol subclasses and bioactive compounds, which promotes complementarity and additivity protective effects. In 2025, a thorough review proved that increasing intake of polyphenols through whole-foods, but not supplements, is associated with significant risk reduction of site-specific cancers, and thus support the notion that health-promoting dietary and lifestyle practices need to be considered as a whole.

## 3.12 Conclusions

Growing evidence highlights the significant role that polyphenols play in cancer prevention, supported by findings from laboratory studies, population research, and emerging clinical investigations. These naturally occurring compounds, found abundantly in plant-derived foods, target multiple phases of carcinogenesis—including initiation, promotion, and progression—through diverse molecular pathways. Their antioxidant capabilities reduce DNA damage and oxidative stress, while anti-inflammatory actions help limit chronic inflammation, both of which are crucial in tumor development. Polyphenols further influence carcinogenesis by regulating detoxification enzymes, controlling cell-cycle checkpoints, triggering apoptosis, and modifying epigenetic patterns. Additionally, they inhibit angiogenesis and metastasis, disrupting tumor growth and spread. The complex interactions between polyphenols and cancer-related genes, as well as the tumor microenvironment, underscore their broad anticancer potential.

Epidemiological data and meta-analyses generally associate increased polyphenol consumption with a lower incidence of common cancers such as breast, colorectal, prostate, and lung cancers. These associations are strengthened by preclinical studies demonstrating efficacy and are beginning to be corroborated by human clinical trials using polyphenol supplements and diets. Nonetheless, issues including variability in bioavailability, metabolism, and differences in study methodologies necessitate further rigorous, well-designed clinical research.

Translating these insights into effective clinical and public-health strategies requires interdisciplinary collaboration among nutritionists, molecular biologists, pharmacologists, and oncologists. Innovations in fields like nutrigenomics, metabolomics, and microbiome science offer promising paths toward individualized polyphenol-based prevention tactics. Moreover, cutting-edge delivery methods, such as nanotechnology, could improve bioavailability barriers.

From a population-health perspective, promoting polyphenol-rich dietary patterns provides a sustainable and accessible approach to reduce cancer risk, complementing current preventive strategies. Enhanced education efforts aimed at both healthcare providers and the public are vital to fully realize the preventive potential of polyphenols.

Ultimately, continued multidisciplinary research and public health initiatives are essential to maximize the impact of polyphenols in cancer prevention and improve global health outcomes.

## References

Aatif M (2023) Current understanding of polyphenols to enhance bioavailability for better therapies. Biomedicines 11(7):2078. https://doi.org/10.3390/biomedicines11072078

Adelipour M, Cheraghzadeh M, Rashidi M (2022) Polyphenols as epigenetic modulators in treating or preventing of cancers. Gene Rep 29:101710. https://doi.org/10.1016/j.genrep.2022.101710

Agarwal A, Sharma V, Tewari R, Koul N, Joseph C, Sen E (2008) Epigallocatechin-3-gallate exhibits anti-tumor effect by perturbing redox homeostasis, modulating the release of pro-inflammatory mediators and decreasing the invasiveness of glioblastoma cells. Mol Med Rep 1(4):511–515

Aghababaei F, Hadidi M (2023) Recent advances in potential health benefits of quercetin. Pharmaceuticals 16(7):1020. https://doi.org/10.3390/ph16071020

Ahmed ZSO, Khan E, Elias N, Elshebiny A, Dou Q (2025) Updated review on natural polyphenols: molecular mechanisms, biological effects, and clinical applications for cancer management. Biomolecules 15(5):629. https://doi.org/10.3390/biom15050629

Alam M, Gulzar M, Akhtar MS, Rashid S, Zulfareen T, Shamsi A, Hassan MI (2024) Epigallocatechin-3-gallate therapeutic potential in human diseases: molecular mechanisms and clinical studies. Mol Biomed 5:73. https://doi.org/10.1186/s43556-024-00240-9

Almanza-Aguilera E, Guiñón-Fort D, Perez-Cornago A, Martínez-Huélamo M, Andrés-Lacueva C, Tjønneland A, Eriksen AK, Katzke V, Bajracharya R, Schulze MB, Masala G, Oliverio A, Tumino R, Manfredi L, Lasheras C, Crous-Bou M, Sánchez MJ, Amiano P, Colorado-Yohar SM, Guevara M, Sonestedt E, Bjartell A, Thysell E, Weiderpass E, Aune D, Aglago EK, Travis RC, Zamora-Ros R (2023) Intake of the Total, classes, and subclasses of (poly) phenols and risk of prostate cancer: a prospective analysis of the EPIC study. Cancers 15(16):4067. https://doi.org/10.3390/cancers15164067

Al-Naqeb G, Kalmpourtzidou A, Giampieri F, De Giuseppe R, Cena H (2024) Genotoxic and antigenotoxic medicinal plant extracts and their main phytochemicals: 'a review.' Front Pharmacol 15:144873. https://doi.org/10.3389/fphar.2024.1448731

Ameer SF, Mohamed MY, Elzubair QA, Sharif EAM, Ibrahim WN (2024) Curcumin as a novel therapeutic candidate for cancer: can this natural compound revolutionise cancer treatment? Front Oncol 14:1438040. https://doi.org/10.3389/fonc.2024.1438040

Angelé-Martínez C, Ameer FS, Raval Y, Huang TTRJ, Anker JN, Brumaghim JL (2022) Polyphenol effects on CuO-nanoparticle-mediated DNA damage, reactive oxygen species generation, and fibroblast cell death. Toxicol Vitro 78:105252. https://doi.org/10.1016/j.tiv.2021.105252

Armutcu F, Akyol S, Ustunsoy S, Turan FF (2015) Therapeutic potential of caffeic acid phenethyl ester and its anti-inflammatory and immuno-modulatory effects (review). Exp Ther Med 9(5):1582–1588. https://doi.org/10.3892/etm.2015.2346

Atanasov AG, Zotchev SB, Dirsch VM, International Natural Product Sciences Taskforce, Supuran CT (2021) Natural products in drug discovery: advances and opportunities. Nat Rev Drug Discov 20(3):200–216. https://doi.org/10.1038/s41573-020-00114-z

Baroni L, Sarni AR, Zuliani C (2021) Plant foods rich in antioxidants and human cognition: a systematic review. Antioxidants 10(5):714. https://doi.org/10.3390/antiox10050714

Berman AY, Motechin RA, Wiesenfeld MY, Holz MK (2017) The therapeutic potential of resveratrol: a review of clinical trials. NPJ Precis Oncol 1(1):35. https://doi.org/10.1038/s41698-017-0038-6

Bertelli A, Biagi M, Corsini M, Baini G, Cappellucci G, Miraldi E (2021) Poly-phenols: from theory to practice. Foods 10(11):2595. https://doi.org/10.3390/foods10112595

Bishayee A (2009) Cancer prevention and treatment with resveratrol: from rodent studies to clinical trials. Cancer Prev Res 2(5):409–418. https://doi.org/10.1158/1940-6207.CAPR-08-0160

Borsoi FT, Neri-Numa IA, de Oliveira WQ, de Araújo FF, Pastore GM (2023) Dietary polyphenols and their relationship to the modulation of non-communicable chronic diseases and epigenetic mechanisms: a mini-review. Food Chem Mol Sci 6:100155. https://doi.org/10.1016/j.fochms.2022.100155

Brattain MG, Strobel-Stevens JD, Fine WD, Webb M, Sarrif AM (1980) Establishment of mouse colonic carcinoma cell lines with different metastatic properties. Cancer Res 40:2142–2146

Briguglio G, Costa C, Pollicino M, Giambò F, Catania S, Fenga C (2020) Polyphenols in cancer prevention: new insights. Int J Funct Nutr 1(2):1–1. https://doi.org/10.3892/ijfn.2020.9

Caban M, Lewandowska U (2021) Inhibiting effects of polyphenols on angiogenesis and epithelial-mesenchymal transition in anterior segment eye diseases. J Funct Foods 87:104761. https://doi.org/10.1016/j.jff.2021.104761

Carreau C, Flouriot G, Bennetau-Pelissero C, Potier M (2008) Enterodiol and enterolactone, two major diet-derived polyphenol metabolites have different impact on ERα transcriptional activation in human breast cancer cells. J Steroid Biochem Mol Biol 110(1–2):176–185. https://doi.org/10.1016/j.jsbmb.2008.03.032
Carter LG, D'Orazio JA, Pearson KJ (2014) Resveratrol and Cancer: Focus on in Vivo Evidence. Endocr Relat Cancer 21(3):R209–R225. https://doi.org/10.1530/ERC-13-0171
Ceci C, Lacal PM, Tentori L, De Martino MG, Miano R, Graziani G (2018) Experimental evidence of the antitumor, antimetastatic and antiangio-genic activity of ellagic acid. Nutrients 10(11):1756. https://doi.org/10.3390/nu10111756
Cháirez-Ramírez MH, de la Cruz-López KG, García-Carrancá A (2021) Polyphenols as antitumor agents targeting key players in cancer-driving signaling pathways. Front Pharmacol 12:710304. https://doi.org/10.3389/fphar.2021.710304
Chen P, Huang HP, Wang Y et al (2019) Curcumin overcome primary gefitinib resistance in non-small-cell lung cancer cells through inducing autophagy-related cell death. J Exp Clin Cancer Res 38:254. https://doi.org/10.1186/s13046-019-1234-8
Cheng K, Chi NN, Liu JD (2019) Green tea extract for treatment of cancers: a systematic review protocol. Med 98(15):e15117. https://doi.org/10.1097/MD.0000000000015117
Cheynier V (2005) Polyphenols in foods are more complex than often thought. Am J Clin Nutr 81(1 Suppl):223S-229S. https://doi.org/10.1093/ajcn/81.1.223S
Chimento A, De Luca A, D'Amico M, De Amicis F, Pezzi V (2023) The involvement of natural polyphenols in molecular mechanisms inducing apoptosis in tumor cells: a promising adjuvant in cancer therapy. Int J Mol Sci 24(2):1680. https://doi.org/10.3390/ijms24021680
Choi YH, Zhang L, Lee WH, Park KY (1998) Genistein-induced G2/M arrest is associated with the inhibition of cyclin B1 and the induction of p21 in human breast carcinoma cells. Int J Oncol 13(2):391–397. https://doi.org/10.3892/ijo.13.2.391
Condezo-Hoyos L, Gazi C, Pérez-Jiménez J (2021) Design of polyphenol-rich diets in clinical trials: a systematic review. Food Res Int 149:110655. https://doi.org/10.1016/j.foodres.2021.110655
Cory H, Passarelli S, Szeto J, Tamez M, Mattei J (2018) The role of polyphenols in human health and food systems: a mini-review. Front Nutr 5:87. https://doi.org/10.3389/fnut.2018.00087
D'Archivio M, Santangelo C, Scazzocchio B, Varì R, Filesi C, Masella R, Giovannini C (2008) Modulatory effects of polyphenols on apoptosis induction: relevance for cancer prevention. Int J Mol Sci 9(3):213–228. https://doi.org/10.3390/ijms9030213
D'Archivio M, Filesi C, Varì R, Scazzocchio B, Masella R (2010) Bioavailability of the polyphenols: status and controversies. Int J Mol Sci 11(4):1321–1342. https://doi.org/10.3390/ijms11041321
Dei Cas M, Ghidoni R (2018) Cancer prevention and therapy with polyphenols: sphingolipid-mediated mechanisms. Nutrients 10(7):940. https://doi.org/10.3390/nu10070940
Di Lorenzo C, Colombo F, Biella S, Stockley C, Restani P (2021) Polyphenols and human health: the role of bioavailability. Nutrients 13(1):273. https://doi.org/10.3390/nu13010273
Dong Y, Wu X, Han L, Bian J, He C, El-Omar E, Gong L, Wang M (2022) The potential roles of dietary anthocyanins in inhibiting vascular endothelial cell senescence and preventing cardiovascular diseases. Nutrients 14(14):2836. https://doi.org/10.3390/nu14142836
Duda-Chodak A, Tarko T (2023) Possible side effects of polyphenols and their interactions with medicines. Molecules 28(6):2536. https://doi.org/10.3390/molecules28062536
Dzah CS, Zhang H, Gobe V, Asante-Donyinah D, Duan Y (2024) Anti- and pro-oxidant properties of polyphenols and their role in modulating glutathione synthesis, activity and cellular redox potential: potential synergies for disease management. Adv Redox Res 11:100099. https://doi.org/10.1016/j.arres.2024.100099
Esteghlal S, Mokhtari MJ, Beyzaei Z (2021) Quercetin can inhibit angiogenesis via the down regulation of MALAT1 and MIAT LncRNAs in human umbilical vein endothelial cells. Int J Prev Med 12(1):59. https://doi.org/10.4103/ijpvm.IJPVM_103_20
Fagundes MDA, Silva ARC, Fernandes GA, Curado MP (2022) Dietary polyphenol intake and gastric cancer: a systematic review and meta-analysis. Cancers 14(23):5878. https://doi.org/10.3390/cancers14235878

Fan Y, Chiu JF, Liu J, Deng Y, Xu C, Zhang J, Li G (2018) Resveratrol induces autophagy-dependent apoptosis in HL-60 cells. BMC Cancer 18(1):581. https://doi.org/10.1186/s12885-018-4504-5
Farhan M (2023) Insights on the role of polyphenols in combating cancer drug resistance. Biomedicines 11(6):1709. https://doi.org/10.3390/biomedicines11061709
Farhan M, Rizvi A (2023) The pharmacological properties of red grape polyphenol resveratrol: clinical trials and obstacles in drug development. Nutrients 15:20. https://doi.org/10.3390/nu15204486
Favari C, de Alvarenga JFR, Sánchez-Martínez L, Tosi N, Mignogna C, Cremonini E, Manach C, Bresciani L, Del Rio D, Mena P (2024) Factors driving the inter-individual variability in the metabolism and bioavailability of (poly) phenolic metabolites: a systematic review of human studies. Redox Biol 71:103095. https://doi.org/10.1016/j.redox.2024.103095
Fike LT, Munro H, Yu D, Dai Q, Shrubsole MJ (2022) Dietary polyphenols and the risk of colorectal cancer in the prospective Southern Community Cohort Study. Am J Clin Nutr 115(4):1155–1165. https://doi.org/10.1093/ajcn/nqac012
Gardeazabal I, Romanos-Nanclares A, Martinez-Gonzalez MA, Sanchez-Bayona R, Vitelli-Storelli F, Gaforio JJ et al (2019) Total polyphenol intake and breast cancer risk in the Seguimiento Universidad de Navarra (SUN) cohort. Brit J Nutr 122(5):542–551. https://doi.org/10.1017/S0007114518003811
Grosso G, Godos J, Lamuela-Raventos R, Ray S, Micek A, Pajak A, Sciacca S, D'Orazio N, Del Rio D, Galvano F (2017) A comprehensive meta-analysis on dietary Flavonoid and Lignan intake and cancer risk: level of evidence and limitations. Mol Nutr Food Res 61(4):1600930. https://doi.org/10.1002/mnfr.201600930
Guo Y, Shi S, Yang N, Tang MX, Duan ZJ, Guo XR, Tang ZH (2023) Comparative assessment of nutritional composition, polyphenol profile and antioxidative properties of wild edible ferns from northeastern China. Food Res Int 163:112237. https://doi.org/10.1016/j.foodres.2022.112237
Guo X, Shen L, Tong Y, Zhang J, Wu G, He Q, Yu S, Ye X, Zou L, Zhang Z, Lian XY (2013) Antitumor activity of caffeic acid 3, 4-dihydroxyphenethyl ester and its pharmacokinetic and metabolic properties. Phytomedicine 20(10):904–912. https://doi.org/10.1016/j.phymed.2013.04.002
Guo X, Tresserra-Rimbau A, Estruch R, Martinez-Gonzalez MA, Medina-Remón A, Castañer O, Corella D, Salas-Salvadó J, Lamuela-Raventós RM (2016) Effects of polyphenol, measured by a biomarker of total polyphenols in urine, on cardiovascular risk factors after a long-term follow-up in the PREDIMED study. Oxid Med Cell Longev 2016, 1:2572606. https://doi.org/10.1155/2016/2572606
Guo K, Feng Y, Zheng X, Sun L, Wasan HS, Ruan S, Shen M (2021) Resveratrol and its analogs: potent agents to reverse epithelial-to-mesenchymal transition in tumors. Front Oncol 11:644134. https://doi.org/10.3389/fonc.2021.644134
Guo H, Xu X, Zhang J, Du Y, Yang X, He Z, Zhao L, Liang T, Guo L (2024) The pivotal role of preclinical animal models in anti-cancer drug discovery and personalised cancer therapy strategies. Pharmaceuticals 17(8):1048. https://doi.org/10.3390/ph17081048
Haş IM, Teleky BE, Vodnar DC, Ştefănescu BE, Tit DM, Niţescu M (2023) Polyphenols and cardiometabolic health: knowledge and concern among Romanian people. Nutrients 15(10):2281. https://doi.org/10.3390/nu15102281
Hashemzaei M, Far AD, Yari A, Heravi RE, Tabrizian K, Taghdisi SM, Sadegh SE, Tsarouhas K, Kouretas D, Tzanakakis G, Nikitovic D, Anisimov NY, Spandidos DA, Tsatsakis AM, Rezaee R (2017) Anticancer and apoptosis-inducing effects of quercetin in vitro and in vivo. Oncol Rep 38(2):819–828. https://doi.org/10.3892/or.2017.5766
Hazafa A, Rehman KU, Jahan N, Jabeen Z (2020) The role of polyphenol (flavonoids) compounds in the treatment of cancer cells. Nutr Cancer 72(3):386–397. https://doi.org/10.1080/01635581.2019.1637006
Heo YJ, Hwa C, Lee GH, Park JM, An JY (2021) Integrative multi-omics approaches in cancer research: from biological networks to clinical subtypes. Mol Cells 44(7):433–443. https://doi.org/10.14348/molcells.2021.0042

Hodges RE, Minich DM (2015) Modulation of metabolic detoxification pathways using foods and food-derived components: a scientific review with clinical application. J Nutr Metab 2015:760689. https://doi.org/10.1155/2015/760689

Hon KW, Naidu R (2024) Synergistic mechanisms of selected polyphenols in overcoming chemoresistance and enhancing chemosensitivity in colorectal cancer. Antioxidants 13(7):7. https://doi.org/10.3390/antiox13070815

Huang Y, Wang W, Jin J (2024) Association between polyphenol subclasses and prostate cancer: a systematic review and meta-analysis of observational studies. Front Nutr 11:1428911. https://doi.org/10.3389/fnut.2024.1428911

Ilbeigi D, Mitra Nourbakhsh M, Shahnaz Khaghani S (2017) Enterolactone reduces telomerase activity and the level of its catalytic subunit in breast cancer cells. Cell J 19(supplement 1):37–43. https://doi.org/10.22074/cellj.2017.4705

Jaganathan SK (2012) Growth inhibition by caffeic acid, one of the phenolic constituents of honey, in HCT 15 colon cancer cells. Sci World J 2012:372345. https://doi.org/10.1100/2012/372345

Jesus A, Mota S, Torres A, Cruz MT, Sousa E, Almeida IF, Cidade H (2023) Antioxidants in sunscreens: which and what for? Antioxidants 12:1. https://doi.org/10.3390/antiox12010138

Juhasz B, Mukherjee S, Das DK (2010) Hormetic response of resveratrol against cardioprotection. Exp Clin Cardiol 15(4):e134–e138

Kang NJ, Shin SH, Lee HJ, Lee KW (2011) Polyphenols as small molecular inhibitors of signaling cascades in carcinogenesis. Pharmacol Therape 130(3):310–324. https://doi.org/10.1016/j.pharmthera.2011.02.004

Karunaweera N, Raju R, Gyengesi E, Münch G (2015) Plant polyphenols as inhibitors of NF-κB induced cytokine production—A potential anti-inflammatory treatment for Alzheimer's disease? Front Mol Neurosci 8:24. https://doi.org/10.3389/fnmol.2015.00024

Kawasaki Y, Sendo T (2021) Three photoinitiators induce breast tumor growth in mouse xenografts with MCF-7 breast cancer cells. Curr Res Toxicol 2:322–328. https://doi.org/10.1016/j.crtox.2021.08.004

Kciuk M, Alam M, Ali N, Rashid S, Głowacka P, Sundaraj R, Celik I, Yahya EB, Dubey A, Zerroug E, Kontek R (2023) Epigallocatechin-3-gallate therapeutic potential in cancer: mechanism of action and clinical implications. Molecules 28(13):5246. https://doi.org/10.3390/molecules28135246

Khan SA, Chatterton RT, Michel N, Bryk M, Lee O, Ivancic D et al (2012) Soy isoflavone supplementation for breast cancer risk reduction: a randomised phase II trial. Cancer Prev Res 5(2):309–319. https://doi.org/10.1158/1940-6207.CAPR-11-0251

Kim KH, Ki MR, Min KH, Pack SP (2023) Advanced delivery system of polyphenols for effective cancer prevention and therapy. Antioxidants 12(5):1048. https://doi.org/10.3390/antiox12051048

Kishino M, Kanehara R, Mori N, Ishihara J, Takachi R, Yamaji T, Iwasaki M, Tsugane S, Sawada N (2025) Dietary polyphenol intake and risk of overall and site-specific cancers: the Japan Public Health Center-based prospective study. J Nutr 155(6):1987–1998. https://doi.org/10.1016/j.tjnut.2025.04.028

Kumar S, Pandey AK (2013) Chemistry and biological activities of flavonoids: an overview. Sci World J 2013:162750. https://doi.org/10.1155/2013/162750

Lee SC, Chan WK, Lee TW, Lam WH, Wang X, Chan TH, Wong YC (2008) Effect of a prodrug of the green tea polyphenol (-)-epigallocatechin-3-gallate on the growth of androgen-independent prostate cancer in vivo. Nutr Cancer 60(4):483–491. https://doi.org/10.1080/01635580801947674

Leong HY, Show PL, Lim MH, Ooi CW, Ling TC (2018) Natural red pigments from plants and their health benefits: a review. Food Rev Int 34(5):463–482. https://doi.org/10.1080/87559129.2017.1326935

Liang YC, Lin-Shiau SY, Chen CF, Lin JK (1999) Inhibition of cyclin-dependent kinases 2 and 4 activities as well as induction of Cdk inhibitors P21 and P27 during growth arrest of human breast carcinoma cells by (-)-epigallocatechin-3-gallate. J Cell Biochem 75(1):1–12. https://doi.org/10.1002/(SICI)1097-4644(19991001)75:1<1::AID-JCB1>3.0.CO;2-N

Liang JL, Yuan HM, Quan C, Chen JQ (2025) Risk factors for gastric cancer: an umbrella review of systematic reviews and meta-analyses. Front Oncol 15:1564575. https://doi.org/10.3389/fonc.2025.1564575

Lim YC, Cha YY (2011) Epigallocatechin-3-gallate induces growth inhibition and apoptosis of human anaplastic thyroid carcinoma cells through suppression of EGFR/ERK pathway and cyclin B1/CDK1 complex. J Surg Oncol 104(7):776–780. https://doi.org/10.1002/jso.21999

Luca SV, Macovei I, Bujor A, Miron A, Skalicka-Woźniak K, Aprotosoaie AC, Trifan A (2020) Bioactivity of dietary polyphenols: the role of metabolites. Crit Rev Food Sci Nutr 60(4):626–659. https://doi.org/10.1080/10408398.2018.1546669

Luo H, Jing H, Zhao H, Zhao Y (2025) Global, regional, and national burden and trends of kidney cancer associated with high BMI from 1990 to 2021: findings from the Global Burden of Disease Study 2021. Plos One 20(3):e0320185. https://doi.org/10.1371/journal.pone.0320185

Maleki Dana P, Sadoughi F, Asemi Z, Yousefi B (2022) The role of polyphenols in overcoming cancer drug resistance: a comprehensive review. Cell Mol Biol Lett 27(1):1. https://doi.org/10.1186/s11658-021-00301-9

Meeran SM, Ahmed A, Tollefsbol TO (2010) Epigenetic targets of bioactive dietary components for cancer prevention and therapy. Clin Epigenetics 1(3):101–116. https://doi.org/10.1007/s13148-010-0011-5

Miki H, Uehara N, Kimura A, Sasaki T, Yuri T, Yoshizawa K, Tsubura A (2012) Resveratrol induces apoptosis via ROS-triggered autophagy in human colon cancer cells. Int J Oncol 40(4):1020–1028. https://doi.org/10.3892/ijo.2012.1325

Molteni A, Brizio-Molteni L, Persky V (1995) In vitro hormonal effects of soybean isoflavones. J Nutr 125(3 Suppl):751S-756S. https://doi.org/10.1093/jn/125.3_Suppl.751S

Monazzah M, Lachenmeier DW (2025) Genotoxicity of coffee, coffee by-products, and coffee bioactive compounds: contradictory evidence from in vitro studies. Toxics 13(5):409. https://doi.org/10.3390/toxics13050409

Most J, Timmers S, Warnke I, Jocken JW, van Boekschoten M, de Groot P, Bendik I, Schrauwen P, Goossens GH, Blaak EE (2016) Combined epigallocatechin-3-gallate and resveratrol supplementation for 12 wk increases mitochondrial capacity and fat oxidation, but not insulin sensitivity, in obese humans: a randomised controlled trial. Am J Clin Nutr 104(1):215–227. https://doi.org/10.3945/ajcn.115.122937

Motevalli M, Stanford FC (2025) Personalised lifestyle interventions for prevention and treatment of obesity-related cancers: a call to action. Cancers 17(8):1255. https://doi.org/10.3390/cancers17081255

Nandakumar D, Dixit R, Pandey M (2025) Cancer prevention through lifestyle modification. Indian J Medl Res 161(2):109–111. https://doi.org/10.25259/IJMR_405_2025

Ngamsamer C, Sirivarasai J, Sutjarit N (2022) The benefits of anthocyanins against obesity-induced inflammation. Biomolecules 12(6):852. https://doi.org/10.3390/biom12060852

Nicastro HL, Trujillo EB, Milner JA (2012) Nutrigenomics and cancer prevention. Curr Nutr Rep 1(1):37–43. https://doi.org/10.1007/s13668-011-0007-6

Niedzwiecki A, Roomi MR, Kalinovsky T, Rath M (2016) Anticancer efficacy of polyphenols and their combinations. Nutr 8(9):552. https://doi.org/10.3390/nu8090552

Ozkan G, Ceyhan T, Çatalkaya G, Rajan L, Ullah H, Daglia M, Capanoglu E (2024) Encapsulated phenolic compounds: clinical efficacy of a novel delivery method. Phytochem Rev 23(3):781–819. https://doi.org/10.1007/s11101-023-09909-5

Patanè P, Laganà P, Devi P, Vig A, Haddad MA, Natalello S, Cava MA, Ameen SM, Hashim HA (2019) Polyphenols and functional foods from the regulatory viewpoint. J AOAC Int 102(5):1373–1377. https://doi.org/10.5740/jaoacint.19-0130

Pavese JM, Farmer RL, Bergan RC (2010) Inhibition of cancer cell invasion and metastasis by genistein. Cancer Metastasis Rev 29(3):465–482. https://doi.org/10.1007/s10555-010-9238-z

Pereira R, Ribeiro D, Silva VRM, Fernandes E (2025) Synthetic flavonoid dimers: synthesis strategies and biological activities. Eur J Med Chem 291:117669. https://doi.org/10.1016/j.ejmech.2025.117669

Petre A (2019) What are polyphenols? Types, benefits, and food sources. Healthline Media LLC, https://www.healthline.com. https://www.healthline.com/nutrition/polyphenols. Accessed 28th Oct 2025

Phan DTA, Kongkaew C, Heinrich M, Dao TCM, Vo TH (2025) From 'traditional'remedies to 'modern'supplements: a systematic review and meta-analysis of pharmaceutical adulteration in weight-loss natural products. Front Pharmacol 16:1594975. https://doi.org/10.3389/fphar.2025.1594975

Picó C, Serra F, Rodríguez AM, Keijer J, Palou A (2019) Biomarkers of nutrition and health: new tools for new approaches. Nutrients 11(5):1092. https://doi.org/10.3390/nu11051092

Prades-Sagarra È, Yaromina A, Dubois LJ (2023) Polyphenols as potential protectors against radiation-induced adverse effects in patients with thoracic cancer. Cancers 15(9):2412. https://doi.org/10.3390/cancers15092412

Quesada-Vázquez S, Eseberri I, Les F, Pérez-Matute P, Herranz-López M, Atgié C, Lopez-Yus M, Aranaz P, Oteo JA, Escoté X, Lorente-Cebrian S, Roche E, Courtois A, López V, Portillo MP, Milagro FI, Carpéné C (2024) Polyphenols and metabolism: from present knowledge to future challenges. J Physiol Biochem 80(3):603–625. https://doi.org/10.1007/s13105-024-01046-7

Rai G, Mishra S, Suman S, Shukla Y (2016) Resveratrol improves the anticancer effects of doxorubicin in vitro and in vivo models: a mechanistic insight. Phytomed 23(3):233–242. https://doi.org/10.1016/j.phymed.2015.12.020

Renaud J, Martinoli MG (2019) Considerations for the use of polyphenols as therapies in neurodegenerative diseases. Int J Mol Sci 20(8):1883. https://doi.org/10.3390/ijms20081883

Reyes-Farias M, Carrasco-Pozo C (2019) The anti-cancer effect of quercetin: molecular implications in cancer metabolism. Int J Mol Sci 20(13):3177. https://doi.org/10.3390/ijms20133177

Rostampour K, Alipour K, Mirjalili F, Forootani B, Yekrang Safakar H, Beigrezaei S, Forbes SC, Salehi-Abargouei A (2025) Dietary flavonoids and lung cancer: a GRADE-assessed systematic review and meta-analysis of observational studies. Nutr Cancer 77(2):164–178. https://doi.org/10.1080/01635581.2024.2417457

Roszkowski S (2023) Application of polyphenols and flavonoids in oncological therapy. Mol 28(10):4080. https://doi.org/10.3390/molecules28104080

Rothwell JA, Knaze V, Zamora-Ros R (2017) Polyphenols: dietary assessment and role in the prevention of cancers. Curr Opin Clin Nutr Metab Care 20(6):512–521. https://doi.org/10.1097/MCO.0000000000000424

Ruhul Amin ARM, Wang D, Nannapaneni S, Lamichhane R, Chen ZC, Shin DM (2021) Combination of resveratrol and green tea epigallocatechin gallate induces synergistic apoptosis and inhibits tumor growth in vivo in head and neck cancer models. Oncol Rep 45(5):87. https://doi.org/10.3892/or.2021.8038

Sakla MS, Shenouda NS, Ansell PJ, Mac-Donald RS, Lubahn DB (2007) Genistein affects HER2 protein concentration, activation, and promoter regulation in BT-474 human breast cancer cells. Endocrine 32(1):69–78. https://doi.org/10.1007/s12020-007-9006-1

Sejbuk M, Mirończuk-Chodakowska I, Karav S, Witkowska AM (2024) Dietary polyphenols, food processing and gut microbiome: recent findings on bioavailability, bioactivity, and gut microbiome interplay. Antiox 13(10):1220. https://doi.org/10.3390/antiox13101220

Sharma E, Attri DC, Sati P, Dhyani P, Szopa A, Sharifi-Rad J, Hano C, Calina D, Cho WC (2022) Recent updates on anticancer mechanisms of polyphenols. Front Cell Develop Biol 10:1005910. https://doi.org/10.3389/fcell.2022.1005910

Shen F, Weber G (1997) Synergistic action of quercetin and genistein in human ovarian carcinoma cells. Oncol Res 9(11–12):597–602

Singaravelan N, Tollefsbol TO (2025) Polyphenol-based prevention and treatment of cancer through epigenetic and combinatorial mechanisms. Nutrients 17(4):616. https://doi.org/10.3390/nu17040616

Singh BN, Shankar S, Srivastava RK (2011) Green tea catechin, epigallocatechin-3-gallate (EGCG): mechanisms, perspectives and clinical applications. Biochem Pharmacol 82(12):1807–1821. https://doi.org/10.1016/j.bcp.2011.07.093

Singla RK, Dubey AK, Garg A, Sharma RK, Fiorino M, Ameen SM, Haddad MA, Al-Hiary M (2019) Natural polyphenols: chemical classification, definition of classes, subcategories, and structures. J AOAC Int 102(5):1397–1400. https://doi.org/10.5740/jaoacint.19-0133

Sova M, Saso L (2020) Natural sources, pharmacokinetics, biological activities and health benefits of hydroxycinnamic acids and their metabolites. Nutrients 12(8):2190. https://doi.org/10.3390/nu12082190

Spencer JPE, Abd El Mohsen MM, Mini-hane AM, Mathers (2008) Biomarkers of the intake of dietary polyphenols: strengths, limitations and application in nutrition research. Brit J Nutr 99(1):12–22. https://doi.org/10.1017/S0007114507798938

Subbaraj GK, Kumar YS, Kulanthaivel L (2021) Antiangiogenic role of natural flavonoids and their molecular mechanism: an update. Egypt J Int Med 33(1):29. https://doi.org/10.1186/s43162-021-00056-x

Talib WH, Awajan D, Alqudah A, Alsawwaf R, Althunibat R, Abu AlRoos M, Al Safadi A, Abu Asab S, Hadi RW, Al Kury LT (2024) Targeting cancer hallmarks with epigallocatechin gallate (EGCG): mechanistic basis and therapeutic targets. Mol 29(6):1373. https://doi.org/10.3390/molecules29061373

Tang SN, Singh C, Nall D, Meeker D, Shankar S, Srivastava RK (2010) The dietary bioflavonoid quercetin synergises with epigallocathechin gallate (EGCG) to inhibit prostate cancer stem cell characteristics, invasion, migration and epithelial-mesenchymal transition. J Mol Signal 5(1):14. https://doi.org/10.1186/1750-2187-5-14

Tsai MC, Chen CC, Tseng TH, Chang YC, Lin YJ, Tsai IN, Wang CC, Wang CJ (2023) Hibiscus anthocyanins extracts induce apoptosis by activating AMP-activated protein kinase in human colorectal cancer cells. Nutrients 15(18):3972. https://doi.org/10.3390/nu15183972

Umesalma S, Nagendraprabhu P, Sudhandiran G (2015) Ellagic acid inhibits proliferation and induced apoptosis via the Akt signaling pathway in HCT-15 colon adenocarcinoma cells. Mol Cell Biochem 399(1):303–313. https://doi.org/10.1007/s11010-014-2257-2

Vladu AF, Ficai D, Ene AG, Ficai A (2022) Combination therapy using polyphe-nols: an efficient way to improve antitumoral activity and reduce resistance. Int J Mol Sci 23(18):10244. https://doi.org/10.3390/ijms231810244

Wan MLY, Co VA, El-Nezami H (2021) Dietary polyphenol impact on gut health and microbiota. Crit Rev Food Sci Nutr 61(4):690–711. https://doi.org/10.1080/10408398.2020.1744512

Wang J, Tang L, Wang JS (2015) Bio-markers of dietary polyphenols in cancer studies: current evidence and beyond. Oxid Med Cell Longev 2015:732302. https://doi.org/10.1155/2015/732302

Wang M, Wang Y, Zhang H (2025) Dietary polyphenols for tumor therapy: bioactivities, nano-therapeutic systems and delivery strategies. Food Funct 16(3):853–866. https://doi.org/10.1039/D4FO04715J

Wiciński M, Erdmann J, Nowacka A, Kuźmiński O, Michalak K, Janowski K, Ohla J, Biernaciak A, Szambelan M, Zabrzyński J (2023) Natural phytochemicals as SIRT activators—Focus on potential biochemical mechanisms. Nutrients 15(16):3578. https://doi.org/10.3390/nu15163578

Wu AH, Lee E, Vigen C (2013) Soy isoflavones and breast cancer. Am Soc Clin Oncol Educ Book 33:102–106. https://doi.org/10.14694/EdBook_AM.2013.33.102

Wu M, Luo Q, Nie R, Yang X, Tang Z, Chen H (2021) Potential implications of polyphenols on aging considering oxidative stress, inflammation, autophagy, and gut microbiota. CritAl Rev Food Sci Nutr 61(13):2175–2193. https://doi.org/10.1080/10408398.2020.1773390

Xu Y, Han M, Huang M, Xu X (2021) Enhanced heat stability and antioxidant activity of myofibrillar protein-dextran conjugate by the covalent adduction of polyphenols. Food Chem 352:129376. https://doi.org/10.1016/j.foodchem.2021.129376

Yahfoufi N, Alsadi N, Jambi M, Matar C (2018) The Immunomodulatory and anti-inflammatory role of polyphenols. Nutrients 10(11):1618. https://doi.org/10.3390/nu10111618

Zamora-Ros R, Cayssials V, Jenab M, Rothwell JA, Fedirko V, Aleksandrova K, Tjønneland A, Kyrø C, Overvad K, Boutron-Ruault MC, Carbonnel F, Mahamat-Saleh Y, Kaaks R, Kühn T, Boeing H, Trichopoulou A, Valanou E, Vasilopoulou E, Masala G, Pala V, Panico S, Tumino R, Ricceri F, Weiderpass E, Lukic M, Sandanger TM, Lasheras C, Agudo A, Sánchez MJ, Amiano P, Navarro C, Ardanaz E, Sonestedt E, Ohlsson B, Nilsson LM, Rutegård M, Bueno-de-Mesquita B, Peeters PH, Khaw KT, Wareham NJ, Bradbury K, Freisling H, Romieu I, Cross AJ, Vineis P, Scalbert A (2018) Dietary intake of total polyphenol and polyphenol classes and the risk of colorectal cancer in the European Prospective Investigation into Cancer and Nutrition (EPIC) cohort. Eur J Epidemiol 33(11):1063–1075. https://doi.org/10.1007/s10654-018-0408-6

Zamora-Ros R, Cayssials V, Franceschi S, Kyrø C, Weiderpass E, Hennings J, Sandström M, Tjønneland A, Olsen A, Overvad K Boutron-Ruault MC, Truong T, Mancini FR, Katzke V, Kühn T, Boeing H, Trichopoulou A, Karakatsani A, Martimianaki G, Palli D, Krogh V, Panico S, Tumino R, Sacerdote C, Lasheras C, Rodríguez-Barranco M, Amiano P, Colorado-Yohar SM, Ardanaz E, Almquist M, Ericson U, Bueno-de-Mesquita BH, Vermeulen R, Schmidt JA, Byrnes G, Scalbert A, Agudo A, Rinaldi S (2020) Polyphenol intake and differentiated thyroid cancer risk in the European prospective investigation into cancer and nutrition (EPIC) cohort. Int J Cancer 146(7):1841–1850. https://doi.org/10.1002/ijc.32589

Zeisel SH (2007) Nutrigenomics and metabolomics will change clinical nutrition and public health practice: insights from studies on dietary requirements for choline2. Am J Clin Nutr 86(3):542–548. https://doi.org/10.1093/ajcn/86.3.542

Zelicha H, Kloting N, Kaplan A, Yaskolka Meir A, Rinott E, Tsaban G, Chassidim Y, Bluher M, Ceglarek U, Isermann B, Stumvoll M, Quayson RN, von Bergen M, Engelmann B, Rolle-Kampczyk UE, Haange SB, Tuohy KM, Diotallevi C, Shelef I, Hu FB, Stampfer MJ, Shai I (2022) The effect of high-polyphenol Mediterranean diet on visceral adiposity: the DIRECT PLUS randomized controlled trial. BMC Med 20(1):327. https://doi.org/10.1186/s12916-022-02525-8.

Zhang L, Angst E, Park JL, Moro A, Dawson DW, Reber HA, Eibl G, Hines J, Go VLW, Lu QY (2010) Quercetin aglycone is bioavailable in murine pancreas and pancreatic xenografts. J Agric Food Chem 58(12):7252–7257. https://doi.org/10.1021/jf101192k

Zhou HB, Chen JM, Cai JT, Du Q, Wu CN (2008) Anticancer activity of genistein on implanted tumor of human SG7901 cells in nude mice. World J Gastroenterol 14(4):627–631. https://doi.org/10.3748/wjg.14.627

Zoi V, Galani V, Lianos GD, Voulgaris S, Kyritsis AP, Alexiou GA (2021) The role of curcumin in cancer treatment. Biomedicines 9(9):1086. https://doi.org/10.3390/biomedicines9091086

# Chapter 4
# Polyphenols in Neuroprotection

**Ashwani Arya, Suchitra Nishal, Manish Dhall, Virender Kumar, and Vineet Mittal**

**Abstract** Polyphenols are naturally occurring bioactive compounds, generally vegetable-derived foods and have been recognised for therapeutic potential due to their potential assistances in management of neurodegenerative illness. These phytochemicals may possess several pharmacological profiles (reduction of oxidative stress, deposition of beta amyloids, neuro-inflammatory responses, mitochondrial dysfunction, aggregation of senile plaques, etc.). Several research findings have shown that these compounds can reduce inflammatory processes by regulating pro-inflammatory cytokines and antioxidant enzymes markers in terms of decreased progression of neurodegenerative disorders. Polyphenols may also target intracellular signalling pathways, ultimately reducing neuroinflammation and neuronal damage. Evidence from several literature findings may suggest the significant role of these polyphenols for neuroprotective effects. This chapter highlights the recent knowledge regarding the therapeutic mechanisms by which polyphenols may improve neuroprotective effects by influencing neuroinflammatory processes, oxidative stress, etc., in the neurodegenerative diseases, along with their significant potential therapeutic value. This chapter summarises available pre-clinical and clinical evidence findings in literature, which may demonstrate the capability of poyphenolic compounds

A. Arya
Department of Pharmaceutical Education and Research, Bhagat Phool Singh Mahila Vishwavidyalaya, Khanpur Kalan, Sonepat 131305, Haryana, India
e-mail: aryaashwani261@gmail.com

S. Nishal · M. Dhall · V. Kumar
College of Pharmacy PGMIS (SDPGIPS), Pt. B. D. Sharma University of Health Sciences, Rohtak, Haryana 124001, India
e-mail: suchitranishal50@gmail.com

M. Dhall
e-mail: dhall_manish123@yahoo.co.in

V. Kumar
e-mail: sachdeva.virender5@gmail.com

V. Mittal (✉)
Department of Pharmaceutical Sciences, Maharshi Dayanand University, Rohtak, Haryana 124001, India
e-mail: drvineet.pharma@mdurohtak.ac.in



R. K. Singla et al. (eds.), *Polyphenols in Health*,
https://doi.org/10.1007/978-3-032-27324-6_4

in terms of slow progression of diseases, also improving cognitive impairments and supporting brain health. Similarly, the chapter discusses the limitations and challenges linked with the transfer of these observations into clinical practice and may highlight future directions in terms of therapeutic potential of polyphenols in neurodegenerative disorders. This work aims to contribute to the development of novel treatment approach when speaking of polyphenols and positive neuroprotective effects, which may expand the recovery and improve health conditions of affected individual by these debilitating conditions worldwide.

**Keywords** Alzheimer's disease · Antioxidant · Beta amyloid · Bioactive compound · Inflammation · Neuroprotection · Polyphenol

## Abbreviations

| | |
|---|---|
| 3-(4,5-dimenthylthiazol-2-yl)-2,5-dimethyltetrazolium bromide | MTT |
| Acetylcholinesterase | AChE |
| Alzheimer's disease | AD |
| Beta amyloid | Aβ |
| Blood–brain barrier | BBB |
| Brain-derived neurotrophic factor | BDNF |
| cAMP-response element-binding protein | CREB |
| Choline acetyltransferase | ChAT |
| Double-blind placebo control study | DBPCS |
| Epigallocatechin gallate | EGCG |
| Epigallocatechin-3-gallate nanoparticle | Nano-EGCG |
| Extracellular-signal-regulated kinase | ERK |
| Food frequency questionnaire | FFQ |
| Hen egg white lysozyme | HEWL |
| Huntington's disease | HD |
| Mini-Mental State Examination Japanese version | MMSE-J |
| Mitogen-activated protein kinase | MAPK |
| Nuclear factor kappa-light-chain-enhancer of activated B cells | NF-kβ |
| Nuclear factor-erythroid 2-related factor 2 | Nrf2 |
| Parkinson's disease | PD |
| Reactive oxygen species | ROS |
| Sirtuin 1 | SIRT1 |
| Superoxide dismutase | SOD |
| Thioflavin T | ThT |
| Total flavonoid content | TFC |
| Total phenolic content | TPC |

## 4.1 Introduction

Polyphenolic compounds represent a heterogeneous group of naturally derived bioactive molecules that have gained growing attention due to their potential role in neuroprotection (Shi et al. 2024). These compounds are generally obtained from herbs and are available as secondary metabolites from defensive and pigmentation mechanisms. Because of the extensive range of chemical structures of these compounds, polyphenols possess several pharmacological potentials such as anti-inflammatory, antioxidant, and neuroprotective effects, which may indicate the beneficial role of these molecules against neurodegenerative illness (Jalouli et al. 2025). Recent studies have demonstrated the significant role of polyphenols in reducing the progression and development of neurodegenerative illness (Shi et al. 2024). They are available in several herb-based food materials vegetables, coffee, fruits, red wine, etc. (Thakur and Kumar 2024). Due to their pharmacological potential such as anti-inflammatory, antioxidant as well as neuroprotective effects, these compounds have been recognised as significant agents for therapeutic interventions towards the management of neurodegenerative illness (Shi et al. 2024). Several findings including clinical, animal experimentations as well as in vitro findings have shown the protective role of these compounds (Figueira et al. 2017a; Song et al. 2024).

Neurodegenerative illness may lead to progressive memory impairment and neuronal loss; these conditions may have notable consequences such as cognitive dysfunctions, movement impairment, and negative effects on social and day-to-day life of individuals (Sing et al. 2024). Lack of treatments as well as the progressive impairment of neurodegenerative illnesses such as Alzheimer's disease (AD), Huntington's disease (HD), and Parkinson's disease (PD) may represent the major challenges to healthcare sector (Caruso et al. 2022; Giri et al. 2024; Gong et al. 2024; Madhubala et al. 2024). The effective therapeutic intervention still remains a challenging approach for reducing the progression of these disorders. Several key factors such as deposition of beta amyloids (Aβ), senile plaques, oxidative stress, neuroinflammation, mitochondrial dysfunctions, and dysfunctions of nuclear factor-erythroid 2-related factor 2 (Nrf2) level may be correlated with neurodegenerative illness. Several key pathological hallmarks of these disorders, necroinflammation, and oxidative stress represent the central key factors associated with progression and development of neurodegenerative illness. Several pro-inflammatory markers including tumour necrosis factors and interleukins have been associated with neuronal damages and development of neurodegeneration (Jalouli et al. 2025). In this regard, naturally available polyphenols have been suggested in the ambit of promising therapeutic approaches against neurodegeneration due to the wide range of multitherapeutic potent pharmacological profiles, including reduction of neuroinflammations, antioxidant action, reduction of Aβ deposition, and accumulation of neurofibrillary tangles in the brain. Several pre-clinical findings mentioned in the available literature show the apparent therapeutic efficacy of several polyphenols such as curcumin, resveratrol, and epigallocatechin gallate for the improvement of cognitive impairments and the decrease of the progression of neurodegenerative illness, including AD (Chen et al. 2025).

Several pathological hallmarks such as oxidative stress, Aβ deposition, senile plaques, neuroinflammation, mitochondrial dysfunction, and Nrf2-related dysfunction may lead to neuronal loss with consequences for the progression of neurodegeneration illness. By targeting these therapeutic targets, polyphenols may show anti-inflammatory effects, antioxidant potential, etc. (Chen et al. 2025; Grabska-Kobyłecka et al. 2023; Jalouliet al. 2025).

The aim of this chapter is to highlight the recent knowledge regarding therapeutic mechanisms by which polyphenols may improve neuroprotective effects. This chapter summarises the available pre-clinical and clinical evidence findings in literature, which may demonstrate the capability of polyphenolic compounds in terms of slow progression of diseases, improvement of cognitive impairments, and positive effects on brain health (Mayer et al. 2024). Similarly, the chapter discusses limitations and challenges linked with the practical translation of these observations into clinical practice, also highlighting future directions in terms of therapeutic potential of polyphenols in neurodegenerative disorders (Socała et al. 2024). This work aims to contribute to the development of novel treatment approaches that may improve outcomes and quality of life for individuals affected by these debilitating conditions worldwide.

## 4.2 Methodology

Polyphenolic compounds ascribed to have therapeutic efficacy in biological systems are presently recognised for neuroprotective effects. The therapeutic potential of polyphenols for neuroprotection has been reviewed in this chapter. For this purpose, several databases-related information including Pubmed, Scopus, Web of Science, etc., were investigated by using key words such as 'Polyphenols + neuroprotection', 'Polyphenols + neurodegenerative diseases', and 'Polyphenols + anti-inflammatory oxidative stress'. The role of polyphenols is elucidated in the following sections when speaking of neuroprotection.

## 4.3 Role of Polyphenols in Neuroprotection

Neurodegenerative ailments are illustrated by gradual loss of neurons and impaired neural function, which lead to cognitive decline, movement difficulties, and a marked reduction in quality life of diseased persons (Singh et al. 2024). Although decades of intensive research have been devoted to the understanding of these disorders, the development of effective therapeutic interventions remains a significant challenge. In this context, polyphenolic natural compounds have gained increasing attention owing to their promising neuroprotective assets (Shi et al. 2024). Recent research has increasingly highlighted the prospective role of polyphenols in slowing progression of neurodegenerative ailments and supporting cognitive health (Jalouli et al. 2025; Shi et al. 2024). These natural compounds are widespread in natural foods such as tea, fruits, coffee, and vegetables along with red wine (Thakur and Kumar 2024). As a result of their anti-inflammatory, strong antioxidant, and neuroprotective effects, polyphenols have drawn significant interest for their possible therapeutic applications in managing neurodegenerative conditions (Shi et al. 2024). The investigation of plant-derived polyphenols as preventive agents against neurodegeneration and other aeging-associated disorders forms a notable part of a broader effort to promote healthy aeging through natural interventions. In vitro experiments, cell culture analyses, and clinical trials have collectively demonstrated the defensive actions of dietary polyphenols over brain health. (Figueira et al. 2017a; Song et al. 2024). Polyphenols may activate the mitogen-activated protein kinase (MAPK)/ extracellular-signal-regulated kinase (ERK) pathway. In addition, they can favour the improvement of memory impairments, the survival of neuronal cells, and the enhancement of synaptic plasticity by means of positive effects on cAMP-response element-binding protein (CREB) and sirtuin 1 (SIRT1), with the consequent increase of brain-derived neurotrophic factor (BDNF). These polyphenols may also be targeted to decrease the nfr-2 level, which may enhance antioxidant enzymes like catalase, glutathione, and superoxide dismutase (SOD) to subside oxidative stress, which may reduce the generation of reactive oxygen species (ROS) and shown protective mechanismd against cellular damages and lipid peroxidation. By targeting the inhibition of nuclear factor kappa-light-chain-enhancers of activated B cells (NF-kβ) signalling expression, polyphenols may decrease the amount of pro-inflammatory markers such as interleukins and tumour necrosis factor-α with consequent decrease of neuroinflammatory responses and the improvement of the homeostasis of neuronal levels (Fiore et al. 2025). Literature findings show the role of polyphenols in neuroprotection as follows.

D'Errico and coworkers extracted 24 phenolic and 25 flavonoids from several Mediterranean forage crops and analysed their activity against enzymes such as acetylcholine esterase and butylcholine esterase, which are essential for the process of neurotransmission (D'Errico et al. 2025). In brief, the work demonstrated that extracts of Mediterranean forage crops such as *Medicago sativa* may be employed as possible alternates for therapeutic AD management.

As per literature findings, the optimised seeds extraction of *Celastrus paniculatus* may enhance total phenolic acid and total flavonoid contents as compared to the conventional one. Green extracts of *C. paniculatus* seeds may improve behavior parameters, enhance oxidative enzymes markers such as SOD, catalase, and gluthione, decrease nitric oxide levels, and inhibit acetylcholinesterase (AChE) enzyme in experimental animal studies (Arya et al. 2022). As per literature findings, the pharmacological potential for AChE inhibitory potential of herbal extracts may be due to enhanced concentration of the polyphenol's herbal extracts. Data interpretation may suggest the significant role of polyphenols for AD management (Melo

Sarkar et al. 2016; Sharififar et al. 2012). Thus, as per these findings, the neuroprotective effects of herbal extract against oxidative stress may be due to improved concentration of polyphenols in plant actives including total phenolic content (TPC) and total flavonoid content (TFC) in green extraction methods.

Filiberti and colleagues extracted polyphenol extracts from potato peel wastes using aqueous ethanol (Filiberti et al. 2025). Potato peel waste polyphenols may show considerable radical scavenging activity in low concentration when exposed to oxidative stress, when speaking of HT-22 neuronal cell line without stress.

Carrotta and coworkers studied the comparative inhibition of Aβ-peptides aggregation by using aqueous extracts of *Moringa oleifera* at room temperature and high temperature on SH- SY5 cell lines. Both extracts were found to be efficient in reducing gallic acid-stimulated Aβ1-42 formation as well as mitochondrial ROS (Carrotta et al. 2025).

Debnath and coworkers used polysuccinimide derivatives to create epigallocatechin-3-gallate nanoparticles (nano-EGCG). By preventing extracellular Aβ aggregation, nano-EGCG effectively shielded neuronal cells from harmful effects. Nano-EGCG was more effective due to its improved chemical stability against degradation, greater interaction with protein aggregates, and successful cell entrance (Debnath et al. 2016).

Han and colleague prepared a 4:6 ratio mixture comprising phlorotannin and fucoidan derived from *Ecklonia cava* and assessed cognitive functions by using passive avoidance, Morris water maze, and Y-maze scans (Han et al. 2021). The mixture was found to exhibit a mitigating impact on Aβ-derived learning and memory failures together with control on expression of choline acetyltransferase (ChAT) and AChE in brain tissue. Resveratrol, epigallocatechin gallate (EGCG), and curcumin are examples of polyphenolic substances that have been shown to decrease activation of microglial and generation of pro-inflammatory cytokines (Liu et al. 2023). Blueberry supplements may enhance the memory in experimental animals, which may be due to increase BDNF levels in brain samples and CREB activation. These flavonoids also increased the level of ERK-I/II phosphorylation for improving long-term memory impairment (Williams et al. 2008). Similarly, blueberry extracts may improve the spatial memory by changing the parameters of hippocampal plasticity such as hippocampal neurogenesis and the activation of the extracellular kinase receptors after the administration of blueberry-rich diet for several weeks (Casadesus et al. 2004).

By preventing Aβ formation, lowering oxidative stress, and regulating microglial activation, curcumin offers neuroprotection (Azzini 2024). Manap and coworkers analysed four natural substances such as curcumin, bacoside A, piperine, and chebulinic acid for confirming cognitive effects through in vitro studies using 3-(4,5-dimenthylthiazol-2-yl)-2,5-dimethyltetrazolium bromide (MTT) assessment, thioflavin T (ThT) assay, and AChE enzyme inhibition assay. In vitro AChE repression revealed that curcumin—together with piperine—exhibited higher AChE inhibition than either alone. This combination maintained cell viability of 85% in SH-SY5Y cell line model, suggesting that these chemicals prevent Aβ-induced neuronal injury (Manap et al. 2019).

Akinyemi and colleagues evaluated impact of curcumin on activity of enzymes acetylcholinesterase and adenosine deaminase using seven groups of rats induced with cadmium for memory loss. Curcumin co-treatment at 12.5 and 25 mg/kg, respectively, improved the cerebral cortex's non-enzymatic antioxidant status and nitric oxide while lowering the amount of malondialdehyde (Akinyemi et al. 2017).

Chen and coworkers discovered that the most significant Aβ-degrading enzyme, neprilysin, can be upregulated by poly-substituted hydroxylcurcuminoids (Chen et al. 2016). The fascaplysin compound, isolated from Fijian sponge *Fascaplysinopsis* sp, showed neuroprotective benefits through stimulating P-glycoprotein expression and inhibiting AChE activity (Manda et al. 2016). Mohammadi and colleagues explored effects of curcumin and diacetylcurcumin (a synthetically derived curcumin) on Aβ fibrillation of hen egg white lysozyme (HEWL). Obtained results indicated that diacetylcurcumin prevented HEWL amyloid fibrillation more successfully than curcumin (Mohammadi et al. 2016).

Martino et al. (2016) investigated the inhibitory activity of curcumin analogues on β-site APP-cleaving enzyme 1 and glycogen synthase kinase-3β. Docking study results showed that the benzyl rings of Aβ 17-based analogues have the best binding affinities since they can only be adorned with tiny substituents, like a fluorine atom. The affinities of curcumin and its derivative for binding on Aβ were also analysed. In contrast to curcumin, the curcumin derivative attenuated the interaction among Aβ and cell membranes, hence tempted better neuroprotection (Qi et al. 2017). Other finding by de Gomes and coworkers evaluated the impact of curcumin-loaded lipid-core nanocapsules on sickness behaviour models and observed behavioural alterations along with reduction in neuroinflammation (de Gomes et al. 2020). With reference to another compound, quercetin, this compound may increase the regulation of complexes of electron transport chain in impaired and normal regulation of neuronal functions in dopaminergic system. Quercetin inhibits the dose-dependent effects in the programmed neuronal loss in nigral neurons, which may possess neuroprotective effects (Karuppagounder et al. 2013). Another polyphenolic extract such as Neumentix has been reported to upgrade the impairment in the neurobehavioural parameters at the dose of 134 mg/kg for two consecutives weeks (Bian et al. 2020).

Polyphenols of lychee seeds may exhibit neuroprotective effects by improving the spatial memory in experimental animals by targeting NLRP-III levels and improve blood–brain barrier (BBB) (Xiong et al. 2021). Resveratrol reversed cognitive deficits in sevoflurane-induced cognitive decline in experimental animals using Morris water maze models at a dose of 100 mg/kg for six consecutive days. Similarly, resveratrol can also reduce the activation of neuronal loss of glial cells at a dose of 30 mg/kg in ischemia-induced neuronal cell loss in experimental animals (Wang et al. 2002). Caffeic acid exhibited neuroprotective effects and also improved amnesia, enhancing neurological scores and working memory at the dose of 60 mg/kg in experimental mice (Pinheiro Fernandes et al. 2014).

Supplemented polyphenols may advance the activity of the brain. Curcumin obtained from *Curcuma longa,* piperine, resveratrol, and eugenol have been reported as reversible antagonistic potential for monoamines oxidase enzyme, which may be linked with degradation of serotonin level (Zhang et al. 2019). Herbal rich-polyphenols dietary intake may improve cognitive dysfunctions and also be ascribed to positively influence pharmacological profiles, including anti-inflammatory, antioxidant, and neuroprotective potential in the management of the neurodegenerative disorders (Grabska-Kobyłecka et al. 2023). Resveratrol and curcumin, which are natural occurring polyphenolic compounds, are reported to decrease the hyperphosphorylation of tau proteins and decrease the oxidative stress, which may indicate their role in neurodegenerative illness as pathological hallmarks associated with this kind of disorders (Chenet al. 2025; Yu et al., 2022). Gallic acid and its derivatives may decrease apoptosis and enhance cell viability, which may mean a significant role in PD prevention (Lu et al. 2006).

De Lorenzi and colleagues investigated how curcumin and its structurally similar analogues, cur6 and cur16, affected the control of Aβ aggregation. Anti-oligomeric action was also observed during in vitro Aβ-42 aggregation process (De Lorenzi et al. 2022). The green extract of *Nardostachys jatamansi* roots may improve polyphenolic profiles including TPC and TFC values if compared to conventional extract. The green extract of *Nardostachys jatamansi* roots may also improve the scavenging inhibition of 2,2-diphenyl-1-picrylhydrazyl (in vitro studies), reduce nitric oxide inhibition, and enhance glutathione, SOD, and catalase levels in experimental animals. These findings may indicate the role of polyphenolic compounds for the neuroprotection against oxidative stress (Arya et al. 2022).

Paramakrishnan and coworkers explored the role of astaxanthin in cerebrovascular damage (induced with injections of streptozotocin) associated with AD in zebrafish. Obtained results indicated that astaxanthin exhibited multitarget activity for therapeutic AD management (Paramakrishnan et al. 2023).

***Arcone and colleagues*** examined the impact of polyphenols derived from *Citrus limon* peel waste on important cholinergic as well as aminergic neurotransmission enzymes, including butyryl cholinesterase and monoamine oxidases, as well as on Aβ 1–40 aggregation along with superoxide dismutase, that influence oxidative stress. These findings showed that the extract displayed multitarget activity. As a result, it can be considered a viable option for managing and preventing neurological conditions (Arcone et al. 2023).

Maiuolo and coworkers carried a study on polyphenolic content obtained from extract of fibres of bergamot (from solid residues) exhibited protective action on the damage induced by treatment with Aβ (Maiuolo et al. 2023). Adhikari-Devkota and colleagues assessed the possible inhibitory effects of extract, fractions, and isolated components from *Citrus* 'Hebesu' on inflammatory reactions caused by lipopolysaccharide using BV-2 murine microglial cells. Citrus "Hebesu" peels may be consumed to help prevent and treat neurological disorders (Adhikari-Devkota et al. 2019).

Narasimhan and coworkers explored scopoletin, an active ingredient derived from *Morinda citrifolia* for attenuation of oxidative stress. It was found to reduce protein aggregation despite lowering oxidative stress, as seen by lower α-synuclein and nitrotyrosine levels. Through the prevention of oxidative damage, scopoletin reduced rotenone-derived apoptosis in SH-SY5Y cells, according to in vitro experiments.

Additionally, scopoletin was reported to enhance neural defence system (Narasimhan et al. 2019).

Another study by Wang and Wang discussed the results of an in vitro oxygen–glucose deprivation/reoxygenation model in primary cortical neurons, based on poncirin (Wang and Wang 2022). Briefly, poncirin inhibited oxidative stress and mitochondrial dysfunction, shielding cortical neurons from oxygen–glucose deprivation/reoxygenation damage.

Kurmi and coworkers looked into the inhibitory effect of polyphenolic stilbene derivatives by employing molecular docking (Kurmi et al. 2025). Related findings suggest that good interaction profiles and high binding affinity can serve as basis for a promising AD therapy. On the other hand, Mishra and colleagues investigated the neuroprotective effect of piperine-incorporated casein micelles for Aβ 1–42-derived neurotoxicity (Mishra et al. 2026). Piperine micelles outperformed free piperine in terms of oxidative stress mitigation, antioxidant activity, and mitochondrial homeostasis restoration. Additionally, cells treated with micelles showed notable improvements in intracellular ROS production, calcium imbalance, mitochondrial superoxide levels, mitochondrial membrane potential, and the expression profile of autophagy and mitophagy markers. Piperine-containing micelles not only increased piperine bioavailability, but also enhanced its neuroprotective benefits. Hence, this analysis offered a viable platform for creating efficient AD therapies by fusing neuroprotective qualities of piperine with improved transport capabilities of casein micelles.

Hou and coworkers created a multipurpose nanocomposite with the ability to scavenge ROS (Hou et al. 2026). Graphene quantum dots were encased in a metal-polyphenol network made of physiologically Co(II)-coordinated EGCG. By reducing oxidative stress and controlling microglial activation, it successfully suppressed Aβ-42 amyloidosis. It has been reported in vivo to be able of passing through BBB and drastically reduce Aβ-42 accumulation in APP/PS1 mice. Furthermore, it may reduce neuroinflammation by scavenging intracellular ROS and controlling microglia's transition.

EGCG has been reported to enhance antioxidant enzymes markers (SOD, glutathione, and catalase) and promote Nrf2 translocation of levels, which may activate the response of transcription antioxidant genes at 50 mg/kg dose (Tang et al. 2024).

Several polyphenolic compounds may exhibit neuroprotective impacts through inhibition of NF-kβ signalling expression and MAPK pathway, which may participate in the reuction of neuroinflammation and enhancement of survival of neuronal cells (Chen et al. 2025). Polyphenols may possess neuroprotective effects and reduce neuroinflammation, which are key factors involved in pathological hallmarks in neurodegenerative illness by decreasing pro-inflammatory markers of tumour necrosis factor, interleukins and inhibiting MAPK and Nf-kβ signalling expression in brain samples (Chen et al. 2025; Singh et al. 2020). Literature described preclinical studies which indicated that resveratrol reduced Aβ deposition, nuroinflammatory responses, and decreased the level of oxidative stress at the dose of 100 mg/kg. This compound may produce neuroprotective effects in experimental animals by the targeting mechanism of inhibition of the Nf-kβ expression (Pasinetti 2012).

Propolis-rich polyphenolic compound at doses of 0.5 mg/kg may produce neuro-protective effects and behavioural estimation in experimental induced PD models. These polyphenols may improve the survival of neuronal cells, also attenuating oxidative stress, and favouring the deposition of synuclein as well as preservation of the neuronal cells in dopaminergic system (Mamashli et al. 2023).

Naringenin may reduce aggregation of Aβ plaques and protect the hyperphosphorylation of tau proteins from neuronal toxicity (D'Egidio et al. 2025). Similarly, this compound may reduce the activation of microglia and the amount of pro-inflammatory cytokines for the generation of neuroinflammation, which may link to degeneration of neurons (Yang et al. 2014). In addition to the reduced oxidative stress along with neuroinflammatory responses, naringenin exhibits critical function in improving cognitive dysfunctions and the dysfunction of mitochondria, which may be implicated in AD (Davinelli et al. 2023). Similarly, this compound may improve memory, decrease oxidative stress, as well as reduce the inflammatory responses in experimental animals at dose 50 mg/kg (Ghofrani et al. 2015; Khan et al. 2012; Ma et al. 2013a).

In general, literature findings reported that polyphenols may be potentially utilised for the development of lead compounds and used a moiety for the possible therapeutic AD intervention. Tables 4.1 and 4.2 present active polyphenolic compounds with respective therapeutic mechanisms linked with neuroprotective effects.

### 4.3.1 Mechanism of Polyphenols for Neuroprotective Effects

Neuroinflammation is a conventional pathological attribute of numerous neurodegenerative forms, including PD, AD, and multiple sclerosis. It is characterised by glial cells activation and increased release of pro-inflammatory cytokines alongside subsequent disruption of normal neuronal functions (Jalouli et al. 2025). These actions are attributable to strong antioxidant anti-inflammatory assets of polyphenols (Sun et al. 2024). Polyphenols may activate the MAPK/ ERK pathway. Moreover, they can favour the improvement of memory impairments, the survival of neuronal cells, and the enhancement of synaptic plasticity by means of positive effects on CREB and SIRT1, with the consequent increase of brain-derived neurotrophic factor (BDNF). These polyphenols may also be targeted to decrease nfr-2 level, which may enhance antioxidant enzymes (catalase, SOD, and glutathione) to lower oxidative stress levels, which may reduce ROS generation and show protective mechanisms against cellular damages and lipid peroxidation. By targeting the inhibition of nuclear factor kappa-light-chain-enhancers of activated B cells (NF-kβ) signalling expression, polyphenols may decrease the amount of pro-inflammatory markers such as interleukins and tumour necrosis factor-α with consequent decrease of neuroinflammatory responses and the improvement of the homeostasis of neuronal levels (Fiore et al. 2025; George et al. 2012).

A key pathway through which polyphenols confer neuroprotection involves the regulation of oxidative stress, which plays a central role in driving neuroinflammatory

**Table 4.1** Role of active polyphenolic compounds in neuroprotection: curcuminoids, resveratrol, quercetin, rutin, catechin, EGCG, kaempferol, and lignans

| Active constituent | Dietary source of polyphenols | Mechanism of action | References |
|---|---|---|---|
| Curcuminoids | Curcumin | Inhibition of NF-κB, Aβ aggregation, Toll-like receptor 4, AChE inhibitions, and alleviation of mitochondrial dysfunction | Qureshi et al. (2017), Jafarisavari et al. (2025), Panaro et al. (2020), Tang (2017), Bagheri et al. (2020) |
| Resveratrol | Grape skin, red wine | Improved mitochondrial functioning; activation of silent information regulator-1; suppression of Aβ aggregation and alteration of τ protein phosphorylation, suppresses p53 activity | Yadav et al. (2022), Zhu et al. (2025), Puranik et al. (2025) |
| Quercetin | Apple, citrus fruits, parsley, tea, onions, sage | Up/down regulation of paraoxonase-2, Nrf2, protein kinase C, c-Jun N-terminal kinase, MAPK signalling cascades, and PI3K/Akt pathways; it may either up-regulate or down-regulate cytokines; Inhibition of Aβ fibrils | Zaplatic et al. (2019), Dos Santos et al. (2026), Islam et al. (2025) |
| Rutin | Buckwheat, cherries, apricots, grapefruit, grapes, oranges etc | Prevention of tau aggregation and cytotoxicity by tau oligomer; reduction of generation of pro-inflammatory cytokines; anti-apoptosis, antioxidant, inhibits p-38 MAPK pathway | Sun et al. (2021), Cordeiro et al. (2021) |
| Catechin, EGCG | Green tea, cocoa beans | Prevention of Aβ aggregation, anti-inflammatory action, AChE inhibition; regulation of signalling pathways for instance, PI3K/Akt, Nrf2/ARE, and MAPK | Yuca et al. (2025), Xicota et al. (2017), Nan et al. (2021), Kopustinskiene et al. (2021) |
| Kaempferol | Green leafy vegetables | Lowering inflammation and oxidative stress; inhibition of Aβ generated neurotoxicity, regulation of cholinergic system | Nejabati et al. (2022), Moalefshahri et al. (2025) |
| Lignans | Wheat, flax, barley, buckwheat, millet, oats, rye, etc | Disruption of the heme oxygenase-1 (HO-1)/NLRP3/nuclear factor E2-related factor 2 (Nrf2) pathway and activation of AMP-activated protein kinase signalling | Zhang et al. (2025), Zhou et al. (2021) |

responses (Tavan et al. 2024). These compounds demonstrate potent antioxidant activity, allowing them to neutralise ROS and to reduce oxidative damage to neuronal cells (Hu et al. 2024; Yahia et al. 2024). By scavenging free radicals, polyphenols help to maintain integrity of neurons and suppress activation of pro-inflammatory signalling pathways. Figure 4.1 depicts the therapeutic mechanism of polyphenols

**Table 4.2** Role of several active polyphenolic compounds in neuroprotection: mangiferin, apigenin, luteolin, genistein, oleuropein, and fisetin

| Active constituent | Dietary source of polyphenols | Mechanism of Action | References |
|---|---|---|---|
| Mangiferin | Mango tree | Attenuated oxidative stress and protection effects from mitochondrial dysfunction as well as regulation of signalling pathways like NLRP3, Nrf2, NF-kB, and MAPK | Shang et al. (2025) |
| Apigenin | Chamomile, parsley, artichokes, and celery | Reduced intensity of NF-κB p65<br>Increase in antioxidant molecules SOD and GSH-px, MARK4 inhibition | Zhang et al. (2025), Hussain et al. (2025) |
| Luteolin | Thyme, parsely, and celery | Augmented antioxidant activity by stimulation of Nrf2/ARE<br>Inhibit NFκB, MAPK/AP-1, and JAK-STAT pathways | Kwon et al. (2017), Mahto et al. (2025) |
| Genistein | Soyabean, peanuts, and legumes | Inhibit Aβ aggregation induced by glucose<br>Regulation of acetyl choline esterase | Gonzalez et al. (2025), Sharma et al. (2025) |
| Oleuropein | Olive oil | Removal of Aβ aggregates of | Fekete et al. (2025), Riquelme et al. (2025) |
| Fisetin | Strawberries, grapes, apple, and onion | Regulation of signalling pathways like ERK and CREB | Nabizadeh et al. (2024) |

for the reduction of inflammatory response and exerts its neuroprotective effects changing signalling pathways of microglia. Polyphenols may decrease the release of pro-inflammatory markers, especially interleukins and tumour necrosis factor, to produce neuroinflammation. Polyphenols also inhibited the Toll-free receptors which may be expressed in microglial cells. In addition to that, these compounds may also inhibit the transcription signalling factors NF-κB), p-38 MAPK pathways. Polyphenols decrease the activation of microglial signalling, and reduction of NF-κB may ultimately lower inflammatory enzymes and the generation of nitric oxide. These pathways are associated with the generation of oxidative stress and neuronal damages. By targeting this multiple therapeutic mechanism, polyphenols may protect the neuron and reduce the production of oxidative stress and neuronal loss. This broad anti-inflammatory mechanism may favour neurons survival and slow progression and development phases in neurodegenerative illness.

Figure 4.2 shows multitherapeutic mechanisms of polyphenols for neuroprotection. This figure may depict that these polyphenols may be targeted as anti-apoptotic effects by decreasing the mitochondrial depolarisation and protecting the dysfunction of mitochondrial function, which may be impaired in neurodegenerative illness. Polyphenols may improve BBB. They may possess anti-inflammatory potential by

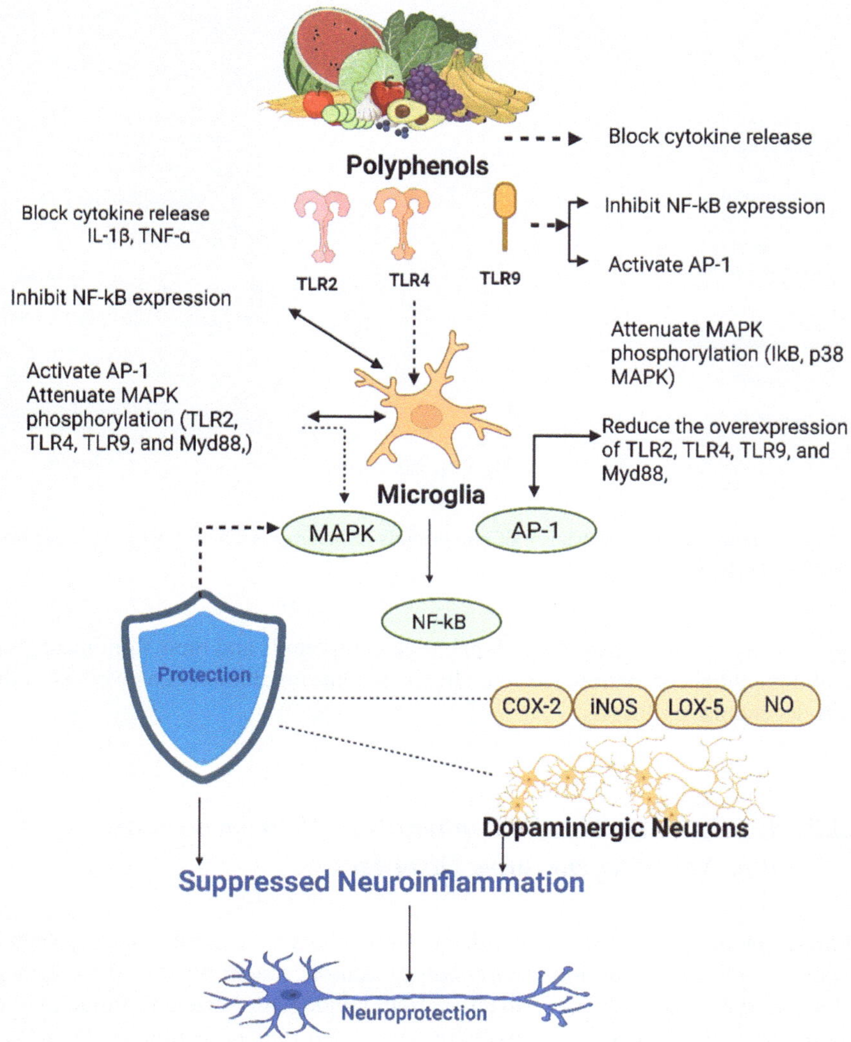


**Fig. 4.1** Targeted therapeutic mechanisms of polyphenols for neuroinflammation. Neuroprotective effects. This picture has been created by the authors

reducing pro-inflammatory markers. These compounds may modulate cell signalling pathways and inhibit the aggregation of proteins, which may be targeted to inhibited Aβ, α-synuclein plasticity, and improved cognitive functions. They also stimulate brain-derived factors, promoting the neurogenesis, and may activate the level of Nrf2 and improve antioxidant enzymes markers. Increasing research findings

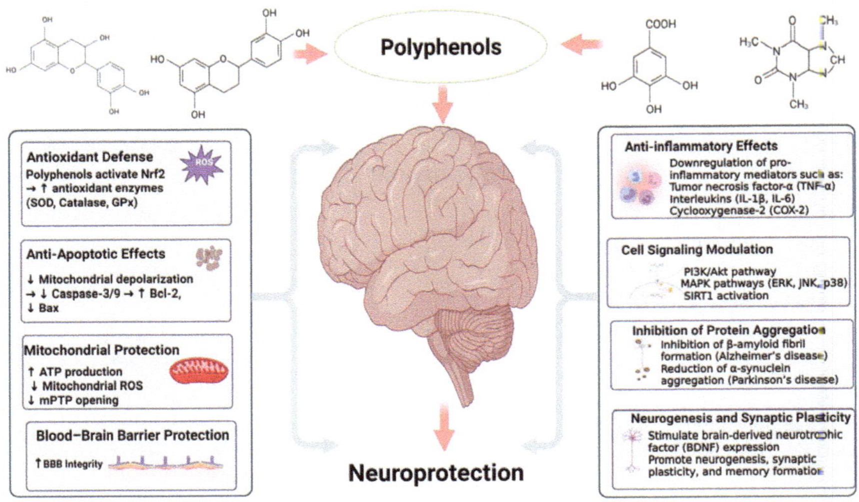


**Fig. 4.2** Multitherapeutic mechanism of polyphenols for neuroprotection. This picture has been created by the authors

suggested a notable interest for polyphenolic compounds and their significant therapeutic potential for neuroprotective effects in neuroinflammation as well as in the neurodegenerative illness.

### *4.3.2 Clinical Studies of Polyphenols in Neuroprotection and Neurodegenerative Disorders*

Clinical evidence reveals that dietary polyphenols exhibit neuroprotective responses and may delay the progression of neurodegenerative disorders through mitochondrial-protective, antioxidant, anti-amyloidogenic, and anti-inflammatory actions. Also, they can attenuate oxidative stress, neuroinflammation, and pathological protein aggregation. Curcumin has been investigated in various clinical studies in AD patients. Although early studies depicted limited cognitive improvement owing to poor bioavailability, later trials using better formulations demonstrated reduced amyloid plaque burden and enhanced memory scores, along with decreased inflammatory biomarkers.

Ringman and coworkers were among the first to investigate curcumin in AD patients, confirming safety but limited cognitive improvement, mainly attributed to poor systemic bioavailability (Ringman et al. 2005). Morillas-Ruiz and colleagues accomplished a multicenter double-blind controlled assessment to measure impact of antioxidant drink using 100 participants divided in: a 52 patients-control group; a second group of early phase-AD patients (24 subjects); and a third group of moderate phase-AD patients (24 subjects). Briefly, it was concluded that antioxidant drinks having polyphenols may decrease the plasmatic concentrations of total homocysteine when consumed on regular basis (Morillas-Ruiz et al 2010).

In Japan, evaluation was done by means of a double-blind, randomised controlled group (29 female and 4 male participants) for studying the impact of drinking green tea on cognitive loss. Green tea powder (2 g/day) with 220.2 mg catechins was given to the participants, and assessment was done using laboratory examination and Mini-Mental State Examination Japanese version (MMSE-J). Significant impact of drinking green tea was not observed on cognitive functions as per the observations of MMSE-J; however, oxidative stress was attenuated in elders (Ide et al. 2015).

In another study, a 12-week randomised controlled trial evaluation of cognitive functions was done after giving cherry juice (200 ml/day) with little anthocyanin content in older (above 70 years) patients having mild to moderate dementia. Vocal, long-term memory, and short-term memory were found to be improved in case of group taking cherry juice (Kent 2017). Green tea consumption for two months was found favourable for improving cognitive functions as per the observation of study on patients suffering from severe AD (Arab 2016).

A study indicating controlled, randomised as well as doubleblinded studies of the coca flavanol for the cross over trial of three periods, demonstrated the improved cognitive dysfunctions at the dose of 520 mg/kg and 994 mg of coca flavanol drinks (Scholey et al. 2009). In addition, a study reported that blueberries and polyphenols extracts obtained from grapes involving the double blinded-randomised and placebo-controlled assessment may improve memory and cognitive dysfunctions in 60 to 70 years-old old patients after the administration of 258 mg of flavanol per day (duration: six months) (Bensalem et al. 2019). Similarly, blueberry supplements comprising anthocyanin during the randomised clinical studies concerning 54 healthy children have indicated the improvement in memory and cognitive impairments as well verbal memory and executive functions of brains (Barfoot et al. 2019). Subsequently, small and coworkers revealed that a bioavailable formulation of curcumin administered to non-demented older adults potentially enhanced memory and attention while reducing amyloid and tau accretion in brain areas associated with AD (Small et al. 2018).

Resveratrol has shown promising results in randomised controlled trials (mild to moderate AD), where long-term supplementation was associated with stabilisation of cognitive decline, reduction of cerebrospinal fluid Aβ40, and attenuation of neuroinflammation. Turner and colleagues reported stabilisation of cognitive decline and altered levels of cerebrospinal fluid Aβ40 in mild to moderate AD patients following long-term resveratrol supplementation (Turner et al. 2015). Supporting these results, Moussa and coworkers demonstarted that resveratrol decreases neuroinflammatory markers and modulates immune responses in individuals with AD (Moussa et al. 2017).

Polyphenols from green tea, particularly EGCG, have been clinically evaluated in PD and cognitive mutilation, depicting modest improvements in motor symptoms, reduced oxidative stress markers, and favourable safety profiles. Green tea polyphenols, mainly EGCG, have been clinically evaluated for neuroprotective potential, with studies conducted by Mandel and Youdim, revealing improvements in oxidative stress parameters and motor signals in PD (Mandel and Youdim 2004).

Flavonoids, namely quercetin and anthocyanins, have depicted cognitive benefits in elderly populations and individuals with mild cognitive impairment, improving memory, executive function, and cerebral blood flow. Clinical trials involving cocoa flavanols and grape polyphenols have further indicated enhanced neurovascular coupling and cognitive performance. Desideri and colleagues studied that elder patients of mild cognitive injury exhibited improvements in executive function and processing speed following high flavanol cocoa administration (Desideri et al. 2012).

Berry-derived anthocyanins have revealed similar effects with Krikorian and coworkers confirming improved verbal memory and learning in older adults after blueberry supplementation (Krikorian et al. 2010). Additionally, flavonoid-rich interventions such as cocoa flavanols and berry-derived anthocyanins have depicted cognitive benefits in elderly populations and individuals with mild cognitive impairment, as evidenced by other researchers (Lamport et al 2011; Whyte and Williams 2015). Collectively, these clinical trials as summarised by Grabska-Kobyłecka and colleagues favour the potential function of polyphenols as adjunctive agents in neuroprotection and neurodegenerative disease treatment, although additional large-scale trials are needed to optimise dosing approaches and bioavailability (Grabska-Kobyłecka et al. 2023). Table 4.3 presents the overview of a few recent human researches concerning polyphenols and correlated neuroprotective effects.

### *4.3.3 Challenges and Limitations of Polyphenols in Neuroprotection and Neurodegenerative Disorders*

Although experimental, pre-clinical, and clinical research findings have emphasised the neuroprotective significance of polyphenols, various limitations restrict their valuable application in therapy of neurodegenerative ailments (Grabska-Kobyłecka et al. 2023; Manach et al. 2004). Their poor bioavailability is one of their main drawbacks. A number of polyphenolic components show rapid metabolic transformation and poor absorption in the intestinal area, which limits their concentration in the systemic circulation and their exposure to the brain (Youdim et al. 2004). Furthermore, BBB restricted permeability may reduce the percentage of active ingredients that reach neuronal tissue, thereby reducing their therapeutic effect. Furthermore, polyphenols' wide structural diversity, chemical instability, and significant conversion into glucuronide and sulphate conjugates make it difficult to predict how they will affect human biology (Del Rio et al. 2013; Manach et al. 2004). Inconsistencies between human trials and experimental models continue to be significant, as neuroprotective responses obtained in pre-clinical findings are not consistently mirrored in clinical practice. This translational research gap may be responsible for reflecting differences in metabolic pathways, disease complexity, and exposure levels among

**Table 4.3** Overview of a few recent human research studies concerning polyphenols and neuroprotective effects

| Study design | Participants | Treatment | Method | Result | Reference |
|---|---|---|---|---|---|
| Cross-sectional study | 1091 including men and women | Intake of flavonoid-rich food | Neuropsychological tests and completed a food frequency questionnaire (FFQ) were used | Expression of flavonoid intake on cognitive functions was not observed | Butchart et al. (2011) |
| Cross-sectional study | 152 including men (61) and women (91) | Equol production status using urine samples | Touch panel-type dementia assessment scale together with mild cognitive impairment screen were used | Soy consumption has been linked to improved cognitive function in equol producers | Igase et al. (2017) |
| 12 weeks, randomised controlled trail, double-blind placebo control study (DBPCS) | 26 including men (13) and women (13) | Blueberry concentrate | 30 ml | Enhanced brain activation and perfusion in regions linked to cognitive function | Bowtell et al. (2017) |
| 16 weeks, randomised controlled trail, DBPCS | 16 including men (9) and women (7) | Blueberry powder | 25 g per day powder | Blood oxygen level-based signal | Boespflug et al. (2018) |
| Randomised 24-week controlled trail, DBPCS | 76 including men (41) and women (25) | Blueberry powder | 25 g per day powder | Cognitive benefits observed in blueberry group | McNamara et al. (2018) |
| 90 days, randomised controlled trail, DBPCS | 37 including men (24) and women (13) | Freeze-dried blueberry | 24 g per day | Recovery in execution function | Miller et al. (2018) |
| Randomised controlled trail, double-blind counter balanced-crossover study | 20 including men (13) and women (07) | Single dose of purple grape juice | 230 ml | Cognitive benefits observed | Haskell-Ramsay et al. (2017) |

(continued)

**Table 4.3** (continued)

| Study design | Participants | Treatment | Method | Result | Reference |
|---|---|---|---|---|---|
| 6.7 years, cohort study | 925 including men (693) and women (232) | Strawberry intake | FFQ assessment | Decreased risk of Alzheimer's disease | Agarwal et al. (2019) |
| 12 weeks, randomised controlled trail, double-blind study | 52 including men (26) and women (26) | Single dose catechin capsule | 336.4 mg catechin | Improvement in attention with single dose and memory improvement with long-term use | Baba et al. (2020) |
| 19.7 years, cohort study | 2801 including men (1457) and women (1344) | Flavonoids | Assessed through semiquantitative FFQ | Low risk for Alzheimer's disease alongside related dementia | Shishtar et al. (2020) |

human participants and animal models (Lamport et al. 2012). Another major limitation is the inconsistency of clinical evidence. Human trials vary greatly in terms of study design, participant characteristics, intervention duration, dosage regimens, and cognitive assessment tools (Vauzour et al. 2012). Another difficulty is revealed by individual differences exhibited by people when asssuming polyphenols. Individual differences in neuroprotective effects result from gut microbiota's crucial role in polyphenol metabolism and bioactivation (Cardona et al. 2013). Additionally, clinical applicability and useful comparison across research studies are limited by the lack of standardised formulations and well-defined dosing strategies (Grabska-Kobyłecka et al. 2023). Even though polyphenols are generally regarded as safe, excessive consumption can result in serious medication-nutrient interactions and gastrointestinal distress, especially in older adults who are taking several medications (Williamson 2017).

Finally, neurodegenerative diseases are multifactorial, complex, progressive conditions that use numerous interrelated practices, incorporating aberrant protein accretion, oxidative stress, impairment of mitochondria, and neuroinflammation. Therefore, polyphenols are best suited as preventive agents or supportive adjuncts rather than as stand-alone therapeutic approaches because they are unlikely to stop the progression of disease on their own (Grabska-Kobyłecka et al. 2023). Future studies should focus on advancing bioavailability, investigating targeted drug delivery approaches, carrying out long-term randomised clinical trials, and implementing tailored approaches based on microbiome and metabolic profiles. To increase the translational significance of polyphenols in the treatment of neurodegenerative diseases, such developments are necessary. Low bioavailability, gastrointestinal tract instability, and rapid metabolism are some of the factors able to limit the therapeutic efficacy of polyphenols, despite their potential to treat neuroinflammation. Additionally, potential for toxicity at elevated dose, lack of clinical data regarding long-term effects, and the existence of variable absorption and distribution patterns among individuals all hinder the use of these medications in neuroinflammatory disorders. Despite their encouraging neuroprotective qualities, their application in clinical settings is hindered by a number of issues. Bioavailability is foremost hurdle that needs to be overcome. Polyphenols frequently have low absorption rates and fast metabolism in human body, which limits their ability to reach and affect specific brain regions (Grabska-Kobyłecka et al. 2023; Nazzaro et al. 2025). The entry of polyphenols into brain is impaired by a selective membrane such as BBB (Carecho et al. 2024; Figueira et al 2017b). The efficacy of polyphenols can be impacted by individual differences in metabolism, digestion, and diet composition, which makes it more challenging to develop a reliable treatment (Martemucci et al. 2024). Honey's composition may be impacted by pollen in addition to nectar from various flower sources. The phytochemical profiles of different flowers may affect the concentration and availability of bioactive chemicals in honey. These compounds, which include phenolic acids, flavonoids, and enzymes, increase its therapeutic potential.

Honey's exclusive floral source, soil composition, and other climate-related factors can affect antioxidant, antibacterial, and anti-inflammatory assets. Additionally, filtration, pasteurisation, and heat exposure are methods of extraction for active components (Antoine et al. 2021; Cucu et al. 2021; Ramly et al. 2021). Therefore, it can be said that changes in climate, as well as how handling and processing methods, can affect their bioavailability and therapeutic efficacy. Furthermore, despite polyphenols' anti-inflammatory and antioxidant qualities, the complexity of neurodegenerative diseases which involve a wide range of pathogenic factors may necessitate more precise and focussed treatments than polyphenols by themselves can offer (Islam et al. 2024).

Future studies need to focus on advancing stability besides bioavailability of polyphenols so as to address aforementioned limitations. A viable opinion is the development of advanced delivery approaches like liposomal formulations or nanoencapsulation that can boost polyphenol absorption along with fasten passage across BBB. An additional approach can be the identification of specific polyphenols that possess better absorption along with targeted actions on neuroinflammatory pathways.

### *4.3.4 Future Perspectives of Polyphenols for Neuroprotection*

Polyphenolic compounds show promise as neuroprotective effects in the neurological disorders by influencing oxidative stress, inflammation, and neurodegeneration through interactions with a variety of biological targets, such as enzymes, receptors, and signalling pathways. However, many drugs are ineffective as a result of their inability to pass through BBB. Innovations in nanotechnology propose excellent opportunities for tailored drug delivery by creating combinations of natural products and nanomedicine for superlative AD treatments (Chen 2024). The inherent disadvantages of pharmaceutical substances can be overcome by nanoparticles via stabilisation and increased solubilisation along with increased stay time of active components (Palmal et al. 2014). A natural bio-based nanocarrier increases polyphenols' bioavailability and is the best choice for encasing, shielding, and transferring them. Natural polymers (polysaccharides/proteins) are apt for delivery and address difficulties like poor oral absorption. These polymers display good biocompatibility, biodegradability, nutritional value, and sustainability. Protein-dependent natural polymers have many useful characteristics, including amphiphilicity, emulsification, foaming, and gelation besides their superior nutritional value. On account of their characteristic molecular arrangement as well as functional assets, they can be transformed to varied nanoscale delivery vehicles (for instance, nanoparticles, nano-emulsions, nanogels, nanofibres, and nanofilms). Consequently, they possess the ability to carry polyphenolic (hydrophilic as well as lipophilic) chemicals (Guan 2022; Rudrapal 2022). Bioavailability of lipophilic polyphenols can be enhanced along with escape from first-pass metabolism by employing lipid derived delivery channels, i.e. nano-emulsions, liposomes, alongside solid lipid nanoparticles (Wang 2022; Zhang 2022). Inorganic hybrid nanoparticles can be created from polyphenols owing to their ability of making covalent and non-covalent bonds. Moreover, nanotechnology creates the opportunity to turn polyphenols into functional materials that are able to regulate as well as stabilise physicochemical interactions among molecules (Fang et al. 2024).

In order to control efficacy along with safety of polyphenols for human use, clinical research is essential (Bolat et al. 2024; Kapoor et al. 2025). Dietary inclusion of these substances may decrease advancement in neuroinflammatory disorders and enrich quality of life. Some polyphenols are proficient in overpassing BBB (Albadrani 2024; Jalouli et al. 2025). Polyphenols possess the ability to control varied signalling pathways associated with neurodegenerative ailments (Ciupei 2024; Nájera-Maldonado et al. 2024). Further research is essential to combat challenges like poor absorption and quick metabolism to get maximum therapeutic efficacy (Abbasi et al. 2024). Hence, clinical trials are the fundamental prerequisite for ascertaining the optimal dose, safety, and efficacy of polyphenols for human use (Chen et al. 2024; Kapoor et al. 2025). Further studies are required to be executed in order to get thorough knowledge of necessary to completely understand the beneficial impacts and clinical appliances of polyphenols.

## 4.4 Conclusion

Polyphenols are valuable bioactive compounds found in several plants having several pharmacological potentials. There are neurodegenerative illnesses characterised by impairment of neuron loss and cognitive decline, which may have consequences to the quality of life of affected communities. In this regard, therapeutic intervention of this illness may be challenging in terms of its effective management. Polyphenols, naturally derived bioactive compounds, have been recognised for their role in neurodegenerative illness for its neuroprotective effects. Recent research has increasingly highlighted the potential role of polyphenols in slowing the progression of neurodegenerative diseases and supporting cognitive health. Increasing research findings suggested that these compounds may explore the significant therapeutic potential for neuroprotective effects in neuroinflammation as well as in the neurodegenerative illness. Due to their neuroprotective potential, minimal toxicity, and ability to contrast processes involved in neurodegeneration, polyphenols are considered promising candidates for the prevention of neurological diseases.

## References

Abbasi A, Hashemi M, Samadi Kafil H, Abbasi Astamal M, Lahouty M, Ghorbani Tajani AG, Hosseini H, Nasirifar SZ (2024) A critical review on the bioavailability promotion of the food bioactive compounds: nano lipid carriers perspective. Pharm Sci 30(3):282–303. https://doi.org/10.34172/ps.2024.11

Adhikari-Devkota A, Kurauchi Y, Yamada T, Katsuki H, Watanabe T, Devkota HP (2019) Anti-neuroinflammatory activities of extract and polymethoxyflavonoids from immature fruit peels of Citrus 'Hebesu'. J Food Biochem 43(6):e12813. https://doi.org/10.1111/jfbc.12813

Agarwal P, Holland TM, Wang Y, Bennett DA, Morris MC (2019) Association of strawberries and anthocyanidin intake with Alzheimer's dementia risk. Nutrients 11(12):3060. https://doi.org/10.3390/nu11123060

Akinyemi AJ, Okonkwo PK, Faboya OA, Onikanni SA, Fadaka A, Olayide I, Akinyemi EO, Oboh G (2017) Curcumin improves episodic memory in cadmium induced memory impairment through inhibition of acetylcholinesterase and adenosine deaminase activities in a rat model. Metab Brain Dis 32(1):87–95. https://doi.org/10.1007/s11011-016-9887-x

Albadrani HM, Chauhan P, Ashique S, Babu MA, Iqbal D, Almutary AG, Abomughaid MM, Kamal M, Paiva-Santos AC, Alsaweed M, Hamed M (2024) Mechanistic insights into the potential role of dietary polyphenols and their nano formulation in the management of Alzheimer's disease. Biomed Pharmacother 174:116376. https://doi.org/10.1016/j.biopha.2024.116376

Antoine CM, Forrest JR (2021) Nesting habitat of ground-nesting bees: a review. Ecol Entomol 46(2):143–159. https://doi.org/10.1111/een.12986

Arab H, Mahjoub S, Hajian-Tilaki K, Moghadasi M (2016) The effect of green tea consumption on oxidative stress markers and cognitive function in patients with Alzheimer's disease: a prospective intervention study. Casp J Int Med 7(3):188–194

Arcone R, D'Errico A, Nasso R, Rullo R, Poli A, Di Donato P, Masullo M (2023) Inhibition of enzymes involved in neurodegenerative disorders and a β 1–40 aggregation by Citrus limon peel polyphenol extract. Molecules 28(17):6332

Arya A, Kaushik D, Almeer R, Bungau SG, Sayed AA, Abdel-Daim MM, Bhatia S, Mittal V (2022) Application of green technologies in design-based extraction of celastrus paniculatus (Jyotishmati) seeds, SEM, GC-MS analysis, and evaluation for memory enhancing potential. Front Nutr 9:871183. https://doi.org/10.3389/fnut.2022.871183

Azzini E, Peña-Corona SI, Hernández-Parra H, Chandran D, Saleena LAK, Sawikr Y, Peluso I, Dhumal S, Kumar M, Leyva-Gómez G, Martorell M, Sharifi-Rad J, Calina D (2024) Neuroprotective and anti-inflammatory effects of curcumin in Alzheimer's disease: Targeting neuroinflammation strategies. Phytother Res 38(6):3169–3189. https://doi.org/10.1002/ptr.8200

Baba Y, Inagaki S, Nakagawa S, Kaneko T, Kobayashi M, Takihara T (2020) Effect of daily intake of green tea catechins on cognitive function in middle-aged and older subjects: a randomised, placebo-controlled study. Molecules 25(18):4265. https://doi.org/10.3390/molecules25184265

Bagheri H, Ghasemi F, Barreto GE, Rafiee R, Sathyapalan T, Sahebkar A (2020) Effects of curcumin on mitochondria in neurodegenerative diseases. Biofactors 46(1):5–20. https://doi.org/10.1002/biof.1566

Barfoot K, May G, Lamport DJ, Ricketts J, Riddell PM, Williams CM (2019) The effects of acute wild blueberry supplementation on the cognition of 7–10-year-old schoolchildren. Eur J Nutr 58:2911–2920. https://doi.org/10.1007/s00394-018-1843-6

Bensalem J, Dudonné S, Etchamendy N, Pellay H, Amadieu C, Gaudout D, Dubreuil S, Paradis ME, Pomerleau S, Capuron L, Hudon C, Layé S, Desjardins Y, Pallet V (2019) Polyphenols from grape and blueberry improve episodic memory in healthy elderly with lower level of memory performance: a bicentric double-blind, randomised, placebo-controlled clinical study. J Gerontol Ser A 74(7):96–1007. https://doi.org/10.1093/gerona/gly166

Bian Y, Yamashita T, Taira Y, Shang J, Tsunoda K, Feng T, Sasaki R, Liu X, Shi X, Tadokoro K, Nomura E, Matsumoto N, Osakada Y, Omote Y, Takemoto M, Hishikawa N, Ohta Y, Abe K (2020) A polyphenolic complex attenuates inflammatory response and blood- brain barrier disruption. Curr Neurovasc Res 17(3):286–293. https://doi.org/10.2174/1567202617666200517105727

Boespflug EL, Eliassen JC, Dudley JA, Shidler MD, Kalt W, Summer SS, Stein AL, Stover AN, Krikorian R (2018) Enhanced neural activation with blueberry supplementation in mild cognitive impairment. Nutr Neurosci 21(4):297–305. https://doi.org/10.1080/1028415X.2017.1287833

Bolat E, Sarıtaş S, Duman H, Eker F, Akdaşçi E, Karav S, Witkowska AM (2024) Polyphenols: secondary metabolites with a biological impression. Nutrients 16(15):2550. https://doi.org/10.3390/nu16152550

Bowtell JL, Aboo-Bakkar Z, Conway ME, Adlam AL, Fulford J (2017) Enhanced task-related brain activation and resting perfusion in healthy older adults after chronic blueberry supplementation. Appl Physiol Nutr Metabol 42(7):773–779. https://doi.org/10.1139/apnm-2016-0550

Butchart C, Kyle J, McNeill G, Corley J, Gow AJ, Starr JM, Deary IJ (2011) Flavonoid intake in relation to cognitive function in later life in the Lothian Birth Cohort 1936. Br J Nutr 106(1):141–148. https://doi.org/10.1017/S0007114510005738

Cardona F, Andrés-Lacueva C, Tulipani S, Tinahones FJ, Queipo-Ortuño MI (2013) Benefits of polyphenols on gut microbiota and implications in human health. J Nutr Biochem 24(8):1415–1422. https://doi.org/10.1016/j.jnutbio.2013.05.001

Carecho R, Marques D, Carregosa D, Masuero D, Garcia-Aloy M, Tramer F, Passamonti S, Vrhovsek U, Ventura MR, Brito MA, Nunes Dos Santos C, Figueira I (2024) Circulating low-molecular-weight (poly) phenol metabolites in the brain: unveiling in vitro and in vivo blood–brain barrier transport. Food Funct 15(15):7812–7827. https://doi.org/10.1039/d4fo01396d

Caruso G, Torrisi SA, Mogavero MP, Currenti W, Castellano S, Godos J, Ferri R, Galvano F, Leggio GM, Grosso G, Caraci F (2022) Polyphenols and neuroprotection: therapeutic implications for cognitive decline. Pharmacol Therap 232:108013. https://doi.org/10.1016/j.pharmthera.2021.108013

Carrotta R, Vilasi S, Costa MA, Librizzi F, Martorana V, Passantino R, Buzzanca C, di Stefano V, Ortore MG, Piccirillo S, Preziuso A (2025) Moringa oleifera extracts as strategic phyto-therapy for Alzheimer's disease. Food Sci Nutr 13(4):e70007. https://doi.org/10.1002/fsn3.70007

Casadesus G, Shukitt-Hale B, Stellwagen HM, Zhu X, Lee HG, Smith MA, Joseph JA (2004) Modulation of hippocampal plasticity and cognitive behavior by short-term blueberry supplementation in aged rats. Nutr Neurosci 7(5–6):309–316. https://doi.org/10.1080/10284150400020482

Chen PT, Chen Z, Hou WC, Yu LC, Chen RPY (2016) Polyhydroxycurcuminoids but not curcumin upregulate neprilysin and can be applied to the prevention of Alzheimer's disease. Sci Rep 6(1):29760. https://doi.org/10.1038/srep29760

Chen X, Walton K, Brodaty H, Chalton K (2024) Polyphenols and diets as current and potential nutrition senotherapeutics in Alzheimer's disease: findings from clinical trials. J Alzheimers Dis 101(s1):S479–S501. https://doi.org/10.3233/JAD-231222

Chen G, Yan S, Chen S, Lin T, Lin X (2025) Polyphenols and Alzheimer's disease: a review on molecular and therapeutic insights with in silico support. Food Sci Nutr 13(9):e70496. https://doi.org/10.1002/fsn3.70496

Ciupei D, Colişar A, Leopold L, Stănilă A, Diaconeasa ZM (2024) Polyphenols: from classification to therapeutic potential and bioavailability. Foods 13(24):4131. https://doi.org/10.3390/foods13244131

Cordeiro LM, Soares MV, da Silva AF, Machado ML, Bicca Obetine Baptista F, da Silveira TL, Arantes LP, Soares FA (2021) Neuroprotective effects of rutin on ASH neurons in Caenorhabditis elegans model of Huntington's disease. Nutr Neurosci 25(11):2288–2301. https://doi.org/10.1080/1028415X.2021.1956254

Cucu AA, Baci GM, Dezsi Ş, Nap ME, Beteg FI, Bonta V, Bobiş O, Caprio E, Dezmirean DS (2021) New approaches on Japanese knotweed (Fallopia japonica) bioactive compounds and their potential of pharmacological and beekeeping activities: challenges and future directions. Plants 10(12):2621. https://doi.org/10.3390/plants10122621

Davinelli S, Medoro A, Ali S, Passarella D, Intrieri M, Scapagnini G (2023) Dietary flavonoids and adult neurogenesis: potential implications for brain aging. Curr Neuropharmacol 21(3):651–668. https://doi.org/10.2174/1570159X21666221031103909

Debnath K, Shekhar S, Kumar V, Jana NR, Jana NR (2016) Efficient inhibition of protein aggregation, disintegration of aggregates, and lowering of cytotoxicity by green tea polyphenol-based self-assembled polymer nanoparticles. ACS Appl Mater Interfaces 8(31): 20309–20318. https://doi.org/10.1021/acsami.6b06853

D'Egidio FE, Qosja F Ammannito F, Topi S, d'Angelo M, Cimini A, Castelli V (2025) Antioxidant and anti-inflammatory defenses in Huntington's disease: roles of NRF2 and PGC-1α, and therapeutic strategies. Life 15(4):577. https://doi.org/10.3390/life15040577

D'Errico A, Reveglia P, Nasso R, Maugeri A, Masullo M, Arcone R, Lecce L, De Vendittis E, Rullo R (2026) Effects of polyphenolic extracts from Mediterranean forage crops on cholinesterases and amyloid aggregation relevant to neurodegenerative diseases. J Appl Pharm Sci 16(03):118–130. http://doi.org/10.7324/JAPS.2026.251960

De Gomes MG, Teixeira FEG, de Carvalho FB, Pacheco CO, da Silva Neto MR, Giacomeli R, Ramalho JB, Dos Santos RB, Domingues WB, Campos VF, Haas SE (2020) Curcumin-loaded lipid-core nanocapsules attenuates the immune challenge LPS-induced in rats: neuroinflammatory and behavioral response in sickness behavior. J Neuroimmunol 345:577270. https://doi.org/10.1016/j.jneuroim.2020.577270

De Lorenzi E, Franceschini D, Contardi C, Di Martino RMC, Seghetti F, Serra M, Bisceglia F, Pagetta A, Zusso M, Belluti F (2022) Modulation of amyloid β-induced microglia activation and neuronal cell death by curcumin and analogues. Int J Mol Sci 23(8):4381. https://doi.org/10.3390/ijms23084381

Del Rio D, Rodriguez-Mateos A, Spencer JPE, Tognolini M, Borges G, Crozier A (2013) Dietary (poly)phenolics in human health: structures, bioavailability, and evidence of protective effects. Antioxid Redox Signal 18(14):1818–1892. https://doi.org/10.1089/ars.2012.4581

Desideri G, Kwik-Uribe C, Grassi D, Necozione S, Ghiadoni L, Mastroiacovo D, Raffaele A, Ferri L, Bocale R, Lechiara MC, Marini C, Ferri C (2012) Benefits in cognitive function, blood pressure, and insulin resistance through cocoa flavanol consumption in elderly subjects with mild cognitive impairment. Hypertension 60(3):794–801. https://doi.org/10.1161/HYPERTENSION AHA.112.193060

Dos Santos MG, Arbo BD, Hort MA (2026) Effects of quercetin and its derivatives in in vivo models of neuro inflammation: a systematic review and meta-analysis. Neural Regen Res 21(5):1783–1792. https://doi.org/10.4103/NRR.NRR-D-24-01175

Fang S, Zhang K, Liu D, Yang Y, Xi H, Xie W, Diao K, Rao Z, Wang D, Yang W (2024) Polyphenol-based polymer nanoparticles for inhibiting amyloid protein aggregation: recent advances and perspectives. Front Nutr 11:1408620. https://doi.org/10.3389/fnut.2024.1408620

Fekete M, Jarecsny T, Lehoczki A, Major D, Fazekas-Pongor V, Csípő T, Lipécz Á, Szappanos Á, Pázmándi EM, Varga P (2025) Mediterranean diet, polyphenols, and neuroprotection: mechanistic insights into resveratrol and oleuropein. Nutrients 17(24):3929. https://doi.org/10.3390/nu17243929

Figueira I, Garcia G, Pimpão RC, Terrasso AP, Costa I, Almeida AF, Tavares L, Pais TF, Pinto P, Ventura MR, Filipe A (2017a) Polyphenols journey through blood-brain barrier towards neuronal protection. Sci Rep 7(1):11456. https://doi.org/10.1038/s41598-017-11512-6

Figueira I, Menezes R, Macedo D, Costa I, Dos Santos CN (2017b) Polyphenols beyond barriers: a glimpse into the brain. Curr Neuropharmacol 15(4):562–594. https://doi.org/10.2174/1570159X14666161026151545

Fiore M, Terracina S, Ferraguti G (2025) Brain neurotrophins and plant polyphenols: a powerful connection. Molecules 30:2657. https://doi.org/10.3390/molecules30122657

Filiberti V, Andreu AB, Silveyra MX (2025) Valorization of potato peel waste as a source of neuroprotective polyphenols: antioxidant and mitochondrial effects in HT-22 neuronal cells. Waste Biomass Valoriz 3:1–4. https://doi.org/10.1007/s12649-025-03390-1

Ghofrani S, Joghataei MT, Mohseni S, Baluchnejadmojarad T, Bagheri M, Khamse S, Roghani M (2015) Naringenin improves learning and memory in an Alzheimer's disease rat model: insights into the underlying mechanisms. Eur J Pharmacol 764:195–201. https://doi.org/10.1016/j.ejphar.2015.07.001

Giri PM, Banerjee A, Ghosal A, Layek B (2024) Neuroinflammation in neurodegenerative disorders: current knowledge and therapeutic implications. Int J Mol Sci 25(7):3995. https://doi.org/10.3390/ijms25073995

Gong G, Ganesan K, Wan Y, Liu Y, Huang Y, Luo Y, Wang X, Zhang Z, Zheng Y (2024) Unveiling the neuroprotective properties of isoflavones: current evidence, molecular mechanisms and future perspectives. Crit Rev Food Sci Nutr 65(16):3112–3148. https://doi.org/10.1080/10408398.2024.2357701

Gonzalez T, Govers RA (2026) A possible role for glucose transporter inhibition in the genistein-induced amelioration of a high glucose C. elegans Alzheimer disease model. Free Radic Biol Med 245:357–360. https://doi.org/10.1016/j.freeradbiomed.2025.12.040

Grabska-Kobyłecka I, Szpakowski P, Król A, Kobyłecki A, Głabiński A, Nowak D (2023) Polyphenols and their impact on the prevention of neurodegenerative diseases and development. Nutrients 15(15):3454. https://doi.org/10.3390/nu15153454

Guan T, Zhang Z, Li X, Cui S, McClements DJ, Wu X, Chen L, Long J, Jiao A, Qiu C, Jin Z (2022) Preparation, characteristics, and advantages of plant protein-based bioactive molecule delivery systems. Foods 11(11):1562. https://doi.org/10.3390/foods11111562

Haskell-Ramsay CF, Stuart RC, Okello EJ, Watson AW (2017) Cognitive and mood improvements following acute supplementation with purple grape juice in healthy young adults. Eur J Nutr 56(8):2621–2631. https://doi.org/10.1007/s00394-017-1454-7

Han HJ, Park SK, Kang JY, Kim JM, Yoo SK, Kim DO, Kim GH, Heo HJ (2021) Mixture of phlorotannin and fucoidan from ecklonia cava prevents the Aβ-induced cognitive decline with mitochondrial and cholinergic activation. Mar Drugs 19(8):434. https://doi.org/10.3390/md19080434

Hou T, Nie D, Ding M, Wang C, Mei K, Lu X, Wang X, Tang S, Wu H, Guan P, Zhu W (2026) Multi-target nanocomposites for Alzheimer's treatment via microenvironment modulation and β-amyloid plaque clearance. J Mater Sci Technol 248:281–294. https://doi.org/10.1016/j.jmst.2025.04.081

Hu B, Ouyang Y, Zhao T, Wang Z, Yan Q, Qian Q, Wang W, Wang S (2024) Antioxidant hydrogels: antioxidant mechanisms, design strategies, and applications in the treatment of oxidative stress-related diseases. Adv Healthc Mater 13(11):e2303817. https://doi.org/10.1002/adhm.202303817

Hussain A, Jairajpuri DS, Anwar S, Choudhury A, Hawwal MF, Firdous A, Alajmi MF, Hassan MI (2025) Apigenin-mediated MARK4 inhibition: a novel approach in advancing Alzheimer's disease therapeutics. Mol Divers 29(6):5745–5756. https://doi.org/10.1007/s11030-025-11104-x

Ide K, Yamada H, Takuma N, Kawasaki Y, Harada S, Nakase J, Ukawa Y, Sagesaka YM (2015) Effects of green tea consumption on cognitive dysfunction in an elderly population: a randomised placebo-controlled study. Nutr J 15(1):49. https://doi.org/10.1186/s12937-016-0168-7

Igase M, Igase K, Tabara Y, Ohyagi Y, Kohara K (2017) Cross-sectional study of equol producer status and cognitive impairment in older adults. Geriatr Gerontol Int 17(11):2103–2108. https://doi.org/10.1111/ggi.13029

Islam F, Roy S, Zehravi M, Paul S, Sutradhar H, Yaidikar L, Kumar BR, Dogiparthi LK, Prema S, Nainu F, Rab SO, Doukani K, Emran TB (2024) Polyphenols targeting MAP kinase signaling pathway in neurological diseases: understanding molecular mechanisms and therapeutic targets. Mol Neurobiol 61(5):2686–2706. https://doi.org/10.1007/s12035-023-03706-z

Islam MR, Al-Imran MI, Zehravi M, Sweilam SH, Mortuza MR, Gupta JK, Shanmugarajan TS, Devi K, Tummala T, Alshehri MA, Rajagopal K (2025) Targeting signaling pathways in neurodegenerative diseases: Quercetin's cellular and molecular mechanisms for neuroprotection. Animal Model Exp Med 8(5):798–818. https://doi.org/10.1002/ame2.12551

Jafarisavari Z, Hasanzadeh E, Tayebi L, Asadpour S (2025) Curcumin Delivery in Regenerative Medicine for Alzheimer's Disease. Regener Eng Transl Med 1–21. https://doi.org/10.1007/s40883-025-00482-1

Jalouli M, Rahman MA, Biswas P, Rahman H, Harrath AH, Lee I-S, Kang S, Choi J, Park MN, Kim B (2025) Targeting natural antioxidant polyphenols to protect neuro inflammation and neurodegenerative diseases: a comprehensive review. Front Pharmacol 16:1492517. https://doi.org/10.3389/fphar.2025.1492517

Kapoor B, Bhardwaj S, Singh R, Sharma A, Dolma Y, Sharma A, Sharma NR, Landi M, Kapoor D (2025) Multifaceted role of plant polyphenols in human health: a comprehensive review. Food Sci Biotechnol 1–5. https://doi.org/10.1007/s10068-025-01991-z

Karuppagounder SS, Madathil SK, Pandey M, Haobam R, Rajamma U, Mohanakumar KP (2013) Quercetin up-regulatesmitochondrial complex-I activity to protect against programmed cell death in rotenone model of Parkinson's disease in rats. Neurosci 236:136–148. https://doi.org/10.1016/j.neuroscience.2013.01.032

Kent K, Charlton K, Roodenrys S, Batterham M, Potter J, Traynor V, Gilbert H, Morgan O, Richards R (2017) Consumption of anthocyanin-rich cherry juice for 12 weeks improves memory and cognition in older adults with mild-to-moderate dementia. Eur J Nutr 56(1):333–341. https://doi.org/10.1007/s00394-015-1083-y

Khan MB, Khan MM, Khan A, Ahmed ME, Ishrat T, Tabassum R, Vaibhav K, Ahmad A, Islam F (2012) Naringenin Ameliorates Alzheimer's disease (AD)-type neurodegeneration with cognitive impairment (AD-TNDCI) caused by the Intracerebroventricular-Streptozotocinin rat model. Neurochem Int 61(7):1081–1093. https://doi.org/10.1016/j.neuint.2012.07.025

Kopustinskiene DM, Bernatoniene J, Jakstas V, Morkuniene R (2021) Catechins, neuroprotection, and brain mitochondria. In: de Oliveira MR (ed) Mitochondrial physiology and vegetal molecules—therapeutic potential of natural compounds on mitochondrial health, pp 455–470. Academic Press, Cambdrige, MA, Oxford, and Amsterdam. https://doi.org/10.1016/B978-0-12-821562-3.00014-9

Krikorian R, Shidler MD, Nash TA, Kalt W, Vinqvist-Tymchuk MR, Shukitt-Hale B, Joseph JA (2010) Blueberry supplementation improves memory in older adults. J Agric Food Chem 58(7):3996–4000. https://doi.org/10.1021/jf9029332

Kurmi SP, Karati D (2025) Inhibitory potential of polyphenolic stilbene derivatives with Glycogen Synthase Kinase-3β (GSK-3β) for Alzheimer's disease: computational and SAR insights. Brain Disorders 17(24):100182. https://doi.org/10.1016/j.dscb.2025.100182

Kwon Y (2017) Luteolin as a potential preventive and therapeutic candidate for Alzheimer's disease. Exp Gerontol 95:39–43. https://doi.org/10.1016/j.exger.2017.05.014

Lamport DJ, Spencer JPE (2012) Issues in the assessment of flavonoid effects on cognitive performance in human subjects. Nutrients 4(12):1486–1503

Lamport DJ, Williams CM, Spencer JPE (2011) Polyphenols and cognition in humans: an overview of current evidence. Nutr Bull 36(4):374–383

Liu W, Cui X, Zhong Y, Ma R, Liu B, Xia Y (2023) Phenolic metabolites as therapeutic in inflammation and neoplasms: molecular pathways explaining their efficacy. Pharmacol Res 193:106812. https://doi.org/10.1016/j.phrs.2023.106812

Lu Z, Nie G, Belton PS, Tang H, Zhao B (2006) Structure–activity relationship analysis of antioxidant ability and neuroprotective effect of Gallic acid derivatives. Neurochem Int 48(4):263–274. https://doi.org/10.1016/j.neuint.2005.10.010

Ma J, Yang WQ, Zha H, Yu HR (2013) Effect of Naringeninon learning and memory ability on model rats with Alzheimer disease. Zhong Yao Cai 36(2):271–276

Madhubala D, Patra A, Khan MR, Mukherjee AK (2024) Phytomedicine for neurodegenerative diseases: the road ahead. Phytother Res 38:2993–3019. https://doi.org/10.1002/ptr.8192

Mahto K, Kuwar OK, Maloo A, Kalia N (2025) Therapeutic potential of luteolin in neurodegenerative disorders: targeting Nrf2, NFκB, MAPK, and JAK-STAT pathways to combat neuroinflammation and apoptosis. Inflammopharmacology 33(9):5011–5021. https://doi.org/10.1007/s10787-025-01846-3

Maiuolo J, Bosco F, Guarnieri L, Nucera S, Ruga S, Oppedisano F, Tucci L, Muscoli C, Palma E, Giuffrè AM, Mollace V (2023) Protective role of an extract waste product from citrus bergamia

in an in vitro model of neurodegeneration. Plants (Basel) 12(11):2126. https://doi.org/10.3390/plants12112126
Mamashli F, Meratan AA, Ghasemi A, Obeidi N, Salmani B, Atarod D, Pirhaghi M, Moosavi-Movahedi F, Mohammad-Zaheri M, Shahsavani MB, Habibi-Kelishomi Z, Goliaei B, Gholami M, Saboury AA (2023) Neuroprotective effect of Propolis Polyphenol-based Nano sheets in cellular and animal models of Rotenone-induced Parkinson's disease. ACS Chem Neurosci 14(5):851–863. https://doi.org/10.1021/acschemneuro.2c00605
Manach C, Scalbert A, Morand C, Rémésy C, Jiménez L (2004) Polyphenols: food sources and bioavailability. Am J Clin Nutr 79(5):727–747. https://doi.org/10.1093/ajcn/79.5.727
Manap AS, Wei Tan AC, Leong WH, Yin Chia AY, Vijayabalan S, Arya A, Wong EH, Rizwan F, Bindal U, Koshy S, Madhavan P (2019) Synergistic effects of curcumin and piperine as potent acetylcholine and amyloidogenic inhibitors with significant neuroprotective activity in SH-SY5Y cells via computational molecular modeling and in vitro assay. Front Aging Neurosci 1:206. https://doi.org/10.3389/fnagi.2019.00206
Manda S, Sharma S, Wani A, Joshi P, Kumar V, Guru SK, Bharate SS, Bhushan S, Vishwakarma RA, Kumar A, Bharate SB (2016) Discovery of a marine-derived bis-indole alkaloid fascaplysin, as a new class of potent P-glycoprotein inducer and establishment of its structure-activity relationship. Eur J Med Chem 107:1–11. https://doi.org/10.1016/j.ejmech.2015.10.049
Mandel S, Youdim MBH (2004) Catechin polyphenols: neurodegeneration and neuroprotection in neurodegenerative diseases. Free Radic Biol Med 37(3):304–317. https://doi.org/10.1016/j.freeradbiomed.2004.04.012
Martemucci G, Khalil M, Di Luca A, Abdallah H, D'Alessandro AG (2024) Comprehensive strategies for metabolic syndrome: how nutrition, dietary polyphenols, physical activity, and lifestyle modifications address diabesity, cardiovascular diseases, and neurodegenerative conditions. Metabolites 14(6):327. https://doi.org/10.3390/metabo14060327
Martino RM, De Simone A, Andrisano V, Bisignano P, Bisi A, Gobbi S, Rampa A, Fato R, Bergamini C, Perez DI, Martinez A, Bottegoni G, Cavalli A, Belluti F (2016) Versatility of the curcumin scaffold: discovery of potent and balanced dual BACE-1 and GSK-3β inhibitors. J Med Chem 59(2):531–544. https://doi.org/10.1021/acs.jmedchem.5b00894
Mayer E, Horn J, Mayer D, Randolph E (2024) Dietary polyphenols to maintain healthier brain measures and cognitive function, as mediated by gut microbiota metabolites. In: Hyland N, Stant C (eds) The gut-brain Axis, Second Edition, pp 341–360. Academic Press, Cambdrige, MA, Oxford, and Amsterdam. https://doi.org/10.1016/B978-0-323-99971-7.00014-X
McNamara RK, Kalt W, Shidler MD, McDonald J, Summer SS, Stein AL, Stover AN, Krikorian R (2018) Cognitive response to fish oil, blueberry, and combined supplementation in older adults with subjective cognitive impairment. Neurobiol Aging 64:147–156. https://doi.org/10.1016/j.neurobiolaging.2017.12.003
Melo Sarkar S, Mazumder SJ, Saha S, Bandyopadhyay U (2016) Management of inflammation by natural polyphenols: a comprehensive mechanistic update. Curr Med Chem 23(16):1657–1695. https://doi.org/10.2174/0929867323666160418115540
Miller MG, Hamilton DA, Joseph JA, Shukitt-Hale B (2018) Dietary blueberry improves cognition among older adults in a randomised, double-blind, placebo-controlled trial. Eur J Nutr 57(3):1169–1180. https://doi.org/10.1007/s00394-017-1400-8
Mishra J, Singh C, Bhatti JS (2025) Neuroprotective role of piperine-encapsulated casein micelles in mitigating intracellular dyshomeostasis in Alzheimer's disease pathology. Mol Neurobiol 63(1):218. https://doi.org/10.1007/s12035-025-05297-3
Moalefshahri R, Hashemy SI, Hosseini H, Sahebkar A (2025) Neuroprotective effect of Kaempferol through modulation of autophagy. Nutr Neurosci 28(12):1463–1479. https://doi.org/10.1080/1028415X.2025.2524702
Mohammadi F, Mahmudian A, Moeeni M, Hassani L (2016) Inhibition of amyloid fibrillation of hen egg-white lysozyme by the natural and synthetic curcuminoids. RSC Adv 6(28):23148–23160. https://doi.org/10.1039/C5RA18992F

Morillas-Ruiz JM, Rubio-Perez JM, Albaladejo MD, Zafrilla P, Parra S, Vidal-Guevara ML (2010) Effect of an antioxidant drink on homocysteine levels in Alzheimer's patients. J Neurol Sci 299(1–2):175–178. https://doi.org/10.1016/j.jns.2010.08.050

Moussa C, Hebron M, Huang X, Ahn J, Rissman RA, Aisen PS, Turner RS (2017) Resveratrol regulates neuro-inflammation and induces adaptive immunity in Alzheimer's disease. J Neuroinflammation 14(1):1. https://doi.org/10.1186/s12974-016-0779-0

Nabizadeh F (2024) Neuroprotective role of Fisetin in Alzheimer's disease: an overview of potential mechanism and clinical findings. Neurol Lett 3:14–25. https://doi.org/10.61186/nl.3.2.14

Nájera-Maldonado JM, Salazar R, Alvarez-Fitz P, Acevedo-Quiroz M, Flores-Alfaro E, Hernández-Sotelo D, Espinoza-Rojo M, Ramírez M (2024) Phenolic compounds of therapeutic interest in neuroprotection. J Xenobiot 14(1):227–246. https://doi.org/10.3390/jox14010014

Nan S, Wang P, Zhang Y, Fan J (2021) Epigallocatechin-3-gallate provides protection against Alzheimer's disease-induced learning and memory impairments in rats. Drug des Devel Ther 15:2013–2024. https://doi.org/10.2147/DDDT.S289473

Narasimhan KKS, Jayakumar D, Velusamy P, Srinivasan A, Mohan T, Ravi DB, Uthamaraman S, Sathyamoorthy YK, Rajasekaran NS, Periandavan K (2019) Morindacitrifolia and its active principle scopoletin mitigate protein aggregation and neuronal apoptosis through augmenting the DJ-1/Nrf2/ARE signaling pathway. Oxid Med Cell Longev 2761041. https://doi.org/10.1155/2019/2761041

Nazzaro F, Coppola F, Fratianni F, Abdalrazeq M, Ombra MN, De Giulio B, Coppola R, Zengin G (2025) Polyphenols bioactive metabolites, and their anti-biofilm and neuroprotective potential. Foods 14(22):3976. https://doi.org/10.3390/foods14223976

Nejabati HR, Roshangar L (2022) Kaempferol as a potential neuroprotector in Alzheimer's disease. J Food Biochem 46(12):e14375. https://doi.org/10.1111/jfbc.14375

Palmal S, Maity AR, Singh BK, Basu S, Jana NR, Jana NR (2014) Inhibition of amyloid fibril growth and dissolution of amyloid fibrils by curcumin–gold nanoparticles. Chemistry 20(20):6184–6191. https://doi.org/10.1002/chem.201400079

Panaro MA, Corrado A, Benameur T, Paolo CF, Cici D, Porro C (2020) The emerging role of curcumin in the modulation of TLR-4 signaling pathway: focus on neuroprotective and anti-rheumatic properties. Int J Mol Sci 21(7):2299. https://doi.org/10.3390/ijms21072299

Paramakrishnan N, Lim KG, Paramaswaran Y, Ali N, Waseem M, Shazly GA, Bin Jardan YA, Muthuraman A (2023) Astaxanthin: a marine drug that ameliorates cerebrovascular-damage-associated Alzheimer's disease in a Zebrafish model via the inhibition of matrix metalloprotease-13. Mar Drugs 21(8):433. https://doi.org/10.3390/md21080433

Pasinetti GM (2012) Novel role of red wine-derived Polyphenols in the prevention of Alzheimer's disease dementia and brain pathology: experimental approaches and clinical implications. Planta Med 78(15):E24. https://doi.org/10.1055/s-0035-1545846

Pinheiro Fernandes FD, Fontenele Menezes AP, de Sousa Neves JC, Fonteles AA, da Silva AT, de Araújo Rodrigues P, Santos do Carmo MR, de Souza CM, de Andrade GM (2014) Caffeic acid protects mice from memory deficits induced by focal cerebral ischemia. Behav Pharmacol 25(7):637–647. https://doi.org/10.1097/FBP.0000000000000076

Puranik N, Kumari M, Tiwari S, Dhakal T, Song M (2025) Resveratrol as a therapeutic agent in Alzheimer's disease: evidence from clinical studies. Nutr 17(15):2557. https://doi.org/10.3390/nu17152557

Qi Z, Wu M, Fu Y, Huang T, Wang T, Sun Y, Feng Z, Li C (2017) Palmitic acid curcumin ester facilitates protection of neuroblastoma against oligomeric Aβ40 insult. Cell Physiol Biochem 44(2):618–633. https://doi.org/10.1159/000485117

Qureshi M, Al-Suhaimi EA, Wahid F, Shehzad O, Shehzad A (2018) Therapeutic potential of curcumin for multiple sclerosis. Neurol Sci 39(2):207–214. https://doi.org/10.1007/s10072-017-3149-5

Ramly NS, Sujanto IS, Abd Ghani A, Huat JT, Alias N, Ngah N (2021) The impact of processing methods on the quality of honey: a review. Malays J Appl Sci 6(1):99–110. https://doi.org/10.37231/myjas.2021.6.1.280

Ringman JM, Frautschy SA, Cole GM, Masterman DL, Cummings JL (2005) A potential role of the curry spice curcumin in Alzheimer's disease. Curr Alzheimer Res 2(2):131–136. https://doi.org/10.2174/1567205053585882
Riquelme CV, Pando ME (2025) Oleuropein: a narrative review on its role in neurodegenerative diseases. Aging Adv 2(2):75–83
Rudrapal M, Mishra AK, Rani L, Sarwa KK, Zothantluanga JH, Khan J, Kamal M, Palai S, Bendale AR, Talele SG, Pathan VT9, Borse LB, Neharkar VS, Sahoo PK (2022) Nanodelivery of dietary polyphenols for therapeutic applications. Molecules 27(24):8706. https://doi.org/10.3390/molecules27248706
Scholey AB, French SJ, Morris PJ, Kennedy DO, Milne AL, Haskell CF (2009) Consumption of cocoa flavanols results in acute improvements in mood and cognitive performance during sustained mental effort. J Psychopharmacol 24(10):1505–1514. https://doi.org/10.1177/0269881109106923
Shang Y, Tian J, Zhang Z, Luo L, Saeheng S, Qiu S, Zhou R, Chen J, Li L (2025) Mangiferin: Sources, anti-inflammatory activities, and molecular mechanisms. J Agric Food Chem 73(43):27145–27160. https://doi.org/10.1021/acs.jafc.5c06234
Sharififar F, Moshafi MH, Shafazand E, Koohpayeh A (2012) Acetyl cholinesterase inhibitory, antioxidant and cytotoxic activity of three dietary medicinal plants. Food Chem 130(1):20–23. https://doi.org/10.1016/j.foodchem.2011.06.034
Sharma D, Singh V, Kumar A, Singh TG (2025) Genistein: a promising ally in combating neurodegenerative disorders. Eur J Pharmacol 991:177273. https://doi.org/10.1016/j.ejphar.2025.177273
Shi R, Gao D, Stoika R, Liu K, Sik A, Jin M (2024) Potential implications of polyphenolic compounds in neurodegenerative diseases. Crit Rev Food Sci Nutr 64(16):5491–5514. https://doi.org/10.1080/10408398.2022.2155106
Shishtar E, Rogers GT, Blumberg JB, Au R, Jacques PF (2020) Long-term dietary flavonoid intake and risk of Alzheimer disease and related dementias in the Framingham offspring cohort. Am J Clin Nutr 112(2):343–353. https://doi.org/10.1093/ajcn/nqaa079
Singh SS, Rai SN, Birla H, Zahra W, Rathore AS, Singh SP (2020) NF-κB-mediated neuroinflammation in Parkinson's disease and potential therapeutic effect of polyphenols. Neurotox Res 37(3):491–507. https://doi.org/10.1007/s12640-019-00147-2. Epub 2019 Dec 10. PMID: 31823227
Singh K, Sethi P, Datta S, Chaudhary JS, Kumar S, Jain D, Gupta JK, Kumar S, Guru A, Panda SP (2024) Advances in gene therapy approaches targeting neuro-inflammation in neurodegenerative diseases. Ageing Res Rev 98:102321. https://doi.org/10.1016/j.arr.2024.102321
Small GW, Siddarth P, Li Z, Miller KJ, Ercoli L, Emerson ND, Martinez J, Wong KP, Liu J, Merrill DA, Chen ST, Henning SM, Satyamurthy N, Huang SC, Heber D, Barrio JR (2018) Memory and brain amyloid and tau effects of a bioavailable form of curcumin in non-demented adults. Am J Geriatr Psychiatry 26(3):266–277. https://doi.org/10.1016/j.jagp.2017.10.010
Socała K, Żmudzka E, Lustyk K, Zagaja M, Brighenti V, Costa AM, Andres-Mach M, Pytka K, Martinelli I, Mandrioli J, Pellati F, Biagini G, Wlaź P (2024) Therapeutic potential of stilbenes in neuropsychiatric and neurological disorders: a comprehensive review of preclinical and clinical evidence. Phytother Res 38(3):1400–1461. https://doi.org/10.1002/ptr.8101
Song J, Zhang Y, Wang H, Wu C, Li Y (2024) The influence of food matrix and processing methods on the bioaccessibility of lutein: a review. J Food Bioact 26:7–23. https://doi.org/10.31665/jfb.2024.18376
Sun XY, Li LJ, Dong QX, Zhu J, Huang YR, Hou SJ, Yu XL, Liu RT (2021) Rutin prevents tau pathology and neuroinflammation in a mouse model of Alzheimer's disease. J Neuroinflammation 18(1):131. https://doi.org/10.1186/s12974-021-02182-3
Sun S, Liu Z, Lin M, Gao N, Wang X (2024) Polyphenols in health and food processing: antibacterial, anti-inflammatory, and antioxidant insights. Front Nutr 11:1456730. https://doi.org/10.3389/fnut.2024.1456730

Tang M, Taghibiglou C (2017) The mechanisms of action of curcumin in Alzheimer's disease. J Alzheimers Dis 58(4):1003–1016. https://doi.org/10.3233/JAD-170188
Tang S, Zhang Y, Botchway BOA, Wang X, Huang M, Liu X (2024) Epigallocatechin-3-Gallate inhibits oxidative stress through the Keap1/Nrf2 signaling pathway to improve Alzheimer disease. Mol Neurobiol 62(3):3493–3507. https://doi.org/10.1007/s12035-024-04498-6
Tavan M, Hanachi P, de la Luz C-G, Segura Carretero A, Mirjalili MH (2024) Natural phenolic compounds with neuroprotective effects. Neurochem Res 49(2):306–326. https://doi.org/10.1007/s11064-023-04046-z
Thakur A, Kumar A (2024) Polyphenols for dyes application. Sci Eng Polyphenols Fundam Ind Scale Appl 157–210. https://doi.org/10.1002/9781394203932.ch7
Turner RS, Thomas RG, Craft S, van Dyck CH, Mintzer J, Reynolds BA, Brewer JB, Rissman RA, Raman R, Aisen PS (2015) A randomised, double-blind, placebo-controlled trial of resveratrol for Alzheimer disease. Neurology 85(16):1383–1391. https://doi.org/10.1212/WNL.0000000000002035
Vauzour D (2012) Dietary polyphenols as modulators of brain functions: biological actions and molecular mechanisms underpinning their beneficial effects. Oxid Med Cell Longev 2012:914273. https://doi.org/10.1155/2012/914273
Wang Q, Xu J, Rottinghaus GE, Simonyi A, Lubahn D, Sun GY, Sun AY (2002) Resveratrol protects against global cerebral ischemic injury in gerbils. Brain Res 958(2):439–447. https://doi.org/10.1016/s0006-8993(02)03543-6
Wang X, Qi Y, Zheng H (2022) Dietary polyphenol, gut microbiota, and health benefits. Antioxid 11(6):1212. https://doi.org/10.3390/antiox11061212
Whyte AR, Williams CM (2015) Effects of flavonoid-rich interventions on cognitive function in older adults. Nutrients 7(12):10416–10446
Williams CM, El Mohsen MA, Vauzour D, Rendeiro C, Butler LT, Ellis JA, Whiteman M, Spencer JP (2008) Blueberry induced changes in spatial working memory correlate with changes in hippocampal CREB phosphorylation and brain-derived neurotrophic factor (BDNF) levels. Free Radic Biol Med 45(3):295–305. https://doi.org/10.1016/j.freeradbiomed.2008.04.008
Williamson G (2017) The role of polyphenols in modern nutrition. Nutr Bull 42(3):226–235. https://doi.org/10.1111/nbu.12278
Xicota L, Rodriguez-Morato J, Dierssen M, de la Torre R (2017) Potential role of (-)-epigallocatechin-3-gallate (EGCG) in the secondary prevention of Alzheimer disease. Curr Drug Targets 18(2):174–195. https://doi.org/10.2174/1389450116666150825113655
Xiong R, Zhou XG, Tang Y, Wu JM, Sun YS, Teng JF, Pan R, Law BY, Zhao Y, Qiu WQ, Wang XL, Liu S, Wang YL, Yu L, Yu CL, Mei QB, Qin DL, Wu AG (2021) Lychee seedpolyphenol protects the blood–brain barrier through inhibiting A_(25–35)-induced NLRP3 inflammasome activation via the AMPK/mTOR/ULK1-mediated autophagy in bEnd.3 cells and APP/PS1 mice. Phytother Res 35(2):954–973. https://doi.org/10.1002/ptr.6849
Yadav E, Yadav P, Khan MMU, Singh H, Verma A (2022) Resveratrol: a potential therapeutic natural polyphenol for neurodegenerative diseases associated with mitochondrial dysfunction. Front Pharmacol 13:922232. https://doi.org/10.3389/fphar.2022.922232
Yahia IBH, Baccouri O, Jalouli M, Boujelbene N, Rahman MA, Harrath AH, Zidi I (2024) The small phyto molecule resveratrol: a promising role in boosting tumor cell chemo sensitivity. Pharmacia 71:1–9. https://doi.org/10.3897/pharmacia.71.e122169
Yang W, Ma J, Liu Z, Lu Y, Hu B, Yu H (2014) Effect of Naringenin on brain insulin signaling and cognitive functions in ICV-STZ induced dementia model of rats. Neurol Sci 35(5):741–751. https://doi.org/10.1007/s10072-013-1594-3
Youdim KA, Qaiser MZ, Begley DJ, Rice-Evans CA, Abbott NJ (2004) Flavonoid permeability across an in situ model of the blood–brain barrier. Free Radic Biol Med 36(5):592–604. https://doi.org/10.1016/j.freeradbiomed.2003.11.023
Yu H, Yamashita T, Hu X, Bian Z, Hu X, Feng T, Tadokoro K, Morihara R, Abe K (2022) Protective and anti-oxidative effects of curcumin and resveratrol on A_-oligomer-induced damage in the SH-SY5Y cell line. J Neurol Sci 441:120356. https://doi.org/10.1016/j.jns.2022.120356

Yuca H, Kopuz A, Karakaya S (2025) Tea Catechins: A useful chemical weapon against Alzheimer's disease. In: Malik J, Mandal SC, Sarup P, Lin GD, Li RW (eds) Diet and Alzheimer's disease: let food be our medicine, pp 161–183. Springer Nature Singapore, Singapore. https://doi.org/10.1007/978-981-95-2736-6_7

Zaplatic E, Bule M, Shah SZ, Uddin MS, Niaz K (2019) Molecular mechanisms underlying protective role of quercetin in attenuating Alzheimer's disease. Life Sci 224:109–119. https://doi.org/10.1016/j.lfs.2019.03.055

Zhang Z, Hamada H, Gerk PM (2019) Selectivity of dietary phenolics for inhibition of human monoamine oxidases A and B. Biomed Res Int 2019:8361858. https://doi.org/10.1155/2019/8361858

Zhang Z, Li X, Sang S, McClements DJ, Chen L, Long J, Jiao A, Jin Z, Qiu C (2022) Polyphenols as plant-based nutraceuticals: health effects, encapsulation, nano-delivery, and application. Foods 11(15):2189. https://doi.org/10.3390/foods11152189

Zhang YQ, Liu ZY, Liu X, Xu WX, Guan W, Li MM, Chen QS, Zhang LL, Kuang HX, Yang BY, Zhang H (2025) Total Lignans from leaves of Schisandra Chinensis improve Alzheimer's disease pathology through the Nrf2/HO-1/NLRP3 pathway. Chem Biodivers 22(12):e00812. https://doi.org/10.1002/cbdv.202500812

Zhou Y, Men L, Sun Y, Wei M, Fan X (2021) Pharmacodynamic effects and molecular mechanisms of lignans from Schisandra chinensis Turcz. (Baill.), a current review. Eur J Pharmacol 892:173796. https://doi.org/10.1016/j.ejphar.2020.173796

Zhu L, Yang M, Fan L, Yan Q, Zhang L, Mu P, Lu F (2025) Interaction between resveratrol and SIRT1: role in neurodegenerative diseases. Naunyn Schmiedebergs Arch Pharmacol 398(1):89–101. https://doi.org/10.1007/s00210-024-03319-w

# Chapter 5
# Polyphenols in Cardiovascular Health: Effects on Blood Pressure and Lipid Profiles with Clinical Evidence Supporting Cardiovascular Benefits

**Vivek P. Chavda, Disha Joshi, Anjali Bedse, Pankti Balar, and Komal Patel**

**Abstract** This chapter offers an in-depth examination of polyphenols and their significant impact on cardiovascular health, focusing primarily on their influence on blood pressure regulation and lipid metabolism. It begins with a review of the biochemical features and metabolic pathways of polyphenols, emphasizing their interactions with cardiovascular tissues and the modulatory role of gut microbiota. The chapter details with the key mechanisms driving their cardioprotective effects, including antioxidant activity, inflammation reduction, and prevention of atherosclerosis. Further, it synthesizes evidence from epidemiological studies and clinical trials that demonstrate the beneficial effects of polyphenols on controlling hypertension and improving lipid profiles. The interplay between polyphenols and cardiovascular medications is also explored, alongside their role in various cardiovascular diseases. Challenges related to bioavailability, standardization, and future research directions are discussed. The goal is to equip healthcare professionals and researchers with knowledge to effectively incorporate polyphenol-based strategies into cardiovascular disease prevention and management.

V. P. Chavda (✉)
Department of Pharmaceutics and Pharmaceutical Technology, LM College of Pharmacy, Ahmedabad, Gujarat, India
e-mail: Vivek7chavda@gmail.com

D. Joshi
Department of Pharmaceutical Technology, L. J. Institute of Pharmacy, LJ University, Ahmedabad, Gujarat, India

A. Bedse
Department of Pharmaceutics, KK Wagh College of Pharmacy, Dr. Baba Saheb Ambedkar Technological University, Nashik, Maharashtra, India
e-mail: apbedse@kkwagh.edu.in

P. Balar
Pharmacy Section, LM College of Pharmacy, Ahmedabad, Gujarat, India

K. Patel
Saraswati Institute of Pharmaceutical Science, Dhanap, Gandhinagar, Gujarat, India

 
R. K. Singla et al. (eds.), *Polyphenols in Health*,
https://doi.org/10.1007/978-3-032-27324-6_5

**Keywords** Antioxidant effect · Blood pressure regulation · Cardiovascular health · Dyslipidemia · Lipid profile modulation · Hypertension · Polyphenol

## Abbreviations

| | |
|---|---|
| ADP | Adenosine diphosphate |
| AMPK | AMP-activated protein kinase |
| ATP | Adenosine triphosphate |
| Ang II | Angiotensin II |
| Bcl-2 | B-cell lymphoma 2 |
| BP | Blood pressure |
| BMI | Body mass index |
| CVD | Cardiovascular disease |
| CGMP | Cyclic guanosine monophosphate |
| COX-2 | Cyclooxygenase-2 |
| DASH | Dietary approaches to stop hypertension |
| EGCG | Epigallocatechin gallate |
| EPIC | European Prospective Investigation into Cancer and Nutrition |
| FMD | Flow-mediated dilatation |
| GPx | Glutathione peroxidase |
| HDL | High-density lipoprotein |
| HDL-C | High-density lipoprotein cholesterol |
| ICAM1 | Intercellular Adhesion Molecule-1 |
| LOX | Lipoxygenase |
| LXR | Liver X receptor |
| LDL-C | Low-density lipoprotein cholesterol |
| LDL | low-density lipoprotein |
| MeDiet | Mediterranean Diet |
| NADPH | Nicotinamide adenine dinucleotide phosphate |
| ENOS | Nitric oxide synthase |
| NO | Nitric oxide |
| Nrf2 | Nuclear factor erythroid 2-related factor 2 |
| NF-κB | Nuclear factor-κB |
| PPARα | Peroxisome proliferator-activated receptor |
| PREDIMED | Prevention with Mediterranean Diet |
| Akt | Protein kinase B |
| RCT | Randomised controlled trial |
| ROS | Reactive oxygen species |
| RAAS | Renin–angiotensin–aldosterone system |
| TG | Triglyceride |
| VCAM1 | Vascular Cell Adhesion Molecule 1 |

## 5.1 Introduction

A variety of secondary metabolites found in plants, polyphenols, are identified by the presence of aromatic rings with one or more hydroxyl groups. They have antioxidant, cardioprotective, and anti-inflammatory properties in humans, and they also help plants with defense, flavor, and pigmentation (Pandey and Rizvi 2009).

### *5.1.1 Classification of Polyphenols*

Polyphenols are broadly classified into four groups. Flavonoids are the largest group, including flavonols (quercetin), flavones (apigenin), flavanones (naringenin), and catechins such as epicatechin (EGCG), anthocyanins (cyanidin, betanin), and isoflavones (genistein) (Panche et al. 2016). Phenolic acids are simple compounds like gallic and caffeic acid (Balasundram et al. 2006). Stilbenes, such as resveratrol, are less common but linked to cardiovascular benefits (Baur and Sinclair 2006). Tannins are large polyphenols, classified into hydrolyzable and condensed forms (proanthocyanidins).

### *5.1.2 Dietary Sources*

Polyphenol-rich plant foods include veggies (broccoli and onions), fruits (berries, apples, and grapes), nuts, seeds, legumes, and whole grains. They are also found in tea, coffee, red wine, and chocolate, as well as in olive oil and spices (Scalbert and Williamson 2000).

### *5.1.3 Cardiovascular Disease*

Cardiovascular diseases (CVD), which include coronary artery disease, stroke, and hypertension, are the leading cause of death around the globe. They account for about 18 million deaths each year. Risk factors for these diseases include hypertension, obesity, diabetes, smoking, and a poor diet. In addition to causing death, CVD also lead to disability and create an economic burden (Martin et al. 2025).

### *5.1.4 Polyphenols and Cardiovascular Protection*

Epidemiological and clinical studies link polyphenol-rich diets with cardiovascular benefits. The Mediterranean diet—rich in fruits, olive oil, and wine—reduces coronary heart disease risk (Domínguez-López et al. 2023). The French paradox highlights low cardiovascular mortality despite high-fat intake, attributed partly to wine-derived resveratrol (Catalgol et al. 2012). In Asia, high green tea consumption is associated with reduced hypertension and stroke. Meta-analyses show that flavonoid-rich foods such as berries, cocoa, and tea improve endothelial function and lower cardiovascular mortality (Rees et al. 2018). These effects arise from the antioxidant, anti-inflammatory, vasodilatory, and lipid-modulating actions of polyphenols.

This chapter describes and categorizes polyphenols. It lists dietary sources and reviews the impact of cardiovascular diseases (CVD). It also presents evidence that supports the heart-protective benefits of polyphenols. This information lays the groundwork for understanding polyphenols as natural dietary approaches for preventing and managing cardiovascular issues.

## 5.2 Biochemistry and Metabolism of Polyphenols Relevant to Cardiovascular Effects

Polyphenols are recognized for their cardioprotective effects, but their activity depends on their chemical form and metabolic processing. Unlike vitamins or minerals, they occur in complex structures that undergo digestion, microbial transformation, and enzymatic modification before absorption. These processes shape their bioavailability, metabolite profile, and tissue-specific actions relevant to heart health (Pham et al. 2021).

### *5.2.1 Absorption, Bioavailability, and Metabolism of Polyphenols*

Polyphenols are mainly ingested as glycosides, esters, or polymers, which are poorly absorbed. Small aglycones and some phenolic acids can cross the intestinal lining directly, but most require enzymatic hydrolysis before absorption. Once absorbed, they undergo extensive Phase I- and especially Phase II-metabolism, where enzymes such as uridine 5′-diphospho-glucuronosyltransferase, sulfotransferase, and catechol-O-methyltransferase convert them into glucuronidated, sulfated, or methylated forms—the primary circulating metabolites, usually present at low micromolar plasma levels (Brglez Mojzer et al. 2016). Conjugation reduces direct antioxidant activity but generates metabolites with distinct biological functions, influencing signaling pathways, transcription factors, and receptors in cardiovascular tissues.

Bioavailability varies, with flavanols like epicatechin showing good absorption, while proanthocyanidins and tannins are poorly absorbed (Scalbert and Williamson 2000).

### 5.2.2 *Role of Gut Microbiota in Polyphenol Biotransformation*

Most dietary polyphenols escape absorption in the small intestine and reach the colon, where gut microbes convert them into smaller, more bioavailable metabolites such as phenolic acids, urolithins, and valerolactones (Selma et al. 2009). For instance, ellagitannins from berries and pomegranates yield urolithins with anti-inflammatory and vasoprotective effects, while flavan-3-ol polymers from cocoa and tea generate phenylpropionic and phenylacetic acids with bioactive potential. Variations in gut microbiota composition influence these metabolite profiles, helping explain individual differences in the cardiovascular benefits of polyphenol-rich diets (Nazzaro et al. 2025).

### 5.2.3 *Circulating Metabolites and Tissue Distribution*

After absorption, polyphenols circulate mainly as conjugated metabolites with plasma half-lives of 1–12 h, influenced by metabolism and enterohepatic recycling. They preferentially distribute to metabolically active organs such as the liver, kidneys, vascular endothelium, and cardiac tissue, where they exert cardioprotective effects (Zhang et al. 2025). Although circulating levels are low compared to in vitro models, regular intake of polyphenol-rich foods ensures continuous exposure. Their effects stem mainly from modulating cell signaling and gene expression rather than direct antioxidant action, meaning even low concentrations can exert significant physiological benefits (Yu et al. 2025).

### 5.2.4 *Interaction of Polyphenols with Cardiovascular Targets*

Polyphenol metabolites affect several processes important for heart health:

(a) *Endothelial function.* Flavanols and resveratrol improve nitric oxide (NO) levels by stimulating endothelial nitric oxide synthase (eNOS) activity. This process leads to vasodilation and better regulation of blood pressure (Zordoky et al. 2014)
(b) *Oxidative stress.* Polyphenols stimulate the body's own antioxidant defenses and lower oxidative changes in low-density lipoproteins (LDL), which is an important step in the development of atherosclerosis (Pandey and Rizvi 2009)

(c) *Inflammation.* They block the nuclear factor-κB (NF-κB) and mitogen-activated protein kinase pathways, which reduces the vascular production of pro-inflammatory cytokines like TNF-α and IL-6 (Guo et al. 2024)
(d) *Lipid metabolism.* Some flavonoids control peroxisome proliferator-activated receptors (PPARα) and boost cholesterol efflux, which helps maintain lipid balance (Guo et al. 2024)
(e) *Platelet function.* Quercetin and catechins lower platelet aggregation and thrombus formation. This helps reduce cardiovascular risk (Pignatelli et al. 2000).

### 5.2.5 Challenges in Translating in Vitro Effects to Clinical Outcomes

Despite strong mechanistic evidence, translating polyphenol research into consistent clinical outcomes remains challenging. In vitro studies often use concentrations far above human plasma levels; on the other side, in vivo researches show that the active forms are mostly conjugated metabolites rather than parent compounds. Moreover, bioavailability is influenced by diet, gut microbiota, genetics, and individual metabolic differences (Del Rio et al. 2012).

Human intervention trials on polyphenols show promising but inconsistent results, largely due to small cohorts, short durations, and difficulties in standardizing intake across diets. Establishing reliable biomarkers of exposure and response and conducting long-term studies are essential to confirm their cardioprotective effects in real-world settings (Del Bo' et al. 2019).

## 5.3 Mechanisms Underlying Polyphenols-Related Cardioprotective Actions

Polyphenols, a broad and varied family of bioactive substances, are well known for their ability to prevent CVD. Their cardioprotective mechanisms are complex and include regulation of mitochondrial homeostasis and cardiomyocyte survival as well as antioxidant, anti-inflammatory, endothelial, anti-atherogenic, and antithrombotic effects (Hedayati et al. 2023).

The antioxidant activity of polyphenols is one of the best-established processes behind their beneficial benefits. Quercetin, resveratrol, and catechins are examples of polyphenols with hydroxylated aromatic structures that have the ability to scavenge reactive oxygen species (ROS) and neutralize free radicals before they cause damage to cellular macromolecules (Behl et al. 2020). Additionally, these substances restrict pro-oxidative processes that produce hydroxyl radicals, such as the Fenton process, by chelating transitional metal ions. Apart from direct scavenging, polyphenols also improve endogenous antioxidant systems by activating nuclear factor erythroid

2-related factor 2 (Nrf2) signaling pathways, which in turn upregulates the production and activity of catalase, superoxide dismutase, and glutathione peroxidase (GPx) (Jomova et al. 2025; Ngo and Duennwald 2022). The consequent rise in cellular redox capacity protects the integrity of endothelial cells and stops oxidative damage to vascular tissues, which is a key factor in early atherogenesis (Behl et al. 2020; Wan et al. 2024).

Vascular tone modulation depends on endothelial homeostasis being maintained. It has been demonstrated that polyphenols increase the activity of endothelial eNOS by phosphorylating it via the AMP-activated protein kinase (AMPK) and phosphatidylinositol-3-kinase/protein kinase B (Akt) pathways. This improves vascular relaxation and reactivity by increasing NO bioavailability. Regular consumption of flavanols and stilbenes raises flow-mediated dilatation (FMD), a measure of endothelial function, according to in vitro and clinical research (Serreli and Deiana 2023; Zippel et al. 2018). By scavenging superoxide radicals and inhibiting the development of nicotinamide adenine dinucleotide phosphate (NADPH) oxidase subunits, which produce ROS, polyphenols help lessen NO breakdown. Therefore, decreased arterial stiffness and systemic blood pressure are a result of enhanced NO-mediated signaling (Behl et al. 2020; Hedayati et al. 2023).

The development of CVD is significantly influenced by chronic inflammation. By altering molecular mediators involved in cytokine generation and immune cell recruitment, polyphenols have potent anti-inflammatory actions. They prevent NF-κB from activating, which regulates the production of pro-inflammatory cytokines (TNF-α, IL-6) and enzymes like lipoxygenase (LOX) and cyclooxygenase-2 (COX-2) (Sharanya et al. 2023). Additionally, polyphenols reduce macrophage polarization toward the pro-inflammatory M1 phenotype and inhibit the Toll-like receptor 4/MyD88 pathway. Polyphenols inhibit the production of adhesion molecules such as intercellular adhesion molecule-1 (ICAM1) and vascular cell adhesion molecule 1 (VCAM1) in endothelial and smooth muscle cells, which facilitate leukocyte entry into vascular walls and prevent vascular remodeling and inflammation (Wan et al. 2024).

By inhibiting oxidative processes in LDL and favourably altering lipid metabolism, polyphenols help to prevent atherogenic damage. They hinder the production of foam cells, a characteristic that characterizes the onset of atherosclerotic plaque, by inhibiting LDL oxidation and lowering its absorption by macrophage scavenger receptors (Wan et al. 2024). Polyphenols such as resveratrol, isoflavones, and curcumin affect the activity of liver X receptors (LXR) and PPARα, improving reverse cholesterol transport and boosting the function of high-density lipoproteins (HDL). Inhibiting enzymes such 3-hydroxy-3-methylglutaryl-CoA reductase, polyphenols also reduce plasma triglycerides and total cholesterol. When taken as a whole, these metabolic changes reduce plaque formation and enhance lipid profiles in both clinical and experimental contexts (Hedayati et al. 2023).

Polyphenols additionally protect the cardiovascular system through their antithrombotic actions. By inhibiting thromboxane $A_2$ production and counteracting adenosine diphosphate (ADP)-induced platelet activation, they prevent platelet aggregation. By interfering with intracellular calcium mobilization in platelets, flavonoids like kaempferol and catechins reduce aggregation and clot formation (Ed Nignpense et al. 2019). Additionally, these substances alter vital enzymes involved in platelet signaling, such as phospholipase A and protein kinase C. These processes work together to reduce thrombogenesis, which enhances polyphenols' overall cardioprotective benefits, particularly in populations at risk for ischemic heart disease (Ludovici et al. 2018).

Recent data emphasizes how crucial mitochondrial pathways are for polyphenol-mediated cardioprotection. By maintaining membrane potential and encouraging biogenesis via PGC-1α/SIRT1/AMPK signaling, polyphenols preserve mitochondrial dynamics. This prevents mitochondrial malfunction brought on by oxidative stress and guarantees effective adenosine triphosphate (ATP) synthesis (Chodari et al. 2021). Additionally, resveratrol and related stilbenes block pro-apoptotic factors and upregulate anti-apoptotic proteins such as B-cell lymphoma 2 (Bcl-2) and survivin to prevent cardiomyocyte death. These mitochondrial actions maintain cardiac contractility and metabolic flexibility, preventing ischemia–reperfusion damage and the advancement of heart failure (Kumar et al. 2017).

Polyphenols are important multimodal agents in cardiovascular protection because of their synergistic effects of antioxidant, anti-inflammatory, lipid-modulating, and endothelial processes. Improvements in blood pressure regulation, lipid metabolism, endothelium reactivity, and glucose management across several polyphenol subclasses are supported by clinical and experimental studies. To identify the best formulations, methods for enhancing bioavailability, and translational consistency across human groups, more research is necessary.

Polyphenols have a broad range of cardioprotective actions via coordinating molecular networks that together preserve vascular integrity, fend off oxidative and inflammatory damage, and preserve homeostasis in the cardiometabolic system. Precision nutrition and pharmaceutical innovation in the prevention and treatment of cardiovascular disease may be advanced by incorporating these mechanistic insights into therapeutic design.

## 5.4 Impact of Polyphenols on Blood Pressure Regulation

### *5.4.1 Epidemiological Evidence. Population Studies Associating Polyphenol Intake with Blood Pressure Changes*

A cross-sectional study of health and living conditions of a representative sample of the population, which resided in Sao Paulo (South-eastern Brazil) between 2008 and 2009, demonstrated a strong and linear inverse correlation between the upper tertile of consumption of some polyphenols types, such as tyrosols, alkylphenols, lignans, stilbenes, and other polyphenols, and the prevalence of hypertension (Miranda et al. 2016).

Epidemiological studies that focus on populations have shown a uniform relationship between increased dietary intake of polyphenols and a decreased blood pressure or lower chance of getting blood pressure. The big cohort studies including an 18-year national study of China stated that people with increased use of polyphenols had a considerably reduced possibility of hypertension development in contrast to people with low consumption (Lin et al. 2023). In the same way, the 'Health, Alcohol and Psychosocial factors In Eastern Europe' study revealed that women in the upper polyphenol intake quartile had a 31%-decreased risk of hypertension compared to these people on the lowest quartile with flavonoid, phenolic acids, and some subgroups offering their protective impacts independently. A multiple polyphenol class synergistic effect was also proposed in the present study (Grosso et al. 2017).

### 5.4.2 Clinical Trials on Polyphenols Showing Antihypertensive Effects

The first clinical evidence revealed that quercetin supplement works to reduce blood pressure in hypertensive subjects. Nonetheless, the blood pressure-reducing effect does not seem to involve systemic oxidative stress reductions (Edwards et al. 2007).

The evidence of several randomized, double-blind, placebo-controlled clinical trials has shown that quercetin supplementation decreases blood pressure in human beings especially individuals with prehypertension or individuals with stage 1-high blood pressure (Larson et al. 2010).

Resveratrol, when taken as a supplement or micronized formulation, has been subject to clinical trials that have demonstrated a significant decrease in systolic and diastolic blood pressure in hypertensive patients (Theodotou et al. 2016).

Godos and colleagues reviewed the effect of resveratrol on vascular health, with varied outcomes in clinical trials on blood pressure and endothelial functional outcomes. Its cardioprotective properties are based on its antioxidant, anti-inflammatory, and antihypertensive action. Resveratrol is metabolized by microbiota in the gut leading to the production of beneficial compounds and shifts the microbiota composition toward cardiovascular health. The individual differences in the composition of the gut microbiota and diet may cause a variability in the outcomes and should be studied further (Godos et al. 2024a).

A novel study has revealed that high levels of cocoa flavanols reduce blood pressure, enhance vascular health, and boost the amount of circulating angiogenic cells in the heart patients. It was found that flavanol-rich foods, including cocoa products, tea, wine, and other fruits and vegetables, all have a cardioprotective effect on patients with heart disease. The review suggests that flavanol-rich chocolate and cocoa products can reduce blood pressure in healthy adults in the short term, but this effect is influenced by factors like baseline blood pressure status. Further clinical trials are needed to establish long-term cardiovascular benefits and safety profiles (Ried et al. 2017).

### 5.4.3 *Cellular and Molecular Mechanisms. Modulation of Renin–Angiotensin–Aldosterone System, Endothelial nO Synthase Activation, and Calcium Channel Regulation*

Renin–angiotensin–aldosterone system (RAAS) is a cascade of hormones which affects the cardiovascular and renal dynamics in a very complicated manner It is mostly mediated through angiotensin II (Ang II) and aldosterone. Ang II mediates vasoconstriction, activates the formation of aldosterone, and increases reabsorption of sodium in the kidneys. The aldosterone connects with the mineralocorticoid receptors that are present in the cytosol and controls transcription of genes. Feedback loops support homeostasis, and the other RAAS axes balance classical action. The dysregulation of these pathways is associated with hypertension, cardiovascular remodeling, and electrolyte disturbance, thus making them major targets of therapies (Triebel and Castrop 2024).

Activation of endothelial eNOS is an essential molecular mechanism in the regulation of vascular homeostasis, the translation of the wide array of extracellular stimuli resembling vasoprotective responses into NO. eNOS is expressively present in endothelial cells under physiological conditions and activated in conjunction with calcium binding to calmodulin and phosphorylation at precise serine residues (e.g., Ser 1177) by kinases like Akt, AMPK, and protein kinaseA. Ligand binding by agonists, such as bradykinin, acetylcholine, or vascular endothelial growth factor to receptors, cause intracellular calcium influx and activation of phosphatidylinositol 3-kinase/Akt pathway, which in turn stimulates the enzyme catalytic activities of eNOS. Excitation of mechanosensory complexes on the endothelial surface by shear stress also triggers eNOS activation (via phosphorylation and protein–protein interaction with caveolin-1, heat shock protein 90, and GTP cyclohydrolase 1) that further promotes its availability as cofactor tetrahydrobiopterin. Activated eNOS uses L-arginine and forms L-citrulline, releasing NO that diffuses into vascular smooth muscle cells to activate soluble guanylate cyclase, increase cyclic guanosine monophosphate (cGMP), and cause body relaxation. In addition to vasodilation, eNOS-generated NO has anti-inflammatory, antithrombotic, and anti-proliferative effects, preventing adhesion of leukocytes, aggregation of platelets, and proliferation of smooth muscle cells. Oxidative stress hampers eNOS activity in such a way it leads to the development of endothelial dysfunction. Such a complex regulation of eNOS activity is an important path to therapy in terms of maintaining vascular integrity and preventing oxidative stress-induced cardiovascular disease (Serreli and Deiana 2023).

Polyphenols can adjust vascular smooth muscle tone is through influencing calcium influx. The eNOS-generated NO activates cGMP-protein kinase G which in turn opens potassium channels leading to membrane hyperpolarization and a consequent decreased calcium influx into smooth muscle cells. The effect of this mechanism is relaxation of the smooth muscle and vasodilation. Also, there are polyphenols that could have a direct action on N- and T-type calcium channels and that may have additional involvement in antihypertensive effects. Collectively, these mechanisms help to describe polyphenols as exerting many aspects of cardiovascular protection, including inhibition of RAAS-mediated vasoconstriction of vascular smooth muscle cells, facilitation of the vasodilatory upregulation of endothelial NO synthesis, and controlling calcium homeostasis within vascular smooth muscle cells. This combined effort contributes to blood pressure reduction with enhancement of vascular activity.

### *5.4.4 Dose-Response Relationships and Effective Interventions*

The best effects on blood pressure are promoted by moderate and regular intake of cocoa flavanols and green tea catechins. Consumptions of approximately 500–900 mg a day of cocoa flavanol decrease systolic and diastolic blood pressure in individuals with prehypertension and hypertension to a significant extent. Likewise, consumption of green tea catechin amounts of 300–400 mg per day is associated with significant changes in blood pressure. Increased doses are not seen to have any further advantage in a proportional manner, and there might be enhancement in occurrence of the side effects like gastrointestinal discomfort. These results indicate the need to define the most favorable and safe intake to achieve cardiovascular benefits with minimal risks of side effects (Davison et al. 2010; Godos et al. 2024b; Ried et al. 2017; Xu et al. 2020).

## 5.5 Effects of Polyphenols on Lipid Profiles

Polyphenols have captured the attention of the scientific community because of their potential cardioprotective effects. Another central level through which polyphenols influence different mechanisms in lipid metabolism and lipoprotein regulation is through lipid metabolism and lipoprotein regulation at the same time an important target of polyphenol action (Iqbal et al. 2023).

The most consistent findings concerning polyphenol effects relate to their effect on serum lipids. Meta-analysis, which included 1,136 participants, was carried out assessing the effect of green tea on lipid profile. Results showed interesting results concerning effects of the consumption of green tea on a regular basis, including the fact that low-density lipoprotein cholesterol (LDL-C) was reduced by approximately 2.19 mg/dL. On the other side, high-density lipoprotein cholesterol (HDL-C) observed changes were not statistically significant. Sub-group and sensitivity analyses showed that the findings were robust against intervention subtypes, doses of green tea catechins, durations of follow-up, health status of participants, and risk of bias of trials. Taken together, the findings of the meta-analysis indicate that the consumption of green tea beverages or extracts is associated with the reduction of serum total cholesterol and LDL-C but not HDL-C concentrations (Zheng et al. 2011).

Similarly, a meta-analysis involving 1,131 participants evaluated the effects of cocoa flavanol supplementation on metabolic and lipid outcomes. The studies selected were of different designs, with a flavanol dose between 166 and 2,110 mg per day and an intervention duration between 2 and 52 weeks. Taken together, cocoa flavanol consumption had beneficial effects on lipid profiles as well as insulin sensitivity. Notably, cocoa flavanol consumption produced a small but significantly increased HDL-C, as well as reduced systemic inflammation and thrombosis Yet, among other cardiometabolic biomarkers considered, no uniform contributions were observed. The study found that cocoa flavanol intake had a positive impact on lipid level, insulin sensitivity, and inflammation in adult individuals, indicating the potential role in improving cardiometabolic health (Lin et al. 2016).

Resveratrol is well known for its powerful antioxidant, anti-inflammatory, and potentially anticancer effects. However, its safety and potential treatment effects in aged persons are not well characterized (Ren et al. 2020). The objective of the trial was to investigate the safety and metabolic effects of resveratrol supplementation in overweight older adults. Thirty-two men and women with an average age of approximately 73 years were recruited and randomly assigned to one of three groups: placebo, a moderate dose of resveratrol (300 mg/day), or a high dose regimen (1000 mg/day) (Anton et al. 2014). The interventions were distributed twice daily as capsules for 90 days. During the study, chemical blood profiles indicated normal physiological values, and there were no differences in adverse event rates between treatment and placebo groups. Notably, both resveratrol doses were associated with a significantly lower glucose at the end of the intervention if compared with placebo, even after adjusting for baseline values ($p < 0.05$). However, when they made intra-group comparisons, they found that glucose levels were not significantly different from baseline in the resveratrol group. These results suggest that short-term supplementation with resveratrol at two (moderate and high) dosages is safe and well tolerated, and does not affect routine biochemical indices in overweight elderly subjects. The observed (although small) attenuation of glucose profile emphasizes resveratrol as a plausible supportive measure for maintaining favorable cardiometabolic health in an elderly population. More research is needed to confirm these initial findings and to elucidate further the therapeutic potential for resveratrol in age-related metabolic dysfunction (Anton et al. 2014).

With reference to quercetin, related supplementation is reported to reduce systolic blood pressure and plasma oxidized LDL concentrations in overweight and obese subjects at high CVD-risk. Quercetin may be a dietary polyphenol with potential cardioprotective effects, although the effect on HDL-cholesterol needs clarification (Egert et al. 2009). Other studies, such as a research on the effects of a Mediterranean dietary pattern on cardiovascular risk and the supplementation of extra-virgin olive oil, showed reliable protective effects (Estruch et al. 2018).

## 5.6 Clinical Evidence Supporting the Cardiovascular Benefits of Polyphenols

The Prevention with Mediterranean Diet (PREDIMED) is a pioneering multicentric, randomized controlled trial carried out in Spain between 2003 and 2011, enrolled to evaluate the long-term effects of the Mediterranean Diet (MeDiet) on the incidence of CVD in men and women with a high risk of cardiovascular disease. Subjects were randomized to MeDiet supplemented with extra-virgin olive oil or nuts, or low-fat diet as a control, and ages 55–80 years. In the PREDIMED trial, significant effects on a major cardiovascular event were shown by a reduction of 30% in the MeDiet groups compared with controls, without increasing weight. The positive changes included reduction of blood pressure. Moreover, subgroup analyses pointed to decreased risk of peripheral arterial disease and atrial fibrillation. Remarkably, PREDIMED demonstrated these benefits in a high-risk, older population that already received standard pharmacotherapy, seizing on dietary change as a powerful residual-risk management maneuver (Ros et al. 2014).

The European Prospective Investigation into Cancer and Nutrition (EPIC) cohort contains >340,000 adults aged 9, and the average follow-up has reached 12.3 years in 7 countries across Europe. Diet quality provides consistent evidence using EPIC analysis, that diet quality, as measured (e.g., using Nutri-Score system) is associated with lower risks of both fatal and non-fatal cardiovascular events, ischemic heart disease, and stroke (Poli and Marangoni 2024).

Recent meta-analyses summarize the results of hundreds of clinical trials and huge cohorts that deal specifically with the aspect of polyphenols (catechin, anthocyanins, curcumin, flavanols, etc.) and cardiovascular health. Diets containing high levels of polyphenols were associated with reduced CVD mortality and all-cause mortality among the aged people (Potì et al. 2019).

Results of meta-analyses highlight a marked heterogeneity of the studies in formulation, source, dose, populations, outcomes, which moderates the capacity to make universally prescriptive clinical suggestions (Wan et al. 2024).

Polyphenol supplementation in clinical trials in heart failures and ischemic heart disease indicates improved vascular markers, lower LDL-C levels, and an improvement of inflammatory markers. As an illustration, a five-year relative decrease in mortality of 37% was recorded in cardiac patients who received high levels of polyphenol-containing nutrition routinely. Resveratrol, anthocyanins, catechins, and stilbenes are commonly used in trials. The postulated mechanisms include radical scavenging, anti-inflammatory, and endothelial-protective effects (Alotaibi et al. 2021).

Neuroprotective polyphenols, including resveratrol, salvianolic acid, and flavanones models have been shown to have neuroprotective effects after ischemic stroke and improve outcomes after stroke through biomodulation of oxidative stress and inflammation pathways. In large cohorts, meta-analysis stipulated that diets that included flavonoids were inversely related with incidence of ischemic stroke. The effects seem to be protective when there is early and long-term consumption in vulnerable groups. Clinical trials doses are extremely varied. Special populations covered are elderly people, individuals with established heart disease, and healthy volunteers. The supplementation periods are between several weeks and years, and most effects are observed in long-term studies (Potì et al. 2019; Cheng et al. 2017).

The majority of clinical studies on polyphenols and blood pressure are randomized controlled trials and crossover trials, as well as large cross-sectional studies and cohort studies. The goal is to measure acute and chronic outcomes of polyphenols (in juices, extracts, or supplements) in weeks to months, using placebo or active control. Cross-sectional and cohort studies examine the prevalence of habitual polyphenol consumption and blood pressure/hypertension with diet questionnaires and biomarker analyses (Godos et al. 2019). Populations under investigation are varied (normotensive, prehypertensive, or in stage 1-hypertension adults; patients with metabolic syndrome, type-2 diabetes; etc.) and also sample sizes (Biesinger et al. 2015). At the same time, there are not standardized doses; also, diet interventions include polyphenol abundant foods (juices, cocoa, berries, and nuts) that reflect common real-world consumption (i.e., a daily portion of berry juice or 1,000 mg extract of grape-seeds). Cohort studies also measure habitual intake by use of food-frequency questionnaires with dose in quintiles or tertiles depending on approximate daily intake (Godos et al. 2019; Hodgson et al. 2014).

Systolic blood pressure, diastolic blood pressure, and hypertension are considered as primary endpoints. Variation in blood pressure, endothelial function, various biomarkers, and adjunct drug effects are observed as secondary endpoints. Changes between baseline and post-intervention are normally compared statistically, and probability values alongside confidence intervals and effect sizes are reported. Other variables concern the use of healthy diets (MeDiet, etc.), physical exercise, or the exclusion of smoking. A number of research studies demonstrate that polyphenols will act synergistically when combined with antihypertensive drugs and will trigger greater diastolic blood pressure lowering and enhance endothelial activity (monitored by NO bioavailability and eNOS activity) (Biesinger et al. 2015; Grosso et al. 2017). It has to be noted that existing clinical data regarding the use of polyphenols in blood pressure management is limited in a number of aspects: small sample sizes (9–70 participants); short-term intervention (2–16 weeks); etc. Some classes of polyphenols such as anthocyanins, lignans, and stilbenes may reduce the risk of hypertension according to observational evidence, but trials of supplementation with individual compounds such as hesperidin, genistein, or resveratrol indicate inconsistent or insignificant effects. In addition, the content of polyphenols in food differs widely due to botanical, environmental, and processing variables, and dietary sources have confounding components such as dietary fiber and healthy fats making the attribution of effects only to polyphenols difficult. Tested doses in animals or in vitro experiments are sometimes higher than the dose that could be administered to a person, and the bioavailability is different in each individual because of gut microbiota and genetics. Often, the clinical endpoints of trials are concentrated on individual blood pressure measurements and not on cardiovascular outcomes or blood pressure fluctuations over time, and the comparative information is limited to the standard antihypertensive drugs. In addition, other researchers are at risk of industry bias, reducing clinical applicability, and other problems. Therefore, as polyphenols are promising as adjuncts in hypertension management, stronger well-powered research with standardized dosage and clinical outcomes in diverse populations is required to solidify their effect on blood pressure management (Grosso et al. 2022; Li et al. 2015; Marino et al. 2020).

## 5.7 Polyphenols in Specific Cardiovascular Conditions

In recent years, the study on dietary polyphenols has moved beyond general cardiovascular risk management to more specific cardiovascular conditions including hypertension, atherosclerosis, heart failure, ischemia, and cardiovascular risks associated with metabolic syndrome or diabetes. This section summarizes the latest evidence on effects of dosage and duration, insights into mechanisms, and comparisons with standard pharmacotherapeutic agents.

Clinical and observational (in vitro and in vivo data) findings indicate that increased consumption of polyphenols may be modestly linked to lower blood pressure levels and a diminished risk of developing hypertension. Notably, a meta-analysis of prospective studies reported that among various flavonoid subclasses, anthocyanins were correlated with approximately an 8% reduction in hypertension risk when comparing individuals with the highest intake to those with the lowest resulting in relative risk of 0.92 and 95% (Godos et al. 2019). A cohort study conducted and data published in July 2023 revealed a significant nonlinear and inverse association between dietary polyphenol intake and hypertension risk. The analysis highlighted that high consumption, particularly of lignans and stilbenes, was correlated with reduced blood pressure levels and a lower incidence of hypertension, emphasizing the protective potential of polyphenol-rich diets (Lin et al. 2023). With reference to dosage and intervention length, clinical trials show considerable heterogeneity. Evidence from a subgroup meta-analysis indicates that hypertensive participants receiving polyphenol extracts, specifically in the range of 500–1,000 mg derived from ellagitannin-rich fruits, experienced marked reductions in systolic blood pressure by approximately 4.95 mmHg (Looi et al. 2022). Some flavonoid supplements led to reduced blood pressure in a randomized trial that included 41 hypertensive patients. However, several other controlled trials did not find a significant benefit with lower dosage regimens or among populations with less severe hypertension (Edwards et al. 2007; Zhang et al. 2023). Several preclinical studies demonstrate that quercetin lowers blood pressure through multiple mechanistic pathways (Kim et al. 2017; Lin et al. 2020; Pereira et al. 2018). This flavonoid has also been shown to lower blood pressure in pregnancy-induced hypertension (Sun et al. 2020).

Polyphenols exert antihypertensive effects through several complementary mechanisms, including modulation of the renin–angiotensin–aldosterone system (Ling et al. 2017), stimulation of endothelial nitric oxide synthesis (Meister et al. 2022), inhibition of vascular smooth muscle contraction (Hameed et al. 2022), reduction of oxidative stress and inflammation (Steven et al. 2019), and regulation of sympathetic nervous system activity via gut–brain signaling (de Brito Alves et al. 2023; Yao et al. 2021).

Beyond controlling blood pressure, the development and stability of atherosclerotic plaques play a pivotal role in cardiovascular risk. Polyphenols contribute to both anti-atherogenic and plaque-stabilizing actions through several interconnected pathways, including antioxidative, anti-inflammatory, and endothelial-protective effects. These mechanisms together limit plaque progression and enhance vascular health. Emerging literature reviews underscore that polyphenols such as 2,4,5-trimethoxycinnamic acid curb the oxidation of low-density lipoprotein (Pieczynska et al. 2020; Serna et al. 2014); in addition, nicotinate-curcumin and other polymethoxyflavones suppress foam cell generation (Chen et al. 2019; Guo et al. 2016), and resveratrol attenuate vascular inflammation (Caiati and Jirillo 2025; Feng et al. 2020; Iqbal et al. 2023; Jiang et al. 2021; Ziółkiewicz et al. 2023). From the clinical perspective, cohort studies have shown that individuals with higher habitual intake of flavonoids experience fewer hospital admissions for atherosclerotic cardiovascular disease, supporting the cardiovascular benefits of a flavonoid-rich diet (Bondonno et al. 2024; Dalgaard et al. 2019). Preclinical research demonstrates that polyphenol consumption leads to decreased lipid accumulation and more robust plaque architecture, such as thicker fibrous caps and smaller necrotic cores (Lian et al. 2018; Zhang et al. 2019). These changes help to slow atherosclerosis progression and promote plaque stability, thereby lowering the likelihood of plaque rupture and the risk of cardiovascular events (Huang et al. 2024).

Heart failure is the end stage of many cardiovascular conditions, characterized by maladaptive remodeling, fibrosis, impaired contractility, and energy metabolic derangements. Recent comprehensive reviews indicate that polyphenols including resveratrol (Zhang et al. 2021), quercetin, kaempferol (Zhang et al. 2019), EGCG (Liu et al. 2018), and other phenolics confer multiple protective effects in animal models of heart failure. These compounds have been shown to limit cardiac hypertrophy, improve ejection fraction, suppress myocardial fibrosis, and elevate antioxidant enzyme activity, with other effects (Gal et al. 2020; Hedayati et al. 2023; Lacerda et al. 2020; Liu et al. 2018; Stankovic et al. 2025; Zhang et al. 2019).

Ischemia reperfusion injury, commonly witnessed in conditions like myocardial infarction or cardiac surgery, is marked by severe oxidative stress and cell death. The heart initially activates adaptive responses within 72 h after ischemia, followed by structural remodeling that leads to lasting changes and potential dysfunction (Cebova and Pechanova 2020). Clinical human data remain very limited but animal studies have proven their well-known therapeutic benefits (Dong et al. 2018; Luo et al. 2019; Wu et al. 2017; Zheng et al. 2020).

Metabolic syndrome and diabetes significantly drive cardiovascular risks like hypertension and heart failure. Polyphenols could be helpful by improving insulin sensitivity, lowering inflammation, reducing fat buildup, and enhancing endothelial function. A recent meta-analysis of 2024 found that polyphenol consumption may decrease the likelihood of metabolic syndrome, although effects vary by polyphenol subclass (Makhtoomi et al. 2024). Another large meta-analysis found that antioxidant polyphenol supplementation improved key cardiometabolic markers such as glucose metabolism, lipid levels, and blood pressure (Khouya et al. 2022; Wan et al. 2024). Compelling research evidence suggests that elevated glucose levels are strongly linked to increased cardiovascular disease risk, even at levels below diabetes thresholds (Kour et al. 2023). A randomized controlled trial published in 2022 demonstrated that sesame oil significantly improves liver enzyme levels and reduces fatty liver severity in women with nonalcoholic fatty liver disease on a calorie-restricted diet (Atefi et al. 2022). Another clinical report found that supplementation with black chokeberry extract led to reductions in blood pressure, blood sugar, and lipid profiles in metabolic syndrome patients, highlighting its potential cardiovascular benefits (Tasic et al. 2021). With the continuous accumulating evidence, it is important to keep in mind that polyphenols generally offer modest improvements for cardiovascular conditions. Polyphenols are best seen as supplements that work as adjunctive rather than replacement to conventional treatments. Still, their mechanistic actions like fighting oxidative stress, reducing inflammation, improving endothelial function, and preventing fibrosis make them promising candidates for integrative heart care.

## 5.8 Interactions of Polyphenols with Cardiovascular Drugs and Nutrients

Certain polyphenols may have synergistic effects with conventional cardiovascular drugs such as by enhancing endothelial function, ameliorating oxidative stress, and suppressing inflammatory pathways. Involved mechanisms may augment the effects of antihypertensives, statins, antiplatelet therapies, and β-blockers. However, the converse effect is also possible involving antagonistic interactions that may arise from pharmacokinetic or pharmacodynamic effects. Literature evidences highlighted that polyphenols (quercetin, catechnis, etc.) can influence drug targets, interact with the renin–angiotensin system, and modify platelet activity, which could affect overall performance of cardiovascular medications (Ali et al. 2023; Arshad et al. 2025; Domínguez-López et al. 2023; Kim et al. 2019; Sleiman et al. 2024; Wisnuwardani 2025). Such type of evidence supports the role of polyphenols as natural RAAS modulators, beneficial in preventing cardiovascular disease. Garlic and *Ginkgo biloba* have been linked to increased anticoagulant effects when taken with warfarin (Ge et al. 2014; Soyata et al. 2020). To prove the given results of in vivo studies, a little amount of robust human trial data are available when speaking of vasodilatory and antioxidant effects of polyphenols in combination with standard pharmacotherapy.

At the pharmacokinetic level, polyphenols can influence the metabolism of cardiovascular drugs (Fraga et al. 2019; Korobkova 2015; Mishra et al. 2024; Mohos et al. 2020; Morais et al. 2022; Tarko and Duda-Chodak 2020). Substantially, effects of polyphenols from whole foods can differ significantly from those observed in supplement trials (Bondonno et al. 2017; Cassidy 2017). Dietary factors can also influence how polyphenols interact with medications; for example, a high-protein meal may alter drug binding and metabolism, which can affect the extent of polyphenol drug interactions (Cianciosi et al. 2021; Duda-Chodak and Tarko 2023).

For healthcare providers working in cardiovascular care, key takeaways include: promoting diets rich in polyphenols sourced from fruits, vegetables, whole grains, tea, and cocoa as part of cardiac health strategies; exercising caution with higher dose of polyphenol supplements in patients on complex cardiovascular drugs by monitoring for potential changes in drug effects or side effects; further considering the timing of supplementation in relation to drug dosing to reduce interaction risks; acknowledging individual differences in polyphenol absorption influenced by gut microbiota and genetics, which supports personalized nutrition approaches; and when advising lifestyle changes, by focusing not only on what foods to eat but also on how diet composition and meal timing can impact both drug and polyphenol metabolism (Iqbal et al. 2023; Lu et al. 2024).

Overall, incorporating polyphenol composed foods is safe and beneficial for most cardiovascular patients, but using higher dose supplements or combining polyphenols with medications requires careful management. Effective integration of polyphenols into cardiovascular care hinges on collaboration among cardiologists, pharmacologists, and nutritionists to maximize benefits while minimizing cardiac risks.

## 5.9 Challenges and Future Directions in Polyphenol Research and Application

Even while polyphenols are becoming more widely acknowledged as bioactive substances with significant health advantages, it is still difficult to convert these benefits into reliable clinical results. One of the biggest challenges to accurate intake assessment and therapeutic use is the variation in polyphenol content across meals and supplements (Gleichenhagen and Schieber 2016). The necessity for standardization of polyphenol content is highlighted by the fact that factors including growing circumstances, extraction techniques, and storage drastically change their concentration and composition (Burkholder-Cooley et al. 2017; Quesada-Vázquez et al. 2024). To improve the repeatability of study findings and guarantee the quality of polyphenol-enriched products utilized in clinical and nutritional contexts, standardized analytical techniques and reference materials should be established.

The bioavailability and effectiveness of polyphenols, which are heavily reliant on host microbiota interactions, metabolism, and digestion, represent another restriction. To differentiate between metabolites that are absorbed and those that are physiologically active in human tissues, better biomarkers of bioavailability are necessary. More accurate identification of substances generated from polyphenols that have physiological effects is becoming possible because to developments in metabolomics and plasma metabolite profiling. When combined with systems biology, this can assist in linking metabolic markers to specific clinical outcomes like vascular function or antioxidant capacity (Qiu et al. 2023; Scott et al. 2022).

A possible remedy for inter-individual heterogeneity in polyphenol response is provided by recent developments in nutrigenomics and precision nutrition. Personalized dietary approaches that maximize polyphenol absorption and utilization are made possible by using genetic and microbiome data (Malcomson and Mathers 2023). Microbiome-dependent metabolism of flavonoids, for example, produces a variety of metabolites with varying health effects on different people, indicating the necessity for individualized dietary recommendations (Bayram et al. 2018).

From a medicinal viewpoint, there is growing interest in creating medications based on polyphenols that target cardiovascular health. By modifying oxidative stress pathways, lipid homeostasis, and endothelial nitric oxide signaling, natural phenolics including resveratrol, quercetin, and catechins have vasoprotective and anti-inflammatory properties (Bayram et al. 2018; Liu et al. 2024). Co-delivery systems, synthetic analogs, and formulations based on nanocarriers are emerging to improve tissue targeting and stability, which is a major step toward clinical translation.

Polyphenol research is changing due to technological advancements including metabolomics, gut microbiome analysis, and integrative systems biology techniques. These resources promote a multifaceted understanding of polyphenol function by enabling researchers to comprehend the intricate relationships among food, metabolism, and disease pathways. However, there are still issues with regulatory harmonization since different foreign standards restrict the clearance of functional foods and nutraceuticals containing polyphenols. To guarantee safety, effectiveness, and labeling procedures in line with public health objectives, more precise regulations are required (Lu et al. 2025).

Ultimately, future polyphenol research must integrate standardization, bioavailability, and personalization frameworks with technological innovation to fully realize therapeutic and preventive potential of polyphenols.

## 5.10 Practical Implications for Clinicians, Nutritionists, and Patients

Polyphenols have emerged as important bioactive compounds that contribute significantly to cardiovascular health, notably through their favorable effects on blood pressure regulation, lipid metabolism, and vascular function (Iqbal et al. 2023). For clinicians and nutritionists, it is crucial to encourage patients to adopt diets rich in diverse polyphenol-containing foods such as berries, grapes, apples, various vegetables, nuts, seeds, green and black teas, cocoa, olive oil, and moderate amounts of red wine. Emphasizing whole-food sources rather than isolated supplements is an important point, as the synergistic effects within these foods enhance cardiovascular benefits and minimize the risk of unwanted interactions (Behl et al. 2020b). These recommendations align well with established heart-healthy dietary models like which naturally supply high polyphenol content alongside balanced nutrients, supporting cardiovascular risk reduction when combined with standard pharmacological therapies (Nagahawatta et al. 2024). Educating patients on practical food choices, portion sizes, and preparation methods that preserve polyphenol content can improve adherence and outcomes. Although polyphenol supplements are widely available, their use warrants caution due to inconsistent bioavailability, dosage variability, and potential interactions with medications; thus, supplementation should be personalized and supervised by healthcare professionals. From a public health perspective, promoting polyphenol-rich diets through policy initiatives, educational programs, and nutrition labeling can contribute to decreased cardiovascular disease incidence at the population level (Oliveira et al. 2021). Ultimately, integrating polyphenol-focused nutritional strategies into clinical care and public health frameworks represents a promising evidence-based approach, with ongoing research and multidisciplinary collaboration essential for optimizing cardiovascular prevention and management outcomes.

## 5.11 Conclusion

Polyphenols contribute significantly to cardiovascular well-being through a variety of mechanisms, including antioxidant effects, enhancement of endothelial function, anti-inflammatory actions, and modulation of lipid metabolism. Their influence on lowering blood pressure and improving cholesterol profiles establishes them as valuable components for the prevention and management of cardiovascular diseases. Both epidemiological data and clinical trials reinforce the protective role of diets abundant in polyphenols against conditions such as hypertension, atherosclerosis, and heart failure.

From a clinical standpoint, incorporating polyphenol-rich foods into lifestyle interventions can synergistically improve outcomes when used alongside conventional pharmacological treatments. Consuming whole foods high in diverse polyphenols—such as fruits, vegetables, nuts, tea, and cocoa—provides complex mixtures that may offer benefits beyond individual supplements. Nonetheless, issues like variability in polyphenol content, limited bioavailability, and individual differences in response present ongoing challenges for clinical application.

Future investigations should focus on large, rigorously designed clinical studies to determine effective dosages, clarify interactions with cardiovascular medications, and identify populations that might benefit in an optimal way. The use of emerging tools like metabolomics and precision nutrition approaches may help tailor polyphenol recommendations based on individual genetic and metabolic profiles. Additionally, developing novel delivery systems that enhance polyphenol stability and absorption could further improve their clinical utility.

Promoting balanced dietary patterns rich in natural polyphenol sources remains a practical and sustainable foundation for reducing cardiovascular risk on both individual and community levels. Continued collaboration between nutrition science, clinical research, and public health efforts is essential to fully harness the heart-protective potential of polyphenols and translate this knowledge into effective prevention strategies.

## References

Ali F, Wang D, Cheng Y, Wu M, Saleem MZ, Wei L, Xie Y, Yan M, Chu J, Yang Y, Shen A, Peng J (2023) Quercetin attenuates angiotensin II-induced proliferation of vascular smooth muscle cells and p53 pathway activation in vitro and in vivo. Biofact 49(4):956–970. https://doi.org/10.1002/biof.1959

Alotaibi BS, Ijaz M, Buabeid M, Kharaba ZJ, Yaseen HS, Murtaza G (2021) Therapeutic effects and safe uses of plant-derived polyphenolic compounds in cardiovascular diseases: a review. Drug des Devel Ther 15:4713–4732. https://doi.org/10.2147/DDDT.S327238

Anton SD, Embry C, Marsiske M, Lu X, Doss H, Leeuwenburgh C, Manini TM (2014) Safety and metabolic outcomes of resveratrol supplementation in older adults: results of a twelve-week, placebo-controlled pilot study. Exp Gerontol 57:181–187. https://doi.org/10.1016/j.exger.2014.05.015

Arshad MT, Ali MKM, Maqsood S, Ikram A, Hossain MS, Aljameel AI, Al-Farga A, Gnedeka KT (2025) Dietary phytochemicals in cardiovascular disease prevention and management: a comprehensive review. Food Sci Nutr 13(9):e70872. https://doi.org/10.1002/fsn3.70872

Atefi M, Entezari MH, Vahedi H, Hassanzadeh A (2022) sesame oil ameliorates alanine aminotransferase, aspartate aminotransferase, and fatty liver grade in women with nonalcoholic fatty liver disease undergoing low-calorie diet: a randomized double-blind controlled trial. Int J Clin Pract 1:4982080. https://doi.org/10.1155/2022/4982080

Balasundram N, Sundram K, Samman S (2006) Phenolic compounds in plants and agri-industrial by-products: antioxidant activity, occurrence, and potential uses. Food Chem 99(1):191–203. https://doi.org/10.1016/j.foodchem.2005.07.042

Baur JA, Sinclair DA (2006) Therapeutic potential of resveratrol: the in vivo evidence. Nat Rev Drug Discov 5(6):493–506. https://doi.org/10.1038/nrd2060

Bayram B, González-Sarrías A, Istas G, Garcia-Aloy M, Morand C, Tuohy K, García-Villalba R, Mena P (2018) Breakthroughs in the health effects of plant food bioactives: a perspective on microbiomics, nutri(epi)genomics, and metabolomics. J Agric Food Chem 66(41):10686–10692. https://doi.org/10.1021/acs.jafc.8b03385

Behl T, Bungau S, Kumar K, Zengin G, Khan F, Kumar A, Kaur R, Venkatachalam T, Tit DM, Vesa CM, Barsan G, Mosteanu DE (2020) Pleotropic effects of polyphenols in cardiovascular system. Biomed Pharmacother 130:110714. https://doi.org/10.1016/j.biopha.2020.110714

Biesinger S, Michaels HA, Quadros AS, Qian Y, Rabovsky AB, Badger RS, Jalili T (2015) A combination of isolated phytochemicals and botanical extracts lowers diastolic blood pressure in a randomized controlled trial of hypertensive subjects. Eur J Clin Nutr 70(1):10–16. https://doi.org/10.1038/ejcn.2015.88

Bondonno NP, Bondonno CP, Blekkenhorst LC, Considine MJ, Maghzal G, Stocker R, Woodman RJ, Ward NC, Hodgson JM, Croft KD (2017) Flavonoid-rich apple improves endothelial function in individuals at risk for cardiovascular disease: a randomized controlled clinical trial. Mol Nutr Food Res 62(3):1700674. https://doi.org/10.1002/mnfr.201700674

Bondonno NP, Parmenter BH, Murray K, Bondonno CP, Blekkenhorst LC, Wood AC, Post WS, Allison MA, Criqui MH, Lewis JR, Hodgson JM (2024) Associations between flavonoid intake and subclinical atherosclerosis: the multi-ethnic study of atherosclerosis. Arterioscler Thromb Vasc Biol 44(11):2347–2359. https://doi.org/10.1161/ATVBAHA.124.321106

Brglez Mojzer E, Knez Hrnčič M, Škerget M, Knez Ž, Bren U (2016) Polyphenols: extraction methods, antioxidative action, bioavailability and anticarcinogenic effects. Mol 21(7):901. https://doi.org/10.3390/molecules21070901

Burkholder-Cooley NM, Rajaram SS, Haddad EH, Oda K, Fraser GE, Jaceldo-Siegl K (2017) Validating polyphenol intake estimates from a food-frequency questionnaire by using repeated 24-h dietary recalls and a unique method-of-triads approach with 2 biomarkers. Am J Clin Nutr 105(3):685–694. https://doi.org/10.3945/ajcn.116.137174

Caiati C, Jirillo E (2025) Cellular and molecular bases for the application of polyphenols in the prevention and treatment of cardiovascular disease. Diseases 13(7):221. https://doi.org/10.3390/diseases13070221

Cassidy A (2017) Berry anthocyanin intake and cardiovascular health. Mol Aspects Med 61:76–82. https://doi.org/10.1016/j.mam.2017.05.002

Catalgol B, Batirel S, Taga Y, Ozer NK (2012) Resveratrol: French paradox revisited. Front Pharmacol 3:141. https://doi.org/10.3389/fphar.2012.00141

Cebova M, Pechanova O (2020) Protective effects of polyphenols against ischemia/reperfusion injury. Mol 25(15):3469. https://doi.org/10.3390/molecules25153469

Chen PY, Li S, Koh YC, Wu JC, Yang MJ, Ho CT, Pan MH (2019) Oolong tea extract and citrus peel polymethoxyflavones reduce transformation of l-carnitine to trimethylamine-N-oxide and decrease vascular inflammation in l-carnitine feeding mice. J Agric Food Chem 67(28):7869–7879. https://doi.org/10.1021/acs.jafc.9b03092

Cheng YC, Sheen JM, Hu WL, Hung YC (2017) Polyphenols and oxidative stress in atherosclerosis-related ischemic heart disease and stroke. Oxid Med Cell Longev 2017:8526438. https://doi.org/10.1155/2017/8526438

Chodari L, Dilsiz Aytemir M, Vahedi P, Alipour M, Vahed SZ, Khatibi SMH, Ahmadian E, Ardalan M, Eftekhari A (2021) Targeting mitochondrial biogenesis with polyphenol compounds. Oxid Med Cell Longev 2021:4946711. https://doi.org/10.1155/2021/4946711

Cianciosi D, Forbes-Hernández TY, Regolo L, Alvarez-Suarez JM, Navarro-Hortal MD, Xiao J, Quiles JL, Battino M, Giampieri F (2021) The reciprocal interaction between polyphenols and other dietary compounds: impact on bioavailability, antioxidant capacity and other physico-chemical and nutritional parameters. Food Chem 375:131904. https://doi.org/10.1016/j.foodchem.2021.131904

Dalgaard F, Bondonno NP, Murray K, Bondonno CP, Lewis JR, Croft KD, Kyrø C, Gislason G, Scalbert A, Cassidy A, Tjønneland A, Overvad K, Hodgson JM (2019) Associations between habitual flavonoid intake and hospital admissions for atherosclerotic cardiovascular disease: a prospective cohort study. Lancet Planet Health 3(11):e450–e459. https://doi.org/10.1016/S2542-5196(19)30212-8

Davison K, Berry NM, Misan G, Coates AM, Buckley JD, Howe PR (2010) Dose-related effects of flavanol-rich cocoa on blood pressure. J Hum Hypertens 24(9):568–576. https://doi.org/10.1038/jhh.2009.105

De Brito Alves JL, Brasil JMA, Maia LA, Da Costa LM, Sampaio KB, De Souza EL (2023) Phenolic compounds in hypertension: targeting gut-brain interactions and endothelial dysfunction. J Funct Foods 104:105531. https://doi.org/10.1016/j.jff.2023.105531

Del Bo' C, Bernardi S, Marino M, Porrini M, Tucci M, Guglielmetti S, Cherubini A, Carrieri B, Kirkup B, Kroon P, Zamora-Ros R, Liberona NH, Andres-Lacueva C, Riso P (2019) Systematic review on polyphenol intake and health outcomes: is there sufficient evidence to define a health-promoting polyphenol-rich dietary pattern? Nutr 11(6):1355. https://doi.org/10.3390/nu11061355

Del Rio D, Rodriguez-Mateos A, Spencer JP, Tognolini M, Borges G, Crozier A (2012) Dietary (poly)phenolics in human health: structures, bioavailability, and evidence of protective effects against chronic diseases. Antioxid Redox Signal 18(14):1818–1892. https://doi.org/10.1089/ars.2012.458

Domínguez-López I, Arancibia-Riveros C, Marhuenda-Muñoz M, Tresserra-Rimbau A, Toledo E, Fitó M, Ros E, Estruch R, Lamuela-Raventós RM (2023) Association of microbiota polyphenols with cardiovascular health in the context of a Mediterranean diet. Food Res Int 165:112499. https://doi.org/10.1016/j.foodres.2023.112499

Dong LY, Chen F, Xu M, Yao LP, Zhang YJ, Zhuang Y (2018) Quercetin attenuates myocardial ischemia-reperfusion injury via downregulation of the HMGB1-TLR4-NF-κB signaling pathway. Am J Transl Res 10(5):1273–1283

Duda-Chodak A, Tarko T (2023) Possible side effects of polyphenols and their interactions with medicines. Mol 28(6):2536. https://doi.org/10.3390/molecules28062536

Ed Nignpense B, Chinkwo KA, Blanchard CL, Santhakumar AB (2019) Polyphenols: modulators of platelet function and platelet microparticle generation? J Mol Sci 21(1):146. https://doi.org/10.3390/ijms21010146

Edwards RL, Lyon T, Litwin SE, Rabovsky A, Symons JD, Jalili T (2007) Quercetin reduces blood pressure in hypertensive subjects. J Nutr 137(11):2405–2411. https://doi.org/10.1093/jn/137.11.2405

Egert S, Bosy-Westphal A, Seiberl J, Kürbitz C, Settler U, Plachta-Danielzik S, Wagner AE, Frank J, Schrezenmeir J, Rimbach G, Wolffram S, Müller MJ (2009) Quercetin reduces systolic blood pressure and plasma oxidised low-density lipoprotein concentrations in overweight subjects with a high-cardiovascular disease risk phenotype: a double-blinded, placebo-controlled cross-over study. Br J Nutr 102(7):1065–1074. https://doi.org/10.1017/S0007114509359127

Estruch R, Ros E, Salas-Salvadó J, Covas MI, Corella D, Arós F, Gómez-Gracia E, Ruiz-Gutiérrez V, Fiol M, Lapetra J, Lamuela-Raventos RM, Serra-Majem L, Pintó X, Basora J, Muñoz MA, Sorlí JV, Martínez JA, Fitó M, Gea A, Hernán MA, Martínez-González MA; PREDIMED Study Investigators (2018) Primary prevention of cardiovascular disease with a Mediterranean diet supplemented with extra-virgin olive oil or nuts. N Engl J Med 378(25):e34. https://doi.org/10.1056/NEJMoa1800389

Feng H, Mou SQ, Li WJ, Zhang N, Zhou ZY, Ding W, Bian ZY, Liao HH (2020) Resveratrol inhibits ischemia-induced myocardial senescence signals and NLRP3 inflammasome activation. Oxid Med Cell Longev 2020:2647807. https://doi.org/10.1155/2020/2647807

Fraga CG, Croft KD, Kennedy DO, Tomás-Barberán FA (2019) The effects of polyphenols and other bioactives on human health. Food Funct 10(2):514–528. https://doi.org/10.1039/c8fo01997e

Gal R, Deres L, Horvath O, Eros K, Sandor B, Urban P, Soos S, Marton Z, Sumegi B, Toth K, Habon T, Halmosi R (2020) Resveratrol improves heart function by moderating inflammatory processes in patients with systolic heart failure. Antioxid 9(11):1108. https://doi.org/10.3390/antiox9111108

Ge B, Zhang Z, Zuo Z (2014) Updates on the clinical evidenced herb-warfarin interactions. Evid Based Complement Alternat Med 2014:957362. https://doi.org/10.1155/2014/957362

Gleichenhagen M, Schieber A (2016) Current challenges in polyphenol analytical chemistry. Curr Opin Food Sci 7:43–49. https://doi.org/10.1016/j.cofs.2015.10.004

Godos J, Romano GL, Gozzo L, Laudani S, Paladino N, Dominguez Azpíroz I, Martínez López NM, Giampieri F, Quiles JL, Battino M, Galvano F, Drago F, Grosso G (2024a) Resveratrol and vascular health: evidence from clinical studies and mechanisms of actions related to its metabolites produced by gut microbiota. Front Pharmacol 15:1368949. https://doi.org/10.3389/fphar.2024.1368949

Godos J, Romano GL, Laudani S, Gozzo L, Guerrera I, Dominguez Azpíroz I, Martínez Diaz R, Quiles JL, Battino M, Drago F, Giampieri F, Galvano F, Grosso G (2024b) Flavan-3-ols and vascular health: clinical evidence and mechanisms of action. Nutr 16(15):2471. https://doi.org/10.3390/nu16152471

Godos J, Vitale M, Micek A, Ray S, Martini D, Del Rio D, Riccardi G, Galvano F, Grosso G (2019) Dietary polyphenol intake, blood pressure, and hypertension: a systematic review and meta-analysis of observational studies. Antioxid 8(6):152. https://doi.org/10.3390/antiox8060152

Grosso G, Godos J, Currenti W, Micek A, Falzone L, Libra M, Giampieri F, Forbes-Hernández TY, Quiles JL, Battino M, La Vignera S, Galvano F (2022) The effect of dietary polyphenols on vascular health and hypertension: current evidence and mechanisms of action. Nutr 14(3):545. https://doi.org/10.3390/nu14030545

Grosso G, Stepaniak U, Micek A, Kozela M, Stefler D, Bobak M, Pajak A (2017) Dietary polyphenol intake and risk of hypertension in the Polish arm of the HAPIEE study. Eur J Nutr 57(4):1535–1544. https://doi.org/10.1007/s00394-017-1438-7

Guo Q, Jin Y, Chen X, Ye X, Shen X, Lin M, Zeng C, Zhou T, Zhang J (2024) NF-κB in biology and targeted therapy: new insights and translational implications. Signal Transduct Target Ther 9(1):53. https://doi.org/10.1038/s41392-024-01757-9

Guo S, Long M, Li X, Zhu S, Zhang M, Yang Z (2016) Curcumin activates autophagy and attenuates oxidative damage in EA.hy926 cells via the Akt/mTOR pathway. Mol Med Rep 13(3):2187–2193. https://doi.org/10.3892/mmr.2016.4796

Hameed S, Ur Rehman A, Massey S, Syed NI, Anwar F, Ahmed D, Ahmad S (2022) Grevillea robusta delayed the progression of experimentally induced hepatic fibrosis and cirrhosis in wistar rats by attenuating the expression of smooth muscle actin, collagen, and TGF-β. Front Pharmacol 13:904584. https://doi.org/10.3389/fphar.2022.904584

Hedayati N, Yaghoobi A, Salami M, Gholinezhad Y, Aghadavood F, Eshraghi R, Aarabi MH, Homayoonfal M, Asemi Z, Mirzaei H, Hajijafari M, Mafi A, Rezaee M (2023) Impact of polyphenols on heart failure and cardiac hypertrophy: clinical effects and molecular mechanisms. Front Cardiovasc Med 10:1174816. https://doi.org/10.3389/fcvm.2023.1174816

Hodgson JM, Croft KD, Woodman RJ, Puddey IB, Bondonno CP, Wu JH, Beilin LJ, Lukoshkova EV, Head GA, Ward NC (2014) Effects of vitamin E, vitamin C and polyphenols on the rate of blood pressure variation: results of two randomised controlled trials. Br J Nutr 112(9):1551–1561. https://doi.org/10.1017/S0007114514002542

Huang M, Xie X, Yuan R, Xin Q, Ma S, Guo H, Miao Y, Hu C, Zhu Y, Cong W (2024) The multi-faceted anti-atherosclerotic properties of herbal flavonoids: a comprehensive review. Pharmacol Res 211:107551. https://doi.org/10.1016/j.phrs.2024.107551

Iqbal I, Wilairatana P, Saqib F, Nasir B, Wahid M, Latif MF, Iqbal A, Naz R, Mubarak MS (2023) Plant polyphenols and their potential benefits on cardiovascular health: a review. Mol 28(17):6403. https://doi.org/10.3390/molecules28176403

Jiang J, Gu X, Wang H, Ding S (2021) Resveratrol improves cardiac function and left ventricular fibrosis after myocardial infarction in rats by inhibiting NLRP3 inflammasome activity and the TGF-β1/SMAD2 signaling pathway. PeerJ 9:e11501. https://doi.org/10.7717/peerj.11501

Jomova K, Alomar SY, Valko R, Liska J, Nepovimova E, Kuca K, Valko M (2025) Flavonoids and their role in oxidative stress, inflammation, and human diseases. Chem Biol Interact 413:111489. https://doi.org/10.1016/j.cbi.2025.111489

Khouya T, Ramchoun M, Hmidani A, Amrani S, Benlyas M, Kasbi Chadli F, Ouguerram K, Alem C (2022) Effect of supplementation with polyphenol extract of Thymus atlanticus on paraoxonase-1 activity, insulin resistance, and lipid profile in high-fat diet-fed hamsters. J Food Biochem 46(9):e14225. https://doi.org/10.1111/jfbc.14225

Kim HR, Kim WK, Ha AW (2019) Effects of phytochemicals on blood pressure and neuroprotection mediated via brain renin-angiotensin system. Nutr 11(11):2761. https://doi.org/10.3390/nu11112761

Kim SG, Kim JR, Choi HC (2017) Quercetin-induced AMP-activated protein kinase activation attenuates vasoconstriction through LKB1-AMPK signaling pathway. J Med Food 21(2):146–153. https://doi.org/10.1089/jmf.2017.4052

Korobkova EA (2015) Effect of natural polyphenols on CYP metabolism: implications for diseases. Chem Res Toxicol 28(7):1359–1390. https://doi.org/10.1021/acs.chemrestox.5b00121

Kour N, Bhagat G, Singh S, Bhatti SS, Arora S, Singh B, Bhatia A (2023) Polyphenols mediated attenuation of diabetes associated cardiovascular complications: a comprehensive review. J Diabetes Metab Disord 23(1):73–99. https://doi.org/10.1007/s40200-023-01326-x

Kumar S, Eroglu E, Stokes JA 3rd, Scissum-Gunn K, Saldanha SN, Singh UP, Manne U, Ponnazhagan S, Mishra MK (2017) Resveratrol induces mitochondria-mediated, caspase-independent apoptosis in murine prostate cancer cells. Oncotarget 8(13):20895–20908. https://doi.org/10.18632/oncotarget.14947

Lacerda D, Türck P, Campos-Carraro C, Hickmann A, Ortiz V, Bianchi S, Belló-Klein A, de Castro AL, Bassani VL, Araujo ASDR (2020) Pterostilbene improves cardiac function in a rat model of right heart failure through modulation of calcium handling proteins and oxidative stress. Appl Physiol Nutr Metab 45(9):987–995. https://doi.org/10.1139/apnm-2019-0864

Larson AJ, Symons JD, Jalili T (2010) Quercetin: a treatment for hypertension? a review of efficacy and mechanisms. Pharm 3(1):237–250. https://doi.org/10.3390/ph3010237

Li SH, Zhao P, Tian HB, Chen LH, Cui LQ (2015) Effect of grape polyphenols on blood pressure: a meta-analysis of randomized controlled trials. PLoS ONE 10(9):e0137665. https://doi.org/10.1371/journal.pone.0137665

Lian N, Tong J, Li W, Wu J, Li Y (2018) Ginkgetin ameliorates experimental atherosclerosis in rats. Biomed Pharmacother 102:510–516. https://doi.org/10.1016/j.biopha.2018.03.107

Lin X, Han T, Fan Y, Wu S, Wang F, Wang C (2020) Quercetin improves vascular endothelial function through promotion of autophagy in hypertensive rats. Life Sci 258:118106. https://doi.org/10.1016/j.lfs.2020.118106

Lin X, Zhang I, Li A, Manson JE, Sesso HD, Wang L, Liu S (2016) Cocoa flavanol intake and biomarkers for cardiometabolic health: a systematic review and meta-analysis of randomized controlled trials. J Nutr 146(11):2325–2333. https://doi.org/10.3945/jn.116.237644

Lin X, Zhao J, Ge S, Lu H, Xiong Q, Guo X, Li L, He S, Wang J, Peng F, Fan Y, Zuo X, Tian C, Ying C (2023) Dietary polyphenol intake and risk of hypertension: an 18-y nationwide cohort study in China. Am J Clin Nutr 118(1):264–272. https://doi.org/10.1016/j.ajcnut.2023.05.001

Ling WC, Liu J, Lau CW, Murugan DD, Mustafa MR, Huang Y (2017) Treatment with salvianolic acid B restores endothelial function in angiotensin II-induced hypertensive mice. Biochem Pharmacol 136:76–85. https://doi.org/10.1016/j.bcp.2017.04.007

Liu L, Zhao W, Liu J, Gan Y, Liu L, Tian J (2018) Epigallocatechin-3 gallate prevents pressure overload-induced heart failure by up-regulating SERCA2a via histone acetylation modification in mice. PLoS ONE 13(10):e0205123. https://doi.org/10.1371/journal.pone.0205123

Liu Y, Luo J, Peng L, Zhang Q, Rong X, Luo Y, Li J (2024) Flavonoids: potential therapeutic agents for cardiovascular disease. Heliyon 10(12):e32563. https://doi.org/10.1016/j.heliyon.2024.e32563

Looi D, Moorthy M, Chaiyakunapruk N, Palanisamy UD (2022) Impact of ellagitannin-rich fruit consumption on blood pressure: a systematic review and meta-analysis of randomized controlled trials. J Funct Foods 99:105320. https://doi.org/10.1016/j.jff.2022.105320
Lu Y, Wang K, Hu L (2025) Advancements in delivery systems for dietary polyphenols in enhancing radioprotection effects: challenges and opportunities. NPJ Sci Food 9(1):51. https://doi.org/10.1038/s41538-025-00419-6
Lu Y, Zhao J, Xin Q, Yuan R, Miao Y, Yang M, Mo H, Chen K, Cong W (2024) Protective effects of oleic acid and polyphenols in extra virgin olive oil on cardiovascular diseases. Food Sci Hum Wellness 13(2):529–540. https://doi.org/10.26599/FSHW.2022.9250047
Ludovici V, Barthelmes J, Nagele MP, Flammer AJ, Sudano I (2018) Polyphenols: anti-platelet nutraceutical? Curr Pharm des 24(2):146–157. https://doi.org/10.2174/13816128236661711109104600
Luo M, Tian R, Yang Z, Peng YY, Lu N (2019) Quercetin suppressed NADPH oxidase-derived oxidative stress via heme oxygenase-1 induction in macrophages. Arch Biochem Biophys 671:69–76. https://doi.org/10.1016/j.abb.2019.06.007
Makhtoomi M, Shateri Z, Mashoufi A, Nouri M, Honarvar B, Keshani P (2024) The association between dietary polyphenol intake and the odds of metabolic syndrome. Sci Rep 14(1):25559. https://doi.org/10.1038/s41598-024-77335-4
Malcomson FC, Mathers JC (2023) Translation of nutrigenomic research for personalised and precision nutrition for cancer prevention and for cancer survivors. Redox Biol 62:102710. https://doi.org/10.1016/j.redox.2023.102710
Marino M, Del Bo' C, Martini D, Porrini M, Riso P (2020) A review of registered clinical trials on dietary (poly)phenols: past efforts and possible future directions. Foods 9(11):1606. https://doi.org/10.3390/foods9111606
Martin SS, Aday AW, Allen NB, Almarzooq ZI, Anderson CAM, Arora P, Avery CL, Baker-Smith CM, Bansal N, Beaton AZ, Commodore-Mensah Y, Currie ME, Elkind MSV, Fan W, Generoso G, Gibbs BB, Heard DG, Hiremath S, Johansen MC, Kazi DS, Ko D, Leppert MH, Magnani JW, Michos ED, Mussolino ME, Parikh NI, Perman SM, Rezk-Hanna M, Roth GA Shah NS, Springer MV, St-Onge MP, Thacker EL, Urbut SM, Van Spall HGC, Voeks JH, Whelton SP, Wong ND, Wong SS, Yaffe K, Palaniappan LP; American Heart Association Council on Epidemiology and Prevention Statistics Committee and Stroke Statistics Committeee (2025) 2025 Heart disease and stroke statistics: a report of us and global data from the American heart association. Circ 151(8):e41–e660. https://doi.org/10.1161/CIR.0000000000001303
Meister ML, Najjar RS, Danh JP, Knapp D, Wanders D, Feresin RG (2022) Berry consumption mitigates the hypertensive effects of a high-fat, high-sucrose diet via attenuation of renal and aortic AT1R expression resulting in improved endothelium-derived NO bioavailability. J Nutr Biochem 112:109225. https://doi.org/10.1016/j.jnutbio.2022.109225
Miranda AM, Steluti J, Fisberg RM, Marchioni DM (2016) Association between polyphenol intake and hypertension in adults and older adults: a population-based study in Brazil. PLoS One 11(10):e0165791. https://doi.org/10.1371/journal.pone.0165791
Mishra AK, Singh R, Rawat H, Kumar V, Jagtap C, Jain A (2024) The influence of food matrix on the stability and bioavailability of phytochemicals: a comprehensive review. Food Hum 2:100202. https://doi.org/10.1016/j.foohum.2023.12.010
Mohos V, Fliszár-Nyúl E, Ungvári O, Kuffa K, Needs PW, Kroon PA, Telbisz Á, Özvegy-Laczka C, Poór M (2020) Inhibitory effects of quercetin and its main methyl, sulfate, and glucuronic acid conjugates on cytochrome P450 Enzymes, and on OATP, BCRP and MRP2 transporters. Nutr 12(8):2306. https://doi.org/10.3390/nu12082306
Morais SL, Gonçalves TF, Delerue-Matos C, Ferrreira-Fernandes H, Pinto GR, Domingues VF, Barroso MF (2022) Cytochrome P450 polymorphisms with impact in cardiovascular drugs metabolisms in European populations. Hum Gene 33:201027. https://doi.org/10.1016/j.humgen.2022.201027

Nagahawatta DP, Liyanage NM, Jayawardena TU, Jeon YJ (2024) Marine polyphenols in cardiovascular health: unraveling structure-activity relationships, mechanisms, and therapeutic implications. Int J Mol Sci 25(15):8419. https://doi.org/10.3390/ijms25158419
Nazzaro F, Coppola F, Fratianni F, Abdalrazeq M, Ombra MN, De Giulio B, Coppola R, Zengin G (2025) Polyphenols bioactive metabolites, and their anti-biofilm and neuroprotective potential. Foods 14(22):3976. https://doi.org/10.3390/foods14223976
Ngo V, Duennwald ML (2022) Nrf2 and oxidative stress: a general overview of mechanisms and implications in human disease. Antioxid 11(12):2345. https://doi.org/10.3390/antiox11122345
Oliveira G, Volino-Souza M, Conte-Junior CA, Alvares TS (2021) Food-derived polyphenol compounds and cardiovascular health: a nano-technological perspective. Food Biosci 41:101033. https://doi.org/10.1016/j.fbio.2021.101033
Panche AN, Diwan AD, Chandra SR (2016) Flavonoids: an overview. J Nutr Sci 5:e47. https://doi.org/10.1017/jns.2016.41
Pandey KB, Rizvi SI (2009) Plant polyphenols as dietary antioxidants in human health and disease. Oxid Med Cell Longev 2(5):270–278. https://doi.org/10.4161/oxim.2.5.9498
Pereira SC, Parente JM, Belo VA, Mendes AS, Gonzaga NA, do Vale GT, Ceron CS, Tanus-Santos JE, Tirapelli CR, Castro MM (2018) Quercetin decreases the activity of matrix metalloproteinase-2 and ameliorates vascular remodeling in renovascular hypertension. Atheroscler 270:146–153. https://doi.org/10.1016/j.atherosclerosis.2018.01.031
Pham VT, Dold S, Rehman A, Bird JK, Steinert RE (2021) Vitamins, the gut microbiome and gastrointestinal health in humans. Nutr Res 95:35–53. https://doi.org/10.1016/j.nutres.2021.09.001
Pieczynska MD, Yang Y, Petrykowski S, Horbanczuk OK, Atanasov AG, Horbanczuk JO (2020) Gut microbiota and its metabolites in atherosclerosis development. Molecules 25(3):594. https://doi.org/10.3390/molecules25030594
Pignatelli P, Pulcinelli FM, Celestini A, Lenti L, Ghiselli A, Gazzaniga PP, Violi F (2000) The flavonoids quercetin and catechin synergistically inhibit platelet function by antagonizing the intracellular production of hydrogen peroxide. Am J Clin Nutr 72(5):1150–1155. https://doi.org/10.1093/ajcn/72.5.1150
Poli A, Marangoni F (2024) Nutri-score and cardiovascular risk: new insights from the EPIC cohorts. Lancet Reg Health Eur 46:101059. https://doi.org/10.1016/j.lanepe.2024.101059
Potì F, Santi D, Spaggiari G, Zimetti F, Zanotti I (2019) Polyphenol health effects on cardiovascular and neurodegenerative disorders: a review and meta-analysis. Int J Mol Sci 20(2):351. https://doi.org/10.3390/ijms20020351
Qiu S, Cai Y, Yao H, Lin C, Xie Y, Tang S, Zhang A (2023) Small molecule metabolites: discovery of biomarkers and therapeutic targets. Signal Transduct Target Ther 8(1):132. https://doi.org/10.1038/s41392-023-01399-3
Quesada-Vázquez S, Eseberri I, Les F, Pérez-Matute P, Herranz-López M, Atgié C, Lopez-Yus M, Aranaz P, Oteo JA, Escoté X, Lorente-Cebrian S, Roche E, Courtois A, López V, Portillo MP, Milagro FI, Carpéné C (2024) Polyphenols and metabolism: from present knowledge to future challenges. J Physiol Biochem 80(3):603–625. https://doi.org/10.1007/s13105-024-01046-7
Rees A, Dodd GF, Spencer JPE (2018) The effects of flavonoids on cardiovascular health: a review of human intervention trials and implications for cerebrovascular function. Nutr 10(12):1852. https://doi.org/10.3390/nu10121852
Ren ZQ, Zheng SY, Sun Z, Luo Y, Wang YT, Yi P, Li YS, Huang C, Xiao WF (2020) Resveratrol: molecular mechanisms, health benefits, and potential adverse effects. MedComm 6(6):e70252. https://doi.org/10.1002/mco2.70252
Ried K, Fakler P, Stocks NP (2017) Effect of cocoa on blood pressure. Cochrane Database Syst Rev 4(4):CD008893. https://doi.org/10.1002/14651858.CD008893.pub3
Ros E, Martínez-González MA, Estruch R, Salas-Salvadó J, Fitó M, Martínez JA, Corella D (2014) Mediterranean diet and cardiovascular health: teachings of the PREDIMED study. Adv Nutr 5(3):330S-S336. https://doi.org/10.3945/an.113.005389

Scalbert A, Williamson G (2000) Dietary intake and bioavailability of polyphenols. J Nutr 130(8S Suppl):2073S-2085S. https://doi.org/10.1093/jn/130.8.2073S
Scott MB, Styring AK, McCullagh JSO (2022) Polyphenols: bioavailability, microbiome interactions and cellular effects on health in humans and animals. Pathog 11(7):770. https://doi.org/10.3390/pathogens11070770
Selma MV, Espín JC, Tomás-Barberán FA (2009) Interaction between phenolics and gut microbiota: role in human health. J Agric Food Chem 57(15):6485–6501. https://doi.org/10.1021/jf902107d
Serna M, Wong-Baeza C, Santiago-Hernández JC, Baeza I, Wong C (2014) Hypocholesterolemic and choleretic effects of three dimethoxycinnamic acids in relation to 2,4,5-trimethoxycinnamic acid in rats fed with a high-cholesterol/cholate diet. Pharmacol Rep 67(3):553–559. https://doi.org/10.1016/j.pharep.2014.12.009
Serreli G, Deiana M (2023) Role of dietary poly-phenols in the activity and expression of nitric oxide synthases: a review. Antioxid 12(1):147. https://doi.org/10.3390/antiox12010147
Sharanya CS, Abhithaj J, Arun KG, Eeda KR, Bhat V, Variyar EJ, Sabu A, Haridas M (2023) Lipoxygenase inhibitory synthetic derivatives of methyl gallate regulate gene expressions of COX-2 and cytokines to reduce animal model arthritis. Sci Rep 13(1):10644. https://doi.org/10.1038/s41598-023-37613-z
Sleiman Y, Reisqs JB, Boutjdir M (2024) Differentiation of sinoatrial-like cardiomyocytes as a biological pacemaker model. Int J Mol Sci 25(17):9155. https://doi.org/10.3390/ijms25179155
Soyata A, Hasanah AN, Rusdiana T (2020) Interaction of warfarin with herbs based on pharmacokinetic and pharmacodynamic parameters. Indones J Pharm 2(2):69–76. https://doi.org/10.24198/idjp.v2i2.27289
Stankovic S, Mutavdzin Krneta S, Djuric D, Milosevic V, Milenkovic D (2025) Plant polyphenols as heart's best friends: from health properties, to cellular effects, to molecular mechanisms of action. Int J Mol Sci 26(3):915. https://doi.org/10.3390/ijms26030915
Steven S, Frenis K, Oelze M, Kalinovic S, Kuntic M, Bayo Jimenez MT, Vujacic-Mirski K, Helmstädter J, Kröller-Schön S, Münzel T, Daiber A (2019) Vascular inflammation and oxidative stress: major triggers for cardiovascular disease. Oxid Med Cell Longev 2019:7092151. https://doi.org/10.1155/2019/7092151
Sun X, Zhang S, Song H (2020) Quercetin attenuates reduced uterine perfusion pressure -induced hypertension in pregnant rats through regulation of endothelin-1 and endothelin-1 type A receptor. Lipids Health Dis 19(1):180. https://doi.org/10.1186/s12944-020-01357-w
Tarko T, Duda-Chodak A (2020) Influence of food matrix on the bioaccessibility of fruit polyphenolic compounds. J Agric Food Chem 68(5):1315–1325. https://doi.org/10.1021/acs.jafc.9b07680
Tasic N, Jakovljevic VLJ, Mitrovic M, Djindjic B, Tasic D, Dragisic D, Citakovic Z, Kovacevic Z, Radoman K, Zivkovic V, Bolevich S, Turnic TN (2021) Black chokeberry Aronia melanocarpa extract reduces blood pressure, glycemia and lipid profile in patients with metabolic syndrome: a prospective controlled trial. Mol Cell Biochem 476(7):2663–2673. https://doi.org/10 1007/s11010-021-04106-4
Theodotou M, Fokianos K, Mouzouridou A, Konstantinou C, Aristotelous A, Prodromou D, Chrysikou A (2016) The effect of resveratrol on hypertension: a clinical trial. Exp Ther Med 13(1):295–301. https://doi.org/10.3892/etm.2016.3958
Triebel H, Castrop H (2024) The renin angiotensin aldosterone system. Eur J Physiol 476(5):705–713. https://doi.org/10.1007/s00424-024-02908-1
Wan S, Luo J, Zhu Y, An P, Luo Y, Xing Q (2024) The effect of antioxidant polyphenol supplementation on cardiometabolic risk factors: a systematic review and meta-analysis. Nutr 16(23) 4206. https://doi.org/10.3390/nu16234206
Wisnuwardani RW (2025) Polyphenols and cardiovascular disease: implications for public health nutrition. In: Preedy VR, Patel VB (eds) Handbook of public health nutrition: international, national, and regional perspectives. Springer Nature Switzerland, Cham, pp 1–21. https://doi.org/10.1007/978-3-031-32047-7_106-1

Wu Y, Xia ZY, Zhao B, Leng Y, Dou J, Meng QT, Lei SQ, Chen ZZ, Zhu J (2017) (-)-Epigallocatechin-3-gallate attenuates myocardial injury induced by ischemia/reperfusion in diabetic rats and in H9c2 cells under hyperglycemic conditions. Int J Mol Med 40(2):389–399. https://doi.org/10.3892/ijmm.2017.3014

Xu R, Yang K, Ding J, Chen G (2020) Effect of green tea supplementation on blood pressure: a systematic review and meta-analysis of randomized controlled trials. Med 99(6):e19047. https://doi.org/10.1097/MD.0000000000019047

Yao Y, Liu T, Yin L, Man S, Ye S, Ma L (2021) Polyphenol-Rich extract from Litchi chinensis seeds alleviates hypertension-induced renal damage in rats. J Agric Food Chem 69(7):2138–2148. https://doi.org/10.1021/acs.jafc.0c07046

Yu W, Zhang Y, Lu Y, Ouyang Z, Peng J, Tu Y, He B (2025) Recent research on the bioactivity of polyphenols derived from edible fungi and their potential in chronic disease prevention. J Funct Foods 124:106627. https://doi.org/10.1016/j.jff.2024.106627

Zhang L, Chen J, Yan L, He Q, Xie H, Chen M (2021) Resveratrol ameliorates cardiac remodeling in a murine model of heart failure with preserved ejection fraction. Front Pharmacol 12:646240. https://doi.org/10.3389/fphar.2021.646240

Zhang L, Guo Z, Wang Y, Geng J, Han S (2019) The protective effect of kaempferol on heart via the regulation of Nrf2, NF-κβ, and PI3K/Akt/GSK-3β signaling pathways in isoproterenol-induced heart failure in diabetic rats. Drug Dev Res 80(3):294–309. https://doi.org/10.1002/ddr.21495

Zhang S, Shen C, Di H, Wang Y, Guan F (2025) Regulatory mechanisms of phenolic acids in metabolic dysfunction-associated steatotic liver disease: a review. Antioxid 14(7):760. https://doi.org/10.3390/antiox14070760

Zhang W, Zheng Y, Yan F, Dong M, Ren Y (2023) Research progress of quercetin in cardiovascular disease. Front Cardiovasc Med 10:1203713. https://doi.org/10.3389/fcvm.2023.1203713

Zheng D, Liu Z, Zhou Y, Hou N, Yan W, Qin Y, Ye Q, Cheng X, Xiao Q, Bao Y, Luo J, Wu X (2020) Urolithin B, a gut microbiota metabolite, protects against myocardial ischemia/reperfusion injury via p62/Keap1/Nrf2 signaling pathway. Pharmacol Res 153:104655. https://doi.org/10.1016/j.phrs.2020.104655

Zheng XX, Xu YL, Li SH, Liu XX, Hui R, Huang XH (2011) Green tea intake lowers fasting serum total and LDL cholesterol in adults: a meta-analysis of 14 randomized controlled trials. Am J Clin Nutr 94(2):601–610. https://doi.org/10.3945/ajcn.110.010926

Ziółkiewicz A, Kasprzak-Drozd K, Rusinek R, Markut-Miotła E, Oniszczuk A (2023) The influence of polyphenols on atherosclerosis development. Int J Mol Sci 24(8):7146. https://doi.org/10.3390/ijms24087146

Zippel N, Loot AE, Stingl H, Randriamboavonjy V, Fleming I, Fisslthaler B (2018) Endothelial AMP-activated kinase α1 phosphorylates eNOS on Thr495 and decreases endothelial NO formation. Int J Mol Sci 19(9):2753. https://doi.org/10.3390/ijms19092753

Zordoky BN, Robertson IM, Dyck JR (2014) Preclinical and clinical evidence for the role of resveratrol in the treatment of cardiovascular diseases. Biochim Biophys 1852(6):1155–1177. https://doi.org/10.1016/j.bbadis.2014.10.016

# Chapter 6
# Polyphenols in Inflammatory Conditions

**Salvatore Parisi, Vito Barone, Giorgia Punzo, D. K. Meera, Sony George, and Suni Mary Varghese**

**Abstract** Polyphenols correspond to a heterogeneous class of organic molecules that have been reported to have beneficial effects on the health of human beings. There are different research areas involving polyphenols. From the medical viewpoint, inflammation can be defined as the resulting defence action of blood cells against foreign substances. Consequently, such a defence strategy can be put in place by the human organism in an automated way, with the aim of protecting the subject from irritating external agents. The speed of recovery can be helped if selected molecules, such as steroidal anti-inflammatory drugs, are available. However, some of these drugs can have unfavourable cardiovascular effects and other problems. Consequently, the use of non-synthetic and natural compounds with similar inhibitory effects, such as polyphenols, would be highly recommended. The situation can also be complicated because of the growing intake of antibiotics by human beings, and also in the ambit of food production industries, from primary production. Anyway, the role of polyphenols as natural antioxidants and anti-inflammatory agents has been repeatedly studied and mentioned. Consequently, a chemical classification of such

S. Parisi (✉) · S. M. Varghese
Department of Nutrition and Food Processing Technology, Lourdes Matha Institute of Hotel Management and Catering Technology, Thiruvananthapuram, Kerala State, India
e-mail: drparisi@inwind.it

S. M. Varghese
e-mail: sunimary@lmihmct.org

V. Barone · G. Punzo
Palermo, Italy
e-mail: vitobarone99@gmail.com

G. Punzo
e-mail: punzogiorgia01@gmail.com

D. K. Meera
Department of Food and Nutrition, Goverment College for Women, Thiruvananthapuram, Kerala State, India
e-mail: meerasunil@gmail.com

S. George
Department of Food Technology, TKM Institute of Technology, Kollam, India
e-mail: sony.g@tkmit.ac.in

 
R. K. Singla et al. (eds.), *Polyphenols in Health*,
https://doi.org/10.1007/978-3-032-27324-6_6

a broad and diversified category should be presented before discussing its medical implications. The chemical categorisation of polyphenols can help when speaking of the comprehension of inflammatory mechanisms and correlated therapeutic applications. The regular consumption of vegetable products containing a good bioavailable amount of phenolic compounds can be recommended with favour, despite the need for additional research in this ambit.

**Keywords** Autoimmune disease · Chronic inflammation · Flavonoid · Heterocyclic ring · Metabolic disorder · Non-flavonoid · Polyphenol

## 6.1 Inflammation Conditions in the Human Being and Natural Antagonists: Polyphenols

Polyphenols correspond to a heterogeneous class of organic molecules that have been reported to have beneficial effects on the health of human beings (Issaoui et al. 2020; Parisi 2019, 2020). In detail, and in accordance with other Chapters in this book (Sects. 6.1–6.3), polyphenols are expected to exhibit the following effects and properties (Arshad et al. 2025; Domingues et al. 2024; Giglio et al. 2018; Hafiz 2025; Kardum et al. 2015; Kupczyk et al. 2025; Liva et al. 2025; Makarewicz et al. 2021; Martiniakova et al. 2025; Mechi et al. 2023; Mendes et al. 2025; Mileo and Miccadei 2016; Rudrapal et al. 2022; Stoner and Mukhtar 1995; Tillman et al. 2025; Wang and Hu 2023):

(a) Enhanced biodigestibility from natural sources
(b) Synergical interaction with gut microbiome
(c) Antioxidant behaviour against oxidative stress
(d) Cardiovascular benefits in terms of restored blood pressure and adequate lipid profiles
(e) Demonstrated advantages against carcinogenesis
(f) Beneficial impact on glucose metabolism and insulin sensitivity
(g) Positive effects on appetite regulation and fat metabolism
(h) Demonstration of positive effects against inflammatory diseases, including arthritis and metabolic syndrome.

Substantially, polyphenols can be considered as valid adjuvant molecules in therapeutic strategies against a variety of human diseases. In fact, the following options have been and still are being explored from the scientific and also economic viewpoints (Arzani et al. 2025; Chen et al. 2025a, b; Kumar et al. 2025; Mantravadi et al. 2025; Narayan et al. 2025; Pan et al. 2025; Puspadewi et al. 2025; Solanki et al. 2025; Uti et al. 2025):

(1) Nano-carrier opportunities (as added compounds in selected pharmaceutical products or novel foods)

(2) Inhibiting agents against peculiar bacteria in the ambit of intestinal and oral health, and when speaking of food preservation with natural strategies.

Finally, it should be noted that the use of polyphenols can offer some drug-nutrient interactions which should be assessed from the health viewpoint. In this ambit, regulatory considerations for health claims should be evaluated when speaking of foods, beverages, and/or pharmaceutical products (Chaachouay 2025; Karabulut 2025; Kyriacou et al. 2025; Nandabaram et al. 2025; Pavić and Rahelić 2025).

As a consequence, there are different research areas involving polyphenols. This Chapter, in particular, discusses the relationship between these molecules and potential applications against inflammatory diseases when speaking of human health. Because of this broad ambit, there are some common points with other Chapters in this book.

### *6.1.1 The Problem: Inflammatory Diseases*

From the medical viewpoint, inflammation can be defined as the resulting defence action of blood cells against foreign substances. Consequently, such a defence strategy can be put in place by the human organism in an automated way, with the aim of protecting the subject from irritating and external agents (Gasmi et al. 2022). In detail, the following attacks may be considered in this ambit, taking into account that the below-mentioned list cannot be fully exhaustive (Abu Rached et al. 2025; Black 2002; Dantzer et al. 2006; Hussain et al. 2016; Zamani-Garmsiri et al. 2022; Pasparakis et al. 2014; Santamarina et al. 2025):

(1) Viruses and other pathogens are able to unleash automated defence answers
(2) Health damages and superficial wounds
(3) Chemical agents able to suddenly cause irritation (food and environmental allergens, etc.)
(4) Radioactive sources
(5) Chronic, autoimmune, and low-grade metabolic disorders: insulin resistance, type-2 diabetes, cardiovascular diseases, obesity, metabolic syndrome, non-alcoholic fatty liver disease
(6) Alcohol abuse
(7) Tobacco use
(8) Low-grade and chronic systemic inflammation, referred to as 'Meta-inflammation', is an established part of the underlying mechanism pertinent to the metabolic disorders, including obesity and peculiar anti-dietary behaviours such as prolonged high-calorie consumption.

In summary, inflammation is a positive response against attacks of different origins. However, being the answer immediate and automated, it could be supposed that the resulting defence is immediately effective, and the final conditions of the attacked subject are also rapidly restored. Actually, most part of the possible situations can differ depending on subjects (age, sex, etc.), and the resulting inflammation can range from simple to severe symptoms. It has to be clarified that these symptoms are the result of peculiar answers to attacks, and the attacked body region(s) or organ(s) have a certain importance. The following severe answers (and correlated symptoms) should be considered (Gasmi et al. 2022; Homey et al. 2006; Yoon and Baek 2005):

(a) Fever against different pathogens (viruses, bacteria, etc.)
(b) Modifications of blood composition
(c) Well-identifiable diseases such as rheumatoid arthritis or bowel disease.

Anyway, the basic problem of inflammatory episodes is the initial contrast against the attacking agent. With reference to inflamed wounds, the healing strategy is basically the deployment of specialised white blood cells against the pathogen(s). This general behaviour may fail partially or totally if (Chalmers et al. 2025; Guo et al. 2025; Harvanová and Duranková 2025; Pichard et al. 2025):

(1) The activation of defences is not complete, or
(2) The activation of defences is excessive if compared with the suffered attack, or
(3) The strategic defence is not completely effective for other causes.

In other terms, the inflammation-related answer concerns the role of cytokines in notable amounts, so that the immune system can be easily and promptly activated (Zhang et al. 2021). With reference to obesity and other low-grade inflammatory situations, the inflammatory process also involves adipokines and the release of some peculiar molecule with the aim of helping infiltration of the obese adipose tissue by white adipose tissue or inflammatory cells (Engin 2017; Landecho et al. 2019; Nani et al. 2021). Should the system fail to define a good inflammatory answer, homeostasis of the attacked subjects would not be restored. As a result, the immune response should last for a long time and in an uncontrolled way, without a guarantee of solving the situation. This process, known as chronic inflammation, concerns the deployment of leukocytes associated with lymphocytes and macrophages, with a resulting local or broad damage to several organs or tissues. In addition, and with reference to chronic inflammation, normal physiological functions cannot be restored. These and other consequences are also related to patients' age and other autoimmune diseases (cardiovascular problems, metabolic syndrome, etc.). Anyway, it is widely accepted that all restoring strategies have to target peculiar inflammatory mechanisms (Bucciantini et al. 2021).

For these reasons, the speed of recovery can be helped if selected molecules, such as steroidal anti-inflammatory drugs, are available. This choice depends, from a purely research viewpoint, on the existence of several pathways linked to arachidonic acid. This molecule has a critical role when speaking of complex biochemical strategies able to control the intensity and temporal extension of suffered pain (Behl et al. 2022; Chen et al. 2025a, b; Das et al. 2025; Ran et al. 2025). In addition, fever, heat, and swelling phenomena concerning the attacked wound or tissue depend strongly on the conversion of arachidonic acid into prostaglandins or leukotrienes. Being steroidal anti-inflammatory drugs able to inhibit cyclooxygenase and the consequent production of inflammatory mediators (prostaglandins), the optimal solution would be the use of similar compounds when speaking of main inflammatory diseases such as rheumatoid and osteo-arthritis, inflammatory bowel disease (also named Crohn's disease), and bronchiectasis. However, it is also known that some of these drugs can have unfavourable cardiovascular effects and other problems (Kim et al. 2025; Li et al. 2025; Omar et al. 2025; Petersen et al. 2025). Consequently, the use of non-synthetic and natural compounds with similar inhibitory effects would be highly recommended (Tumbarski et al. 2025).

It should be also noted that inflammatory symptoms can be strictly linked with oxidative stress (Martín and Ramos 2016) when chronic diseases are observed (Bucciantini et al. 2021), because of the notable production of reactive oxygen species with consequent damage of cellular compounds: the genetic structure named deoxyribonucleic acid (DNA), and also simple proteins and fat molecules (lipid peroxidation) (Hussain et al. 2016). In fact, oxidised proteins can cause the release of inflammatory signal compounds (Abdelmawgood et al. 2025; Berlett and Stadtman 1997; Hernández-Ruiz et al. 2025; Jomova et al. 2025; Xiao et al. 2025). Other reported symptoms concern neurological functions: neuroinflammation can be easily linked to cell death and neurodegeneration (Elsheikh et al. 2025; Popa-Wagner et al. 2013; Selvaraj et al. 2025; Trofin et al. 2025). Anyway, oxidative stress can be considered as a phenomenon which is linked to the presence of oxygen- or nitrogen-containing free radicals (examples: superoxide anions; hydrogen peroxides; etc.). These reactive molecules of different origins are actually produced by normal intracellular systems; consequently, their presence should be expected because the accumulation of similar molecules is the cause for the subsequent production of dangerous compounds able to kill living cells. As a consequence, the organism continuously produces enzymes and other molecules able to neutralise the presence of reactive species, with the desired result: the equilibrium of free radicals and the reparation of damaged targets (oxidised protein chains, glycated molecules, cellular tissues, DNA, nucleic acids, and also permeable cell membranes) (Gasmi et al. 2022; Hussain et al. 2016; Kuczyńska-Wiśnik et al. 2010; Jones 2006; Niess et al. 1999; Popa-Wagner et al. 2013; Storz and Imlayt 1999; Tian et al. 2020). Substantially, certain biologically active molecules such as xanthine oxidase and nicotinamide adenine dinucleotide phosphate (NADPH) oxidase can augment the amount of reactive species (and they should be inhibited), while other enzymes with clear antioxidant functions (glutathione peroxidase, catalase, superoxide dismutase, etc.) would be enhanced when speaking of good contrast to oxidative stress (Mishra et al. 2013; Yahfoufi et al. 2018; Yoon and Baek 2005).

The situation can also be complicated because of the growing intake of antibiotics by human beings, and also in the ambit of food production industries, from primary production. In fact, antibiotics can easily act as a collateral cause of production of similar reactive species (Grant and Hung 2013; Guillouzo and Guguen-Guillouzo 2020; Kuczyńska-Wiśnik et al. 2010; Wright, 2007).

Anyway, the role of polyphenols as natural antioxidants and anti-inflammatory agents has been repeatedly studied and mentioned. Consequently, a chemical classification of such a broad and diversified category should be presented before discussing its medical implications.

### 6.1.2 *Polyphenols from a General Viewpoint*

From the chemical viewpoint, polyphenols can be subdivided into two macro-categories (Behl et al. 2022; Ganesan and Xu 2017; Magrone et al. 2019; Matacchione et al. 2020; Nani et al. 2021; Zhang et al. 2021) (Fig. 6.1):

(a) Flavonoids
(b) Non-flavonoids.

The complexity of polyphenols (more than 8,000 different compounds mentioned so far) is linked to the great diversification in the above-mentioned simple classification. In fact, 'flavonoids' should be further subdivided into seven sub-categories as follows (Baruah 2011; Feng et al. 2017; Mateen et al. 2016; Panche et al. 2016; Petti and Scully 2009):

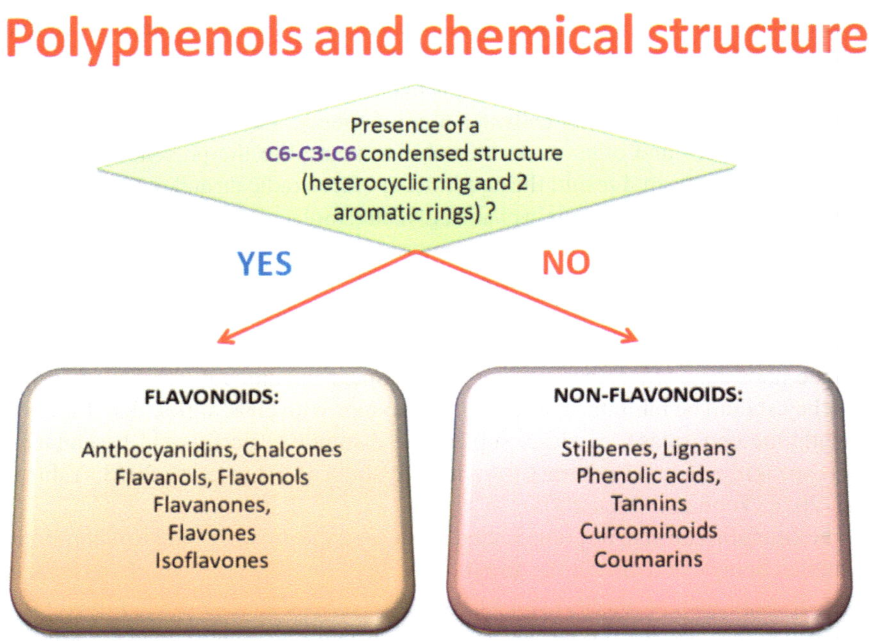


**Fig. 6.1** Subdivision of polyphenols in two categories depending on their basic chemical structure. This picture has been created by the authors

(1) Anthocyanidins
(2) Chalcones
(3) Flavanols
(4) Flavonols
(5) Flavanones
(6) Flavones
(7) Isoflavones.

In addition, the 'non-flavonoids' macrocategory should be discriminated in various sub-groups (the list may not be fully exhaustive) (Amio et al. 2016; Baruah 2011; Gasmi et al. 2022; Mateen et al. 2016; Zhang et al. 2021):

(1) Stilbenes
(2) Lignans
(3) Phenolic acids (hydroxybenzoic acids: syringic, gentisic acids, etc.; and hydroxycinnamic acids: ferulic, *p*-coumaric, caffeic acids, etc.)
(4) Tannins (procyanidins and ellagitannins)
(5) Curcominoids
(6) Coumarins.

From the structural viewpoint and with reference to the above-mentioned classification, flavonoids and non-flavonoid skeletons can be discussed.

Flavonoids have a clear $C_6$–$C_3$–$C_6$ condensed structure, where a heterocyclic ring with oxygen atom(s) links two aromatic rings. Sometimes, the central ring could have four carbon atoms, or the structure could be a xanthone system ($C_4$–$C_1$–$C_1$) (Feng et al. 2017). The classification into different sub-categories can be shown graphically in Figs. 6.2, 6.3, 6.4, 6.5, 6.6 and 6.7. A more adequate description can be graphically shown in Figs. 6.8, 6.9, and 6.10, where the general $C_6$–$C_3$–$C_6$ structure is explained as the union of three A, B, and C rings (Fig. 6.8): the C-ring is the heterocyclic structure (where oxygen is in position 2), and the A-ring is on the left and completely conjugated with the C-ring. On the other side, the B-ring is only connected to the C-ring without condensation. Figure 6.9 shows a 'chromone' structure where the C-ring has a carboxyl group in position 4, and the B-ring does not exist. Finally, Fig. 6.10 shows a 2-phenylchromone structure where the chromone skeleton has the B-ring in position 2 on the heterocyclic ring. This system is the main structure for flavonoids.

With reference to the heterogeneous class of non-flavonoids, the main difference with flavonoids is the absence of the above-mentioned $C_6$–$C_3$–$C_6$ structure. In fact, their skeleton can be assumed to belong to the broad group of polyphenols because of the presence of aromatic rings with associated hydroxyl groups. Apart from this main feature, the skeleton can be roughly linear or ramified, meaning that there is a notable variety of non-flavonoid structures. For this reason, the subdivision into stilbenes, lignans, phenolic acids, tannins, curcuminoids, coumarins, etc., can be justified (Rentzsch et al. 2009; Zhang et al. 2021).

In details:

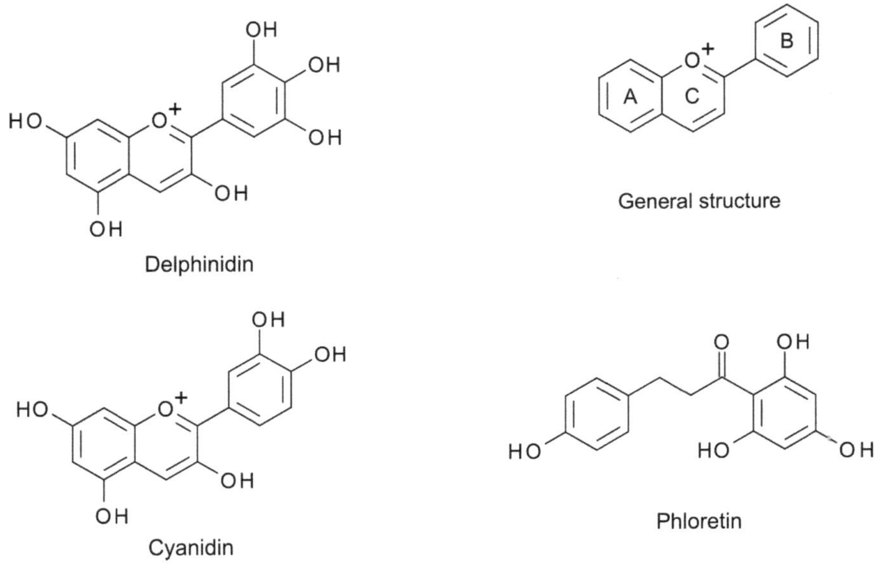


**Fig. 6.2** Anthocyanins and chalcones are based on a similar three-ring structure (top left of the figure). Chemically speaking, anthocyanins represent the aglycones of anthocyanins, with some hydroxyl and/or methoxyl groups on A, B, and C-rings (Fig. 6.8). Delphinidin and cyanidin are good examples. On the other side, chalcones such as phloretin do not have the classical B-ring (only an aliphatic chain). BKchem version 0.13.0, 2009 (http://bkchem.zirael.org/index.html), has also been used for drawing this structure. This picture has been created by the authors

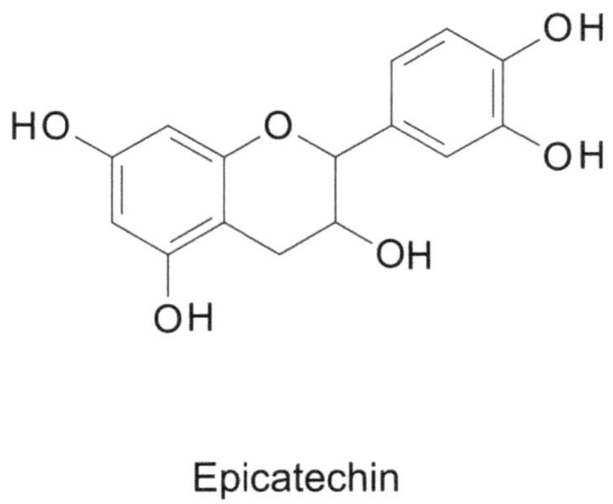


**Fig. 6.3** Chemically speaking, flavanols can be considered as a derivative structure of flavans with one additional hydroxyl group on the heterocyclic (central) ring. This Figure shows the chemical structure of epicatechin. BKchem version 0.13.0, 2009 (http://bkchem.zirael.org/index.html), has also been used for drawing this structure. This picture has been created by the authors

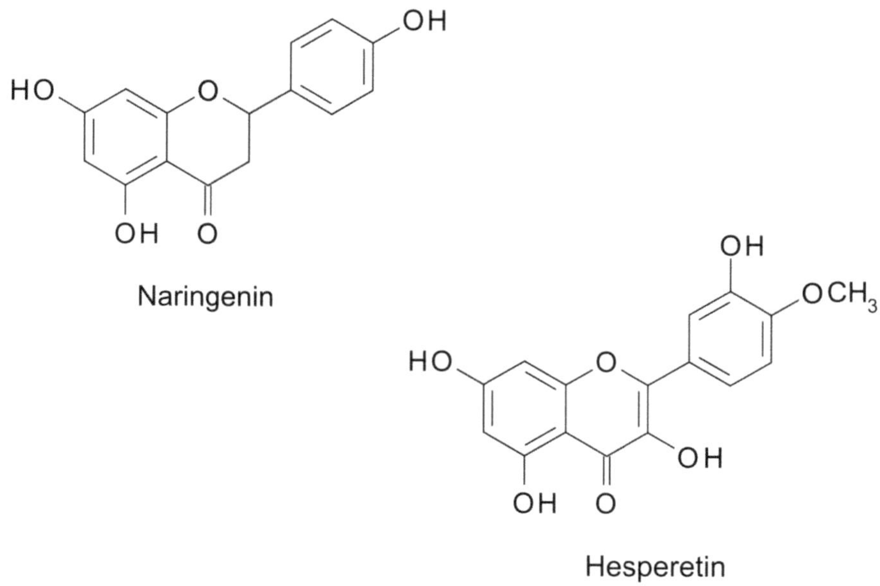


**Fig. 6.4** Chemically speaking, flavonols can be considered similar to flavanols (Fig. 6.3) with the presence of a ketone group in the heterocyclic (central) ring. In other words, it is a 2-phenylchromone structure (Fig. 6.10). Naringenin and hesperetin are good examples. BKchem version 0.13.0, 2009 (http://bkchem.zirael.org/index.html), has also been used for drawing this structure. This picture has been created by the authors

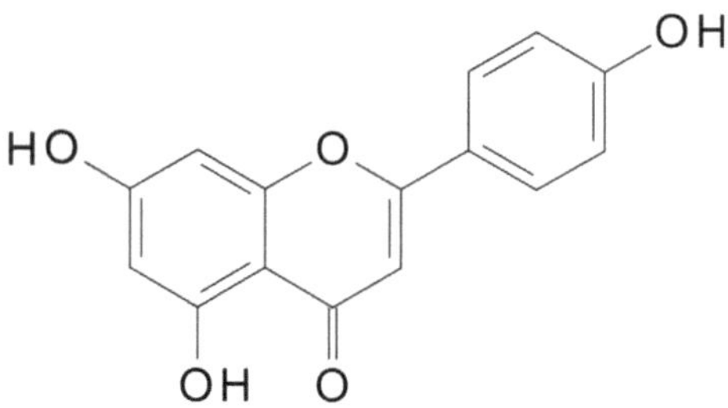


Apigenin

**Fig. 6.5** Apigenin is a good example of flavones. Chemically speaking, the structure shows two aromatic rings connected by a $C_3$-pyran ring with a ketone group and another unsaturated bond. In other words, it is a 2-phenylchromone structure (Fig. 6.10). BKchem version 0.13.0, 2009 (http://bkchem.zirael.org/index.html), has also been used for drawing this structure. This picture has been created by the authors

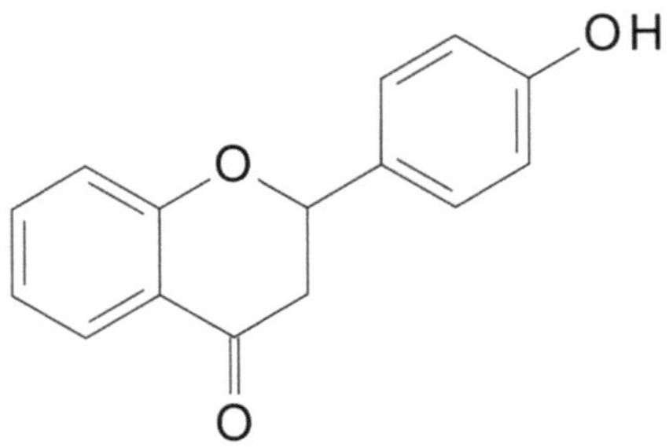

Flavanones

**Fig. 6.6** A basic structure for flavanones. Different from flavones (Fig. 6.3), the heterocyclic ring does not contain unsaturated C–C bonds (no conjugation with the carboxyl group). In other words, it is slightly different from a common 2-phenylchromone structure (Fig. 6.10). BKchem version 0.13.0, 2009 (http://bkchem.zirael.org/index.html), has also been used for drawing this structure. This picture has been created by the authors

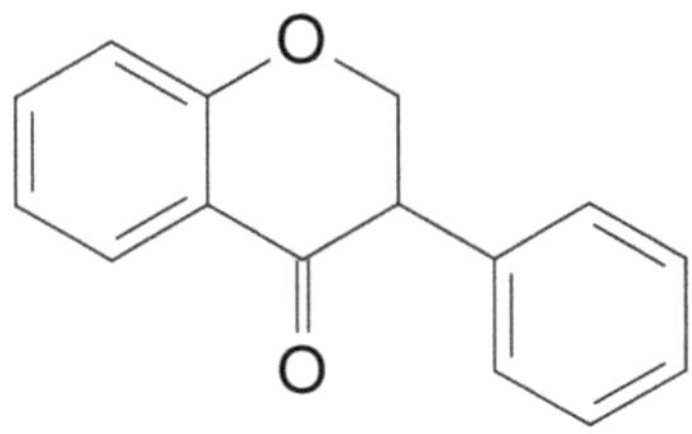

Isoflavones

**Fig. 6.7** A general structure for isoflavones. These compounds are similar to flavones (Fig. 6.3) except for the link of the aromatic ring next to the carboxyl group on the heterocyclic ring instead of the oxygen atom (as in flavones). In other words, it is a 3-phenylchromone structure. BKchem version 0.13.0, 2009 (http://bkchem.zirael.org/index.html), has also been used for drawing this structure. This picture has been created by the authors

(a) Stilbenes (Fig. 6.11) have a $C_6$–$C_2$–$C_6$ structure as follows: a central C–C unsaturated chain which is connected with two aromatic rings, one per each part. As a consequence, these molecules can exist as a couple of stereoisomers. One of the most well-known stilbenes is resveratrol (Roupe et al. 2006; Teka et al. 2022)
(b) Lignans generally show a 2,3-dibenzylbutane skeleton with two phenolic rings. There are many different versions ascribed to this sub-category (Khoo and Falk 2014; Park, 2014)
(c) The category of phenolic acids (hydroxybenzoic acids: syringic, gentisic acids, etc.; and hydroxycinnamic acids; ferulic, *p*-coumaric, caffeic acids, etc.) has the

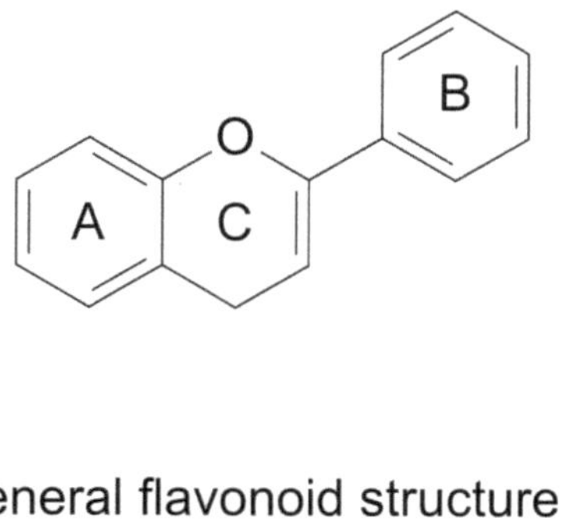


**Fig. 6.8** A general flavonoid structure is explained as the union of three A, B, and C rings. The C-ring is the heterocyclic structure (where oxygen is in position 2), and A-ring is on the left and completely conjugated with C-ring. On the other side, the B-ring is only connected to C-ring without condensation. BKchem version 0.13.0, 2009 (http://bkchem.zirael.org/index.html), has also been used for drawing this structure. This picture has been created by the authors

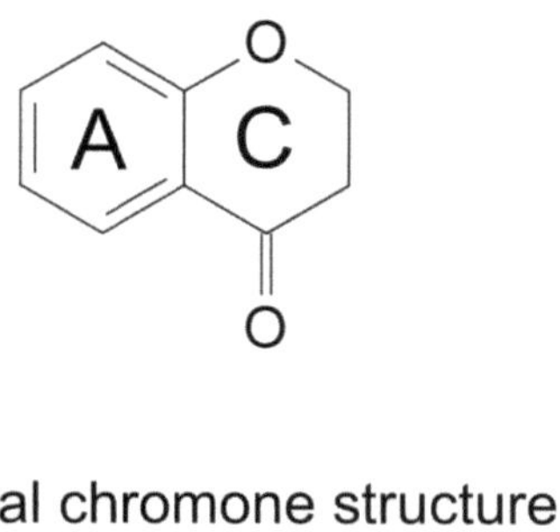


**Fig. 6.9** A 'chromone' structure where the C-ring has a carboxyl group in position 4, and the B-ring does not exist. BKchem version 0.13.0, 2009 (http://bkchem.zirael.org/index.html), has also been used for drawing this structure. This picture has been created by the authors

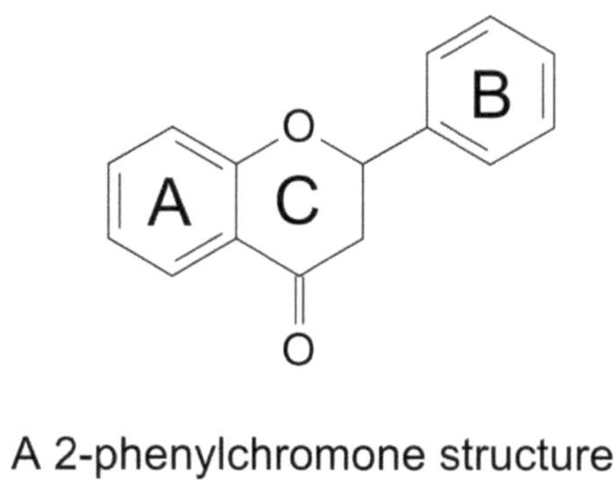


**Fig. 6.10** A 2-phenylchromone structure where the chromone skeleton has the B-ring in position 2 on the heterocyclic ring. BKchem version 0.13.0, 2009 (http://bkchem.zirael.org/index.html), has also been used for drawing this structure. This picture has been created by the authors

following features (de Córdova and Medina 2014; Neilson et al. 2017; Ratnavathi 2019): one aromatic ring, a carboxylic group, and one or more than one hydroxyl and/or methoxyl groups.

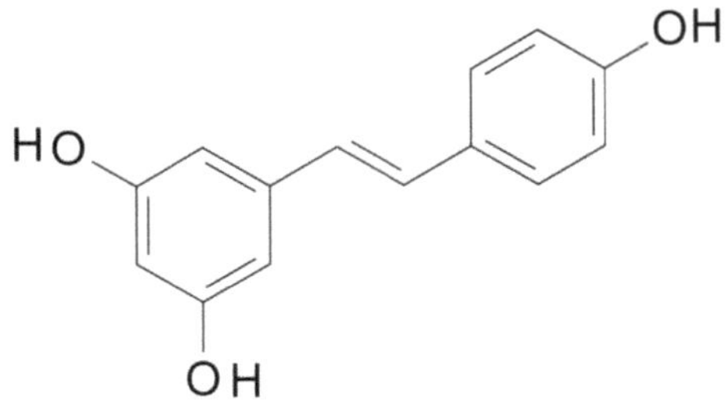

Resveratrol

**Fig. 6.11** A simplified stilbene structure: resveratrol. BKchem version 0.13.0, 2009 (http://bkchem.zirael.org/index.html), has also been used for drawing this structure. This picture has been created by the authors

(d) Tannins (condensed and hydrolysable tannins) have a very challenging structure. Hydrolysable tannins correspond to a matrix containing a glucose molecule and hydroxyl groups linked to an esterified gallic acid, able to form insoluble clusters with other compounds. Condensed (non-hydrolisable) tannins are larger than hydrolysable tannins (Abdullahi et al. 2022; Bouyahya et al. 2021; Govindarajan et al. 2016)

(e) Finally, curcuminoids have a diarylheptanoid structure (Feng and Zhang 2023; Subramani et al. 2017), while coumarins are based on an aromatic benzene ring condensed with a α-pyrone ring (de Elguea-Culebras et al. 2022; Yang et al. 2023).

The above-mentioned polyphenols can have interesting anti-inflammatory properties, which can be explained by means of peculiar mechanisms. Their positive features are useful against several inflammatory diseases, including arthritis and metabolic syndrome.

### *6.1.3 Polyphenols as Effective Antinflammatory Agents and Adjuvants. Mechanisms of Action Against Arthritis and Metabolic Syndrome*

From the medical viewpoint, inflammation can be contrasted with the aim of polyphenols because of their undeniable antioxidant nature. Diseases such as arthritis and metabolic syndrome can be treated in this way (Coletro et al. 2021; Ilari et al. 2024; Sirše 2022). Being studied phenolics compounds >8,000 different molecules so far, and also taking into account that their origin is substantially vegetable, the amount of reported research is really remarkable. In addition, the subdivision of phenolic compounds into at least 13 different sub-categories may be challenging enough when speaking of inflammatory mechanisms and correlated therapeutic applications.

One peculiar example—which can be used to show the usefulness and the variety of active roles for polyphenols in a therapeutic ambit—is offered by polyphenols found in extra-virgin olive oil (EVOO). It has been reported that this product and the original olive contain a good amount of phenolics acids, lignans, and flavonoids at least, and a good amount in EVOO could reach or exceed 60 mg per 100 g of oil (Bucciantini et al. 2021; Dinda et al. 2011; Servili et al. 2009). With reference to EVOO-related polyphenols, the following antinflammatory mechanisms—substantially, the ability to contrast different cellular mechanisms by means of pure antioxidant behaviour—have been reported so far (Bulotta et al. 2013; Carluccio et al. 2007; Cicerale et al. 2012; Gavahian et al. 2019; Marcelino et al. 2019; Tihana et al. 2019):

(1) Polyphenols such as oleuropein aglycone (derived from the glucose-bound form of asecoiridoid) have the possibility, because of a catecholic skeleton, of interfering with peroxidation in several tissues by scavenging action against peroxyl active radical particles. The resulting reaction product would be very stable, with a substantial delay of degenerative processes in place
(2) At the same time, the mentioned oleuropein aglycone can chelate metals such as copper, with the ability to interfere with the oxidation of low-density lipoproteins favoured by copper sulphate
(3) Generally speaking, polyphenols may contrast and delay catalase inactivation by hypochlorous acid. The result is the contrast to lipoprotein oxidation and consequent atherosclerosis
(4) Being polyphenols reported to have antibacterial activity against several pathogens able to live in the human intestine, resulting in a good selection of beneficial intestinal microorganisms, which may improve the production of short fatty acid chains with important effects on the modulation of gene expression through epigenetic mechanisms. Also, blood glucose can be easily regulated by means of these short fatty acids.

These non-exhaustive examples can give an idea of the action of polyphenols on enzyme activity, when specifically speaking of direct action against enzymatic compounds responsible for chronic diseases such as arthritis. The common key action is the modulation of the production and action mechanisms of inflammatory mediators in a favourable way toward the end of the inflammatory response (Behl et al. 2022; Oliviero et al. 2018).

As an example, the NF-kB signalling pathway is responsible for the regulation and control of the expression of different genes involved in the inflammatory response, such as pro-inflammatory cytokines. The alteration of several steps in this pathway by polyphenols gives the possibility of contrasting enzymatic activities such as phosphorylation, also reducing the production of reactive oxygen species and the consequent stimulation of this pathway.

Similar situations have been reported when speaking of the inhibitory action of an EVOO polyphenol, quercetin, on the mitogen-activated protein kinase pathway, with consequent suppression of the production of pro-inflammatory mediators (Behl et al. 2022).

With concern to metabolic syndrome—the concomitant occurrence of metabolic dysfunctions with resulting diabetes, cardiovascular disease, and obesity (Chiva-Blanch and Badimon 2017; Gasmi et al. 2022)—one of the best known action mechanisms ascribed to polyphenols, and citrus flavonoids in particular, is the ability to modulate lipid metabolism in a favourable way. In contrast to lipid peroxidation, aldose reductase inhibition, and activation of insulin receptors have also been observed and studied when speaking specifically of quercetin, flavonoids, and epicatechin, respectively, with interesting effects when speaking of diabetes treatments (Ganugapati et al. 2011; Letan 1966; Madeswaran et al. 2012). Inhibition of cholesterol absorption in blood plasma has also been reported by phytosterols from olives (Belitz et al. 2009; Haddad et al. 2020). Other reported effects concern the boosting effect on endothelial function by means of the activation of nitric oxide synthase, paraoxonase 1, and protein kinase B, and the effective inhibition of α-amylase and pancreatic lipase in obese patients by tannin extracts, with good weight reduction results (Huxley and Neil 2003; Lagana et al. 2020; Mulvihill et al. 2016; Oliveira et al. 2015; Panche et al. 2016).

On these bases, it can be assumed that the regular consumption of vegetable products containing a good bioavailable amount of phenolic compounds can be recommended with favour, despite the need for additional research in this ambit.

## Abbreviations

| | |
|---|---|
| DNA | Deoxyribonucleic acid |
| EVOO | Extra-virgin olive oil |
| NADPH | Nicotinamide adenine dinucleotide phosphate |

## References

Abdelmawgood IA, Kotb MA, Hassan HS, Mahana NA, Rochdi AM, Sayed NH, Elsafoury RH, Saber AM, Youssef MN, Waheeb NG, Al-Rifai MWA, Badr AM, Abdelkader AE (2025) Gentisic acid attenuates ovalbumin-induced airway inflammation, oxidative stress, and ferroptosis through the modulation of Nrf2/HO-1 and NF-κB signaling pathways. Int Immunopharm 146:113764. https://doi.org/10.1016/j.intimp.2024.113764

Abdullahi A, Tijjani A, Abubakar AI, Khairulmazmi A, Ismail MR (2022) Plant biomolecule antimicrobials: an alternative control measures for food security and safety. In: Mandal SC, Nayak AK, Dhara AK (eds) Herbal biomolecules in healthcare applications. Academic Press, Cambridge, MA, pp 381–406. https://doi.org/10.1016/B978-0-323-85852-6.00024-X

Abu Rached N, Dietrich JW, Ocker L, Stockfleth E, Haven Y, Myszkowski D, Bechara FG (2025) Endotyping insulin-glucose homeostasis in hidradenitis suppurativa: the impact of diabetes mellitus and inflammation. J Clin Med 14(7):2145. https://doi.org/10.3390/jcm14072145

Amiot MJ, Riva C, Vinet A (2016) Effects of dietary polyphenols on metabolic syndrome features in humans: a systematic review. Obes Rev 17(7):573–586. https://doi.org/10.1111/obr.12409

Arshad MT, Maqsood S, Ikram A, Abdullahi MA (2025) Synergistic effects of carrot (Daucus carota L.) root and papaya (Carica papaya L.) extracts in combating inflammation and enhancing immune functions—A new perspective. Food Sci Nutr 13(6):e70313. https://doi.org/10.1002/fsn3.70313

Arzani V, Soleimani M, Fritsch T, Jacob UM, Calabrese V, Arzani A (2025) Plant polyphenols, terpenes, and terpenoids in oral health. Open Med 20(1):20251183. https://doi.org/10.1515/med-2025-1183

Baruah JB (2011) Chemistry of phenolic compounds: state of the art. Nova Science Publishers, Hauppauge, NY, USA

Behl T, Mehta K, Sehgal A, Singh S, Sharma N, Ahmadi A, Arora S, Bungau S (2022) Exploring the role of polyphenols in rheumatoid arthritis. Crit Rev Food Sci Nutr 62(19):5372–5393. https://doi.org/10.1080/10408398.2021.1924613

Belitz HD, Grosch W, Schieberle P (2009) Food chemistry, 4th revised and extended. Springer-Verlag, Berlin, Heidelberg, Germany

Berlett BS, Stadtman ER (1997) Protein oxidation in aging, disease, and oxidative stress. J Biol Chem 272(33):20313–20316

Black PH (2002) Stress and the inflammatory response: a review of neurogenic inflammation. Brain Behav Immun 16(6):622–653. https://doi.org/10.1016/S0889-1591(02)00021-1

Bouyahya A, El Omari N, Elmenyiy N, Guaouguaou FE, Balahbib A, Belmehdi O, Salhi N, Imtara H, Mrabti HN, El-Shazly M, Bakri Y (2021) Moroccan antidiabetic medicinal plants: Ethnobotanical studies, phytochemical bioactive compounds, preclinical investigations, toxicological validations and clinical evidences; challenges, guidance and perspectives for future management of diabetes worldwide. Trends Food Sci Technol 115:147–254. https://doi.org/10.1016/j.tifs.2021.03.032

Bucciantini M, Leri M, Nardiello P, Casamenti F, Stefani M (2021) Olive polyphenols: antioxidant and anti-inflammatory properties. Antioxidants 10(7):1044. https://doi.org/10.3390/antiox10071044

Bulotta S, Oliverio M, Russo D, Procopio A (2013) Biological activity of oleuropein and its derivatives. In: Ramawat KG, Mérillon JM (eds) Natural products. Springer, Berlin/Heidelberg, Germany, pp 3605–3638

Carluccio MA, Massaro M, Scoditti E, de Caterina R (2007) Vasculoprotective potential of olive oil components. Mol Nutr Food Res 51(10):1225–1234. https://doi.org/10.1002/mnfr.200600305

Chaachouay N (2025) Synergy, additive effects, and antagonism of drugs with plant bioactive compounds. Drugs Drug Candidates 4(1):4. https://doi.org/10.3390/ddc4010004

Chalmers JD, Metersky M, Aliberti S, Morgan L, Fucile S, Lauterio M, McDonald PP (2025) Neutrophilic inflammation in bronchiectasis. Eur Respir Rev 34(176):240179. https://doi.org/10.1183/16000617.0179-2024

Chen S, Xie Y, Guo D, Li T, Tan Z, Ran X, Wen X (2025a) Effects of Salmonella Typhimurium infection on intestinal flora and intestinal tissue arachidonic acid metabolism in Wenchang chickens. Front Microbiol 16:1514115. https://doi.org/10.3389/fmicb.2025.1514115

Chen Z, Xiao C, Zhang J, Jian S, Li P, Lin J, He C, Chen Z, Qi Y, Shi J, Chen Q, Chen J, Bo H (2025b) The impact of diet on the colonization of beneficial microbes from an ecological perspective. J Agric Food Chem 73(17):10069–10092. https://doi.org/10.1021/acs.jafc.5c02086

Chiva-Blanch G, Badimon L (2017) Effects of polyphenol intake on metabolic syndrome: current evidences from human trials. Oxid Med Cell Longev 1:5812401. https://doi.org/10.1155/2017/5812401

Cicerale S, Lucas LJ, Keast RS (2012) Antimicrobial, antioxidant and anti-inflammatory phenolic activities in extra virgin olive oil. Curr Opin Biotechnol 23(2):129–135. https://doi.org/10.1016/j.copbio.2011.09.006

Coletro HN, Diniz AP, Guimarães NS, Carraro JCC, Mendonça RDD, Meireles AL (2021) Polyphenols for improvement of inflammation and symptoms in rheumatic diseases: systematic review. Sao Paulo Med J 139:615–623. https://doi.org/10.1590/1516-3180.2020.0766.R1.22042021

Dantzer R, Castanon N, Lestage J, Moreau M, Capuron L (2006) Inflammation, sickness behaviour and depression. In: Steptoe A (ed) Depression and physical illness. Cambridge University Press, Cambridge, UK, pp 265–279. https://doi.org/10.1017/CBO9780511544293.013

Das UN, Hacimüftüoglu A, Akpinar E, Gul M, Abd El-Aty AM (2025) Crosstalk between renin and arachidonic acid (and its metabolites). Lipids Health Dis 24(1):52. https://doi.org/10.1186/s12944-025-02463-3

de Córdova MF, Medina AR (2014) Analytical methods for determination of polyphenols in beer. In: Preedy V (ed) Processing and impact on antioxidants in beverages. Academic Press, Cambridge, MA, pp 289–299. https://doi.org/10.1016/B978-0-12-404738-9.00029-5

de Elguea-Culebras GO, Bravo EM, Sánchez-Vioque R (2022) Potential sources and methodologies for the recovery of phenolic compounds from distillation residues of Mediterranean aromatic plants. An approach to the valuation of by-products of the essential oil market—A review. Ind Crops Prod 175:114261. https://doi.org/10.1016/j.indcrop.2021.114261

Dinda B, Debnath S, Banik R (2011) Naturally occurring iridoids and secoiridoids. An updated review, part 4. Chem Pharm Bull 59:803–833. https://doi.org/10.1248/cpb.59.803

Domingues YO, Lemes GA, de Oliveira FL, de Souza TR, Silva B, Bento JAC, Morzelle MC (2024) In vitro simulated gastrointestinal digestion and bioaccessibility of phenolic compounds and antioxidants of soursop (Annona muricata L.) peel and pulp. Lwt 208:116694. https://doi.org/10.1016/j.lwt.2024.116694

Elsheikh AA, Abd-Almotaleb NA, Ahmed MM, Khayal EES (2025) IONPs-induced neurotoxicity via cascade of neuro-oxidative stress, parthanatos-mediated cell death, neuro-inflammation and neurodegenerative changes: ameliorating effect of rosemary methanolic extract. Toxicol Rep 14:101935. https://doi.org/10.1016/j.toxrep.2025.101935

Engin A (2017) The pathogenesis of obesity-associated adipose tissue inflammation. In: Engin A (ed) Obesity and lipotoxicity. Advances in experimental medicine and biology, vol 960. Springer, Cham, pp 221–245. https://doi.org/10.1007/978-3-319-48382-5_9

Feng J, Zhang Y (2023) The potential benefits of polyphenols for corneal diseases. Biomed Pharmacother 169:115862. https://doi.org/10.1016/j.biopha.2023.115862

Feng W, Hao Z, Li M (2017) Isolation and structure identification of flavonoids. In: Justino GC (ed) Flavonoids, from biosynthesis to human health. Intech Open, Rijeka, Croatia, pp 17–43. https://doi.org/10.5772/67810

Ganesan K, Xu B (2017) A critical review on polyphenols and health benefits of black soybeans. Nutrients 9(5):455. https://doi.org/10.3390/nu9050455

Ganugapati J, Mukkavalli S, Sahithi A (2011) Docking studies of green tea flavonoids as insulin mimetics. Int J Comp App 30(4):48–52

Gasmi A, Mujawdiya PK, Noor S, Lysiuk R, Darmohray R, Piscopo S, Lenchyk L, Antonyak H, Dehtiarova K, Shanaida M, Polishchuk A, Shanaida V, Peana M, Bjørklund G (2022) Polyphenols in metabolic diseases. Molecules 27(19):6280. https://doi.org/10.3390/molecules27196280

Gavahian M, Khaneghah AM, Lorenzo JM, Munekata PES, Garcia-Mantrana I, Collado MC, Meléndez-Martínez AJ, Barba FJ (2019) Health benefits of olive oil and its components: impacts on gut microbiota antioxidant activities, and prevention of noncommunicable diseases. Trends Food Sci Technol 88:220–227. https://doi.org/10.1016/j.tifs.2019.03.008

Giglio RV, Patti AM, Cicero AF, Lippi G, Rizzo M, Toth PP, Banach M (2018) Polyphenols: potential use in the prevention and treatment of cardiovascular diseases. Curr Pharm Des 24(2):239–258. https://doi.org/10.2174/1381612824666180130112652

Govindarajan RK, Revathi S, Rameshkumar N, Krishnan M, Kayalvizhi N (2016) Microbial tannase: current perspectives and biotechnological advances. Biocatal Agric Biotechnol 6:168–175. https://doi.org/10.1016/j.bcab.2016.03.011

Grant SS, Hung DT (2013) Persistent bacterial infections, antibiotic tolerance, and the oxidative stress response. Virulence 4(4):273–283. https://doi.org/10.4161/viru.23987

Guillouzo A, Guguen-Guillouzo C (2020) Antibiotics-induced oxidative stress. Curr Opin Toxicol 20–21:23–28. https://doi.org/10.1016/j.cotox.2020.03.004

Guo D, Liu C, Zhu H, Cheng Y, Huo X, Guo Y, Qian H (2025) Food-induced adverse reactions: a review of physiological food quality control, mucosal defense mechanisms, and gastrointestinal physiology. Toxics 13(1):61. https://doi.org/10.3390/toxics13010061

Haddad MA, El-Qudah J, Abu-Romman S, Obeidat M, Iommi C, Jaradat DSM (2020) Phenolics in Mediterranean and Middle East important fruits. J AOAC Int 103(4):930–934. https://doi.org/10.1093/jaocint/qsz027

Hafiz MS (2025) Anti-diabetic potentials of coffee polyphenols: a narrative review. Curr Nutr Food Sci 21(2):202–211. https://doi.org/10.2174/0115734013292395240219091340

Harvanová G, Duranková S (2025) Inflammatory process: factors inducing inflammation, forms and manifestations of inflammation, immunological significance of the inflammatory reaction. Alergol Polska-Pol J Allergol 12(1):54–61. https://doi.org/10.5114/pja.2025.147674

Hernández-Ruiz RG, Olivares-Ochoa XC, Salinas-Varela Y, Guajardo-Espinoza D, Roldán-Flores LG, Rivera-Leon EA, López-Quintero A (2025) Phenolic compounds and anthocyanins in legumes and their impact on inflammation, oxidative stress, and metabolism: comprehensive review. Molecules 30(1):174. https://doi.org/10.3390/molecules30010174

Homey B, Steinhoff M, Ruzicka T, Leung DY (2006) Cytokines and chemokines orchestrate atopic skin inflammation. J Allergy Clin Immunol 118(1):178–189. https://doi.org/10.1016/j.jaci.2006.03.047

Hussain T, Tan B, Yin Y, Blachier F, Tossou MC, Rahu N (2016) Oxidative stress and inflammation: what polyphenols can do for us?. Oxidative Med Cell Longev 1:7432797. https://doi.org/10.1155/2016/7432797

Huxley RR, Neil HAW (2003) The relation between dietary flavonol intake and coronary heart disease mortality: a meta-analysis of prospective cohort studies. Eur J Clin Nutr 57(8):904–908. https://doi.org/10.1038/sj.ejcn.1601624

Ilari S, Nucera S, Morabito L, Caminiti R, Mazza V, Ritorto G, Passacatini LC, Macrì R, Scarano F, Serra M, Scali E, Maiuolo J, Oppedisano F, Palma E, Muscoli S, Proietti S, Tomino C, Mollace V, Muscoli C (2024) A systematic review of the effect of polyphenols on alterations of the intestinal microbiota and shared bacterial profiles between metabolic syndrome and acne. Nutrients 16(21):3591. https://doi.org/10.3390/nu16213591

Issaoui M, Delgado AM, Iommi C, Chammem N (2020) Polyphenols and the mediterranean diet. Springer International Publishing, Cham

Jomova K, Alomar SY, Valko R, Liska J, Nepovimova E, Kuca K, Valko M (2025) Flavonoids and their role in oxidative stress, inflammation, and human diseases. Chemico-Biol Interact 413:111489. https://doi.org/10.1016/j.cbi.2025.111489

Jones DP (2006) Redefining oxidative stress. Antioxid Redox Signal 8(9–10):1865–1879. https://doi.org/10.1089/ars.2006.8.1865

Karabulut G (2025) Tea polyphenols: their role in human health and drug interactions. Dissertation, University of Padova, Italy

Kardum N, Milovanović B, Šavikin K, Zdunić G, Mutavdžin S, Gligorijević T, Spasić S (2015) Beneficial effects of polyphenol-rich chokeberry juice consumption on blood pressure level and lipid status in hypertensive subjects. J Med Food 18(11):1231–1238. https://doi.org/10.1089/jmf.2014.0171

Khoo C, Falk M (2014) Polyphenols in the prevention and treatment of vascular and cardiac disease, and cancer. In: Watson RR, Preedy VR, Zibadi S (eds) Polyphenols in human health and disease, 1st edn. Elsevier Inc., Amsterdam, The Netherlands, pp 1049–1065

Kim JW, Yoon JS, Park S, Kim H, Kim BY, Lee H, Park SH, Kim SK, Choe JY (2025) Risk of cardiovascular disease associated with long-term use of non-steroidal anti-inflammatory drugs in ankylosing spondylitis. Rheumatology 64(1):261–267. https://doi.org/10.1093/rheumatology/kead684

Kuczyńska-Wiśnik D, Matuszewska E, Furmanek-Blaszk B, Leszczyńska D, Grudowska A, Szczepaniak P, Laskowska E (2010) Antibiotics promoting oxidative stress inhibit formation of Escherichia coli biofilm via indole signalling. Res Microbiol 161(10):847–853. https://doi.org/10.1016/j.resmic.2010.09.012

Kumar S, Mukherjee R, Gaur P, Leal É, Lyu X, Ahmad S, Puri P, Chang CM, Raj VS, Pandey RP (2025) Unveiling roles of beneficial gut bacteria and optimal diets for health. Front Microbiol 16:1527755. https://doi.org/10.3389/fmicb.2025.1527755
Kupczyk D, Bilski R, Szeleszczuk Ł, Mądra-Gackowska K, Studzińska R (2025) The role of diet in modulating inflammation and oxidative stress in rheumatoid arthritis, ankylosing spondylitis, and psoriatic arthritis. Nutrients 17(9):1603. https://doi.org/10.3390/nu17091603
Kyriacou NM, Gross AS, McLachlan AJ (2025) Green tea catechins as perpetrators of drug pharmacokinetic interactions. Clin Pharmacol Ther 118(1):45–61. https://doi.org/10.1002/cpt.3547
Lagana P, Coniglio MA, Fiorino M, Delgado AM, Chammen N, Issaoui M, Gamzza ME, Iommi C, Soraci L, Haddad MA, Delia S (2020) Phenolic substances in foods and anticarcinogenic properties: a public health perspective. J AOAC Int 103(4):935–939. https://doi.org/10.1093/jaocint/qsz028
Landecho MF, Tuero C, Valentí V, Bilbao I, de la Higuera M, Frühbeck G (2019) Relevance of leptin and other adipokines in obesity-associated cardiovascular risk. Nutrients 11(11):2664. https://doi.org/10.3390/nu11112664
Letan A (1966) The relation of structure to antioxidant activity of quercetin and some of its derivatives. J Food Sci 31(4):395–399. https://doi.org/10.1111/j.1365-2621.1966.tb01897.x
Li W, Ma Y, Yang L (2025) Systematic review and meta-analysis: no evidence that low-dose non-steroidal anti-inflammatory drugs (NSAIDs) reduce the risk of pancreatitis after endoscopic retrograde cholangiopancreatography (ERCP). BMC Gastroenterol 25(1):83. https://doi.org/10.1186/s12876-025-03690-9
Liva K, Panagiotopoulos AA, Foscolou A, Amerikanou C, Vitali A, Zioulis S, Argyri K, Panoutsopoulos GI, Kaliora AC, Gioxari A (2025) High polyphenol extra virgin olive oil and metabolically unhealthy obesity: a scoping review of preclinical data and clinical trials. Clin Pract 15(3):54. https://doi.org/10.3390/clinpract15030054
Madeswaran A, Umamaheswari M, Asokkumar K, Sivashanmugam S, Subhadradevi V, Jagannath P (2012) In-silico docking studies of cyclooxygenase inhibitory activity of commercially available flavonoids. Asian J Pharm Life Sci 2(2):174–181
Magrone T, Magrone M, Russo MA, Jirillo E (2019) Recent advances on the anti-inflammatory and antioxidant properties of red grape polyphenols: in vitro and in vivo studies. Antioxidants 9(1):35. https://doi.org/10.3390/antiox9010035
Makarewicz M, Drożdż I, Tarko T, Duda-Chodak A (2021) The interactions between polyphenols and microorganisms, especially gut microbiota. Antioxidants 10(2):188. https://doi.org/10.3390/antiox10020188
Mantravadi PK, Kovi BS, Reddy SR, Namasivayam GP, Kalesh K, Parthasarathy A (2025) Probing and manipulating the gut microbiome with chemistry and chemical tools. Gut Microbiome 6:e6. https://doi.org/10.1017/gmb.2025.4
Marcelino G, Hiane PA, Freitas KC, Santana LF, Pott A, Donadon JR, Guimarães RCA (2019) Effects of olive oil and its minor components on cardiovascular diseases, inflammation, and gut microbiota. Nutrients 11(8):1826. https://doi.org/10.3390/nu11081826
Martín MA, Ramos S (2016) Cocoa polyphenols in oxidative stress: potential health implications. J Funct Foods 27:570–588. https://doi.org/10.1016/j.jff.2016.10.008
Martiniakova M, Sarocka A, Penzes N, Biro R, Kovacova V, Mondockova V, Sevcikova A, Ciernikova S, Omelka R (2025) Protective role of dietary polyphenols in the management and treatment of type 2 diabetes mellitus. Nutrients 17(2):275. https://doi.org/10.3390/nu17020275
Matacchione G, Gurău F, Baldoni S, Prattichizzo F, Silvestrini A, Giuliani A, Pugnaloni A, Espinosa E, Amenta F, Bonafè M, Procopio AD, Rippo MR, Olivieri F, Sabbatinelli J (2020) Pleiotropic effects of polyphenols on glucose and lipid metabolism: focus on clinical trials. Ageing Res Rev 61:101074. https://doi.org/10.1016/j.arr.2020.101074
Mateen S, Moin S, Zafar A, Khan AQ (2016) Redox signalling in rheumatoid arthritis and the preventive role of polyphenols. Clin Chim Acta 463:4–10. https://doi.org/10.1016/j.cca.2016.10.007

Mechi D, Baccouri B, Martín-Vertedor D, Abaza L (2023) Bioavailability of phenolic compounds in Californian-Style table olives with Tunisian aqueous olive leaf extracts. Molecules 28(2):707. https://doi.org/10.3390/molecules28020707

Mendes I, Ribeiro MGC, de Souza LF, Rosa CDOB, Hermsdorff HHM, Bressan J (2025) Effect of polyphenol supplementation on adiposity: a systematic review of randomized clinical trials. Curr Nutr Rep 14(1):36. https://doi.org/10.1007/s13668-025-00626-3

Mileo AM, Miccadei S (2016) Polyphenols as modulator of oxidative stress in cancer disease: new therapeutic strategies. Oxid Med Cell Longev 1:6475624. https://doi.org/10.1155/2016/6475624

Mishra A, Sharma AK, Kumar S, Saxena AK, Pandey AK (2013) Bauhinia Variegata leaf extracts exhibit considerable antibacterial, antioxidant, and anticancer activities. Biomed Res Int 1:915436. https://doi.org/10.1155/2013/915436

Mulvihill E, Burke A, Huff M (2016) Citrus flavonoids as regulators of lipoprotein metabolism and atherosclerosis. Ann Rev Nutr 36:275–299. https://doi.org/10.1146/annurev-nutr-071715-050718

Nandabaram A, Nelson VK, Mayasa V (2025) Assessment of possible food-drug interactions of pearl millet diet on gliclazide in diabetic rats. Texila Int J Pub Health 13(1):614–628. https://doi.org/10.21522/TIJPH.2013.13.01.Art058

Nani A, Murtaza B, Sayed Khan A, Khan NA, Hichami A (2021) Antioxidant and anti-inflammatory potential of polyphenols contained in Mediterranean diet in obesity: molecular mechanisms. Molecules 26(4):985. https://doi.org/10.3390/molecules26040985

Narayan S, Nagpal K, Kumar P (2025) Polyphenol-conjugated polysaccharide nanoplatforms for enhanced therapeutic efficacy. Expert Opin Drug Deliv 22(9):1237–1255. https://doi.org/10.1080/17425247.2025.2514714

Neilson AP, Goodrich KM, Ferruzzi MG (2017) Bioavailability and metabolism of bioactive compounds from foods. In: Coulston AM, Boushey CJ, Ferruzzi MG, Delahanty LM (eds) Nutrition in the prevention and treatment of disease. Academic Press, Cambridge, MA, pp 301–319. https://doi.org/10.1016/B978-0-12-802928-2.00015-1

Niess AM, Dickhuth HH, Northoff H, Fehrenbach E (1999) Free radicals and oxidative stress in exercise—Immunological aspects. Exerc Immunol Rev 5:22–56

Oliveira R, Gonçalves G, Inácio F, Koehnlein EA, De Souza CGM, Bracht A, Peralta RM (2015) Inhibition of pancreatic lipase and triacylglycerol intestinal absorption by a pinhão coat (Araucaria angustifolia) extract rich in condensed tannin. Nutrients 7(7):5601–5614. https://doi.org/10.3390/nu7075242

Oliviero F, Scanu A, Zamudio-Cuevas Y, Punzi L, Spinella P (2018) Anti-inflammatory effects of polyphenols in arthritis. J Sci Food Agric 98(5):1653–1659. https://doi.org/10.1002/jsfa.8664

Omar A, Omar H, Mohamad A, Haroon S (2025) Non-steroidal anti-inflammatory drugs consumption and awareness about risks during pregnancy in AL-Bayda City, Libya. Alq J Med Appl Sci 8(1):119–128. https://doi.org/10.54361/ajmas.258119

Pan X, Duan Y, Liu S, Wang Y, Li Q, Jiang F, Li Y, Huang Z, Su L, Li X, Liu M, Zhou X, Tang H (2025) All-in-one: harnessing multifunctional natural polysaccharide spray hydrogel loaded with polyphenol-metal nanoparticles for fruit preservation. Food Chem 470:142638. https://doi.org/10.1016/j.foodchem.2024.142638

Panche AN, Diwan AD, Chandra SR (2016) Flavonoids: an overview. J Nutr Sci 5:e47. https://doi.org/10.1017/jns.2016.41

Parisi S (2019) Analysis of major phenolic compounds in foods and their health effects. AOAC J 102(5):1354–1355. https://doi.org/10.5740/jaoacint.19-0127

Parisi S (2020) Characterization of major phenolic compounds in selected foods by the technological and health promotion viewpoints. J AOAC 103(4):904–905. https://doi.org/10.1093/jaoacint/qsaa011

Park J (2014) Flaxseed secoisolariciresinol diglucoside and visceral obesity, a closer look at its chemical properties, absorption, metabolism, bioavailability, and effects on visceral fat, lipid profile, systemic inflammation, and hypertension. In: Watson RR (ed) Nutrition in the prevention

and treatment of abdominal obesity book, 1st edn. Academic Press, San Diego, CA, USA, pp 317–328
Pasparakis M, Haase I, Nestle FO (2014) Mechanisms regulating skin immunity and inflammation. Nat Rev Immunol 14(5):289–301. https://doi.org/10.1038/nri3646
Pavić E, Rahelić V (2025) Food-drug interactions. In: Banić M, Erceg D (eds) Drug interactions in gastroenterology: a clinical guide. Humana, Cham, Switzerland, pp 245–263. https://doi.org/10.1007/978-3-031-69800-2_15
Petersen SR, Bonnesen K, Grove EL, Pedersen L, Schmidt M (2025) Bleeding risk using non-steroidal anti-inflammatory drugs with anticoagulants after venous thromboembolism: a nationwide Danish study. Eur Heart J 46(1):58–68. https://doi.org/10.1093/eurheartj/ehae736
Petti S, Scully C (2009) Polyphenols, oral health and disease: a review. J Dent 37(6):413–423. https://doi.org/10.1016/j.jdent.2009.02.003
Pichard V, Guilbaud M, Devaux M, Jaulin N, Journou M, Cospolite M, Garcia A, Ferry N, Michalak-provost S, Gernoux G, Adjali O (2025) Incomplete elimination of viral genomes is associated with chronic inflammation in nonhuman primate livers after AAV-mediated gene transfer. Gene Ther 32(3):287–298, 1–12.https://doi.org/10.1038/s41434-025-00514-z
Popa-Wagner A, Mitran S, Sivanesan S, Chang E, Buga AM (2013) ROS and brain diseases: the good, the bad, and the ugly. Oxid Med Cell Longev 1:963520. https://doi.org/10.1155/2013/963520
Puspadewi R, Milanda T, Muhaimin M, Chaerunisaa AY (2025) Nanoparticle-encapsulated plant polyphenols and flavonoids as an enhanced delivery system for anti-acne therapy. Pharmaceuticals 18(2):209. https://doi.org/10.3390/ph18020209
Ran J, Wang Q, Lu T, Pang X, Liao S, He X (2025) Integrating metabolomics and network analyses to explore mechanisms of Geum japonicum var. chinense against pulmonary fibrosis: involvement of arachidonic acid metabolic pathway. Int J Mol Sci 26(4):1462. https://doi.org/10.3390/ijms26041462
Ratnavathi CV (2019) Grain structure, quality, and nutrition. In: C. Aruna C, Visarada KBRS, Venkatesh Bhat B, Tonapi VA (eds) Breeding sorghum for diverse end uses. Woodhead Publishing, Duxford, Cambridge, and Kidlington, UK, pp 193–207. https://doi.org/10.1016/B978-0-08-101879-8.00012-7
Rentzsch M, Wilkens A, Winterhalter P (2009) Non-flavonoid phenolic compounds. In: Moreno Arribas MV, Polo MC (eds) Wine chemistry and biochemistry. Springer, New York, NY, USA, pp 509–527
Roupe KA, Remsberg CM, Yáñez JA, Davies NM (2006) Pharmacometrics of stilbenes: seguing towards the clinic. Curr Clin Pharmacol 1(1):81–101. https://doi.org/10.2174/157488406775268246
Rudrapal M, Khairnar SJ, Khan J, Dukhyil AB, Ansari MA, Alomary MN, Fahad M. Alshabrmi FM, Palai S, Deb PK, Devi R (2022) Dietary polyphenols and their role in oxidative stress-induced human diseases: insights into protective effects, antioxidant potentials and mechanism (s) of action. Front Pharmacol 13:806470. https://doi.org/10.3389/fphar.2022.806470
Santamarina AB, Moreira RG, Mennitti LV, Ferreira YAM, Jucá A, Prado CM, Pisani LP (2025) Low-carbohydrate diet enriched with omega-3 and omega-9 fatty acids modulates inflammation and lipid metabolism in the liver and white adipose tissue of a mouse model of obesity. Nutr Metab Cardiovasc Dis 35(8):103932. https://doi.org/10.1016/j.numecd.2025.103932
Selvaraj NR, Nandan D, Nair BG, Nair VA, Venugopal P, Aradhya R (2025) Oxidative stress and redox imbalance: common mechanisms in cancer stem cells and neurodegenerative diseases. Cells 14(7):511. https://doi.org/10.3390/cells14070511
Servili M, Esposto S, Fabiani R, Urbani S, Taticchi A, Mariucci F, Selvaggini R, Montedoro GF (2009) Phenolic compounds in olive oil: antioxidant, health and organoleptic activities according to their chemical structure. Inflammopharmacology 17:76–84. https://doi.org/10.1007/s10787-008-8014-y
Sirše M (2022) Effect of dietary polyphenols on osteoarthritis—Molecular mechanisms. Life 12(3):436. https://doi.org/10.3390/life12030436

Solanki R, Rawat L, Tabasum S, Pal S, Patel S, Sabarwal A (2025) A comprehensive review of anti-cancer mechanisms of polyphenol honokiol and nano carrier-based approaches to enhance its therapeutic potential. Phytochem Rev 24(6):1–27. https://doi.org/10.1007/s11101-025-10090-0

Stoner GD, Mukhtar H (1995) Polyphenols as cancer chemopreventive agents. J Cell Biochem 59(S22):169–180. https://doi.org/10.1002/jcb.240590822

Storz G, Imlayt JA (1999) Oxidative stress. Curr Opin Microbiol 2(2):188–194. https://doi.org/10.1016/S1369-5274(99)80033-2

Subramani PA, Panati K, Lebaka VR, Reddy DD, Narala VR (2017) Nanostructures for curcumin delivery: possibilities and challenges. In: Grumezescu AM (ed) Nano- and microscale drug delivery systems: design and fabrication. Elsevier, Inc., Amsterdam, The Netherlands, pp 393–418. https://doi.org/10.1016/B978-0-323-52727-9.00021-2

Teka T, Zhang L, Ge X, Li Y, Han L, Yan X (2022) Stilbenes: source plants, chemistry, biosynthesis, pharmacology, application and problems related to their clinical application-a comprehensive review. Phytochemistry 197:113128. https://doi.org/10.1016/j.phytochem.2022.113128

Tian C, Hao L, Yi W, Ding S, Xu F (2020) Polyphenols, oxidative stress, and metabolic syndrome. Oxid Med Cell Longev 2020:7398453. https://doi.org/10.1155/2020/7398453

Tihana Ž, Abdelkebir R, Alcantara C, Carmen M, García-Pérez JV, Meléndez-Martínez AJ, Re A, Lorenzo JM, Barba FJ (2019) From extraction of valuable compounds to health promoting benefits of olive leaves through bioaccessibility, bioavailability andimpact on gut microbiota. Trends Food Sci Technol 83:63–77. https://doi.org/10.1016/j.tifs.2018.11.005

Tillman L, Rieres JM, Ahjem E, Bishop-Guest F, McGrath M, Hatrick H, Pranjol MZI (2025) Thinking outside the therapeutic box: the potential of polyphenols in preventing chemotherapy-induced endothelial dysfunction. Cells 14(8):566. https://doi.org/10.3390/cells14080566

Trofin DM, Sardaru DP, Trofin D, Onu I, Tutu A, Onu A et al (2025) Oxidative stress in brain function. Antioxidants 14(3):297. https://doi.org/10.3390/antiox14030297

Tumbarski Y, Ivanov I, Todorova M, Apostolova S, Tzoneva R, Nikolova K (2025) Phenolic content, antioxidant activity and in vitro anti-inflammatory and antitumor potential of selected bulgarian propolis samples. Biomedicines 13(2):334. https://doi.org/10.3390/biomedicines13020334

Uti DE, Alum EU, Atangwho IJ, Ugwu OPC, Egbung GE, Aja PM (2025) Lipid-based nano-carriers for the delivery of anti-obesity natural compounds: advances in targeted delivery and precision therapeutics. J Nanobiotechnol 23(1):1–37. https://doi.org/10.1186/s12951-025-03412-z

Wang K, Hu S (2023) The synergistic effects of polyphenols and intestinal microbiota on osteoporosis. Front Immunol 14:1285621. https://doi.org/10.3389/fimmu.2023.1285621

Wright GD (2007) On the road to bacterial cell death. Cell 130(5):781–783

Xiao CL, Lai HT, Zhou JJ, Liu WY, Zhao M, Zhao K (2025) Nrf2 signaling pathway: focus on oxidative stress in spinal cord injury. Mol Neurobiol 62(2):2230–2249. https://doi.org/10.1007/s12035-024-04394-z

Yahfoufi N, Alsadi N, Jambi M, Matar C (2018) The immunomodulatory and anti-inflammatory role of polyphenols. Nutrients 10(11):1618. https://doi.org/10.3390/nu10111618

Yang Q, Liu Y, Guo Y, Jiang Y, Wen L, Yang B (2023) New insights of fig (Ficus carica L.) as a potential function food. Trends Food Sci Technol 140:104146. https://doi.org/10.1016/j.tifs.2023.104146

Yoon JH, Baek SJ (2005) Molecular targets of dietary polyphenols with anti-inflammatory properties. Yonsei Med J 46(5):585–596

Zamani-Garmsiri F, Emamgholipour S, Rahmani Fard S, Ghasempour G, Jahangard Ahvazi R, Meshkani R (2022) Polyphenols: potential anti-inflammatory agents for treatment of metabolic disorders. Phytother Res 36(1):415–432. https://doi.org/10.1002/ptr.7329

Zhang S, Xu M, Zhang W, Liu C, Chen S (2021) Natural polyphenols in metabolic syndrome: protective mechanisms and clinical applications. Int J Mol Sci 22(11):6110. https://doi.org/10.3390/ijms22116110

# Chapter 7
# Nanotechnology in Polyphenol Delivery Systems: Advances in Bioavailability Through Nanocarriers and Potential Therapeutic Applications

**Vivek P. Chavda, Dixa Vaghela, Arya Vyas, and Salvatore Parisi**

**Abstract** This chapter thoroughly examines the use of nanotechnology-based delivery systems to address the limited bioavailability and stability of polyphenols, enhancing their therapeutic efficacy. It begins with an overview of polyphenol structures and their inherent challenges in conventional delivery, such as poor solubility and rapid metabolism. The discussion then focuses on diverse nanocarrier types—including lipid-based, polymeric, inorganic, and biomimetic systems—and their roles in improving absorption, protecting bioactive compounds, and enabling targeted release. Covered key therapeutic areas include cancer, cardiovascular, neurodegenerative, metabolic, inflammatory, antimicrobial, and antiviral applications. The chapter further reviews recent innovations, safety and regulatory considerations, and emerging directions such as personalized nanomedicine and artificial intelligence-assisted formulation design. Overall, the chapter highlights the potential of nanotechnology to significantly optimize polyphenol-based therapies by overcoming pharmacokinetic barriers and advancing clinical applications across multiple disease domains.

**Keywords** Antioxidant and anti-inflammatory therapy · Chronic disease prevention and treatment · Nanoformulation and bioavailability · Polymeric and lipid nanoparticle · Polyphenol-loaded nanocarrier · Targeted and controlled drug delivery · Translational and clinical application

V. P. Chavda (✉) · D. Vaghela · A. Vyas
Department of Pharmaceutics and Pharmaceutical Technology, L M College of Pharmacy, Ahmedabad, Gujarat, India
e-mail: Vivek7chavda@gmail.com

S. Parisi
Department of Nutrition and Food Processing Technology, Lourdes Matha Institute of Hotel Management and Catering Technology, Thiruvananthapuram, Kerala State, India
e-mail: drparisi@inwind.it

 
R. K. Singla et al. (eds.), *Polyphenols in Health*,
https://doi.org/10.1007/978-3-032-27324-6_7

## Abbreviations

| | |
|---|---|
| AD | Alzheimer's disease |
| AMPK | AMP-activated protein kinase |
| BCS | Biopharmaceutical Classification System |
| BBB | Blood–brain barrier |
| EE | Encapsulation efficiency |
| EGCG | Epigallocatechin gallate |
| FDA | Food and Drug Administration |
| GIF | Gastrointestinal fluid |
| GIT | Gastrointestinal tract |
| AuNP-BP | Gold NP with polyphenolic compounds derived from *Bacopa procumbens* |
| GMP | Good manufacturing practice |
| OH | Hydroxyl |
| IVIVC | In vitro–in vivo correlation |
| NP | Nanoparticle |
| NLC | Nanostructured lipid carrier |
| NF-κB | Nuclear factor kappa-light-chain-enhancer of activated B cells |
| PD | Parkinson's disease |
| PK | Pharmacokinetic |
| PK/PD | Pharmacokinetic-pharmacodynamic |
| PF | PepFect |
| PLA | Poly (D,L-lactide) |
| PLGA | Poly(lactic-co-glycolic acid) |
| PDI | Polydispersity index |
| PEG | Polyethylene glycol |
| ROS | Reactive oxygen species |
| SEDDS | Self-emulsifying drug delivery system |
| SLN | Solid lipid nanoparticle |

## 7.1 Introduction

Polyphenols constitute a vast and varied class of naturally occurring compounds characterized by multiple phenolic structures. These bioactive molecules are abundant in plant-derived foods such as fruits, vegetables, tea, coffee, wine, nuts, and cocoa products. Their chemical families include flavonoids, phenolic acids, stilbenes, lignans, and tannins (Lin et al. 2016). Extensive research has demonstrated their beneficial effects on human health, chiefly due to their antioxidant and anti-inflammatory properties, alongside their roles in protecting cardiovascular health and potentially reducing risks of cancer and neurodegenerative diseases (Brglez Mojzer et al. 2016).

However, leveraging polyphenols therapeutically is often hampered by their inherently poor solubility in water, susceptibility to chemical degradation, rapid metabolism, and overall limited bioavailability when administered orally. These pharmacokinetic constraints significantly limit their effective bioactive concentrations in target tissues, thus impeding clinical efficacy (Aatif 2023).

Nanotechnology offers innovative strategies to surmount these obstacles by encapsulating polyphenols within nanoscale delivery systems that shield them from degradation, enhance aqueous solubility, and promote targeted, controlled release. Various nanocarrier platforms—including liposomes, polymeric nanoparticles, nanoemulsions, and phytosomes—have demonstrated the ability to improve polyphenol absorption and extend systemic circulation, thereby amplifying their therapeutic effects (Karnwal et al. 2024).

This chapter comprehensively reviews the chemistry of polyphenols alongside advances in nanotechnology-based delivery methods. We discuss the prevailing challenges in polyphenol administration, delve into diverse nanocarrier formulations and their bioavailability enhancement mechanisms, and explore current therapeutic applications and emerging trends. The intent is to equip researchers and clinicians with insights into optimizing polyphenol use via nanotechnology to foster improved disease management outcomes.

## 7.2 Challenges in Polyphenol Delivery: Bioavailability and Stability Issues

Due to their possible health advantages, polyphenols have attracted considerable interest. Nevertheless, their low solubility and absorption in the gastrointestinal tract limit polyphenols' bioavailability. Furthermore, because of their many hydroxyl (OH) groups, polyphenols can break down when exposed to light, heat, and alkaline conditions (Oliver et al. 2016). Numerous studies have demonstrated that polyphenols are advantageous for intestinal health and possess beneficial qualities, including reducing inflammation and providing protection against cancer by regulating gut flora. This reasons explains the frequent inclusion of polyphenols in functional diet menus designed to confer health benefits through consumption. With the increase in global population and the growing recognition of the necessity for healthy living, the market demand for polyphenols is also rising. The utilization of polyphenols in nutraceuticals is limited, however, due to the fact that many polyphenolic compounds exhibit poor oral bioavailability (Farhan et al. 2023; Aravind et al. 2021).

Numerous factors, including gastric residence time, permeability, and solubility within the gut, influence the bioavailability of polyphenols after oral administration. Moreover, conditions experienced during food processing and storage (such as temperature, oxygen levels, and light exposure) or within the gastrointestinal tract (including pH levels, enzymes, and the presence of other nutrients) may affect polyphenol stability. The inadequate bioavailability of polyphenols is due in part to their poor solubility in water and low dissolution rates (Amidon et al. 1995). Most polyphenols fall into class II (low solubility, high permeability) and class IV (low solubility, low permeability) according to the Biopharmaceutical Classification System (BCS) classification, which restricts their activity and potential health benefits. For class II and class IV substances, enhancing the drug's solubility and dissolution rate in gastrointestinal fluid may improve its bioavailability. Different methods can be used to improve the solubility of polyphenols. Methods for enhancing solubility comprise inclusion complexes (Ntuli et al. 2022), micronisation (Lee et al. 2025), solid dispersion (Li et al. 2013), nanosuspension (Yang et al. 2020), solid lipid nanoparticles (Kipp 2004), liposomes (Mignet et al. 2013), self-emulsifying drug delivery systems (SEDDS) (Balata et al. 2016), and gel-based systems (Kaur and Kaur 2014).

Polyphenols' bioavailability is closely connected to the varying pH levels and enzymatic activities found in the gastrointestinal fluids (GIF) throughout the digestive tract. The fate of polyphenols that are administered orally is greatly affected by the stomach's acidic environment (pH 1.2), the small intestine's more neutral conditions (pH 6.8), and various metabolizing enzymes found in the gastrointestinal tract (GIT). As a result, many polyphenols can degrade in the stomach's acidic environment and undergo enzymatic breakdown in the gastrointestinal tract, leading to diminished bioavailability (Imam et al. 2021). The chemical structure of polyphenols and the type of sugars in their glycosylated form have a significant impact on their intestinal absorption (Bertelli et al. 2021),

Usually, polyphenols are glycosylated, and the sugars linked to them are typically released before absorption. After ingestion, various dietary components—such as dietary fiber, lipids, proteins, and digestible carbohydrates—can influence the availability of polyphenols for absorption. Moreover, polyphenols' capacity to attach to proteins constitutes a major constraint on their absorption. Complex polyphenols are decomposed into simpler molecules by various enzymes during digestion. Also present are insoluble complexes formed from phenolic moieties and indigestible polysaccharides, protein (Morzel et al. 2022), and other dense polyphenols, such as condensed tannins and lignins (Ajayi et al. 2021). Polyphenol bioavailability shows variation. Isoflavones exhibit the greatest bioavailability, with phenolic acids, flavanols, flavanones, flavonols, anthocyanins, and proanthocyanidins following in that order (Di Lorenzo et al. 2021). Hydrophobic compounds like curcumin and naringenin have low absorption rates because of their rapid metabolism and excretion (Anand et al. 2007). It is known that anthocyanins are unstable, especially during food processing and when exposed to gastrointestinal diseases, which can restrict their absorption (Di Lorenzo et al. 2021). Those polyphenols that are not taken up by the small intestine are passed on to the colon, where they are transformed into other bioactive phenolic metabolites by the colonic microflora. Before they enter the bloodstream, these metabolites are structurally modified, primarily in the liver (El Monfalouti and Eddine Kartah 2024).

## 7.3 Nanotechnology: Principles and Advantages for Polyphenol Delivery

Nanotechnology has numerous applications in the food sector, encompassing areas such as food safety and quality, targeted delivery of essential compounds with enhanced bioavailability, new product development, management of technological processes, packaging solutions, and sensory enhancements like modifications to texture and flavor. Although the initial focus was on developing nanodrugs and nanopharmaceuticals, the use of nanotechnology in food production and agriculture is growing quickly (Milinčić et al. 2019).

The medicinal potential of dietary polyphenols has been investigated, both individually and in combination. These polyphenols have demonstrated potential therapeutic properties both in vitro and in vivo. Although they have advantageous effects, several traits such as low solubility, inadequate permeability, lack of stability, and quick compound release pose limitations on investigating their clinical uses (Rostami et al. 2019). Other elements, including ineffective systemic management, stability issues, and low bioavailability, further restrict their therapeutic clinical advantages. One possible solution to this challenge is the nanoformulation of polyphenols, which has yielded some interesting results (Tabrez et al. 2020). The nanoencapsulation of polyphenols could enhance circulation, improve localization, increase efficacy, and reduce the risk of multi-drug resistance. Evidence suggests that nanocarriers are better materials for encapsulating phenolic chemicals and enhancing their bioavailability (Yang et al. 2020). A diverse array of nanomaterials organic, inorganic, or organic/inorganic hybrids have been developed, each possessing distinctive physical and chemical properties. Nanomaterials, with sizes between 1 and 100 nm, possess relatively large surface areas and are designed for various biomedical applications, including disease treatment, imaging, and sensing (Chaudhari et al. 2024). Nevertheless, the acute toxicity and long-term negative effects of these engineered nanomaterials present significant health concerns. Consequently, it is crucial to create new kinds of nanomaterials that exhibit high performance and biocompatibility (Guo et al. 2021).

The current strategy relies heavily on nanotechnology, and its application in herbal medicinal systems, especially drug delivery systems, is anticipated to expand rapidly. Herbal drugs used in innovative delivery systems can enhance the pharmacological profile and address the challenges associated with medicinal or herbal plants in the future (Khan et al. 2021). Therefore, herbal nanocarriers will help to confront deadly diseases like cancer, hypertension, diabetes, and other diseases. Recent studies and advancements in nanotechnology have demonstrated the ability to deliver herbal therapeutic molecules to their targeted locations for the treatment of various critical diseases (Egbuna et al. 2022; Rauta et al. 2019). In recent years, various delivery methods based on herbal medicine have been effectively employed alongside the application of nanotechnology in health and environmental fields. Nanopolymers can be used in various ways in the fields of health and environmental science. It also leads to serious health concerns. The improved biological properties, including toxicities, of nanoformulations—such as nanoemulsion, nanoencapsulation, polymeric nanoparticles (NP), nanoliposomes, and solid–liquid NP—are attributed to their physicochemical properties like shape, size, and surface characteristics. Although dietary polyphenols have a good protective function, their poor bioavailability and metabolic instability result in inefficient delivery into the body. Such problems can be overcome through the nanoformulation of polyphenols, which can be delivered efficiently for cancer treatment and for neurological and cardiovascular diseases (Anand et al. 2022; Squillaro et al. 2018).

## 7.4 Types of Nanocarriers Used for Polyphenol Delivery

To tackle the challenge posed by the limited bioavailability of polyphenols, nanotechnology has shown promise for improving polyphenol absorption, especially when speaking of nano-based carriers. Due to their tiny dimensions, NP can move through small capillaries and cells, which helps achieve maximal polyphenol accumulation and sustained release (Nedovic et al. 2011). Research is underway regarding the use of lipid-based materials, dendrimers, and polymeric NP. Due to their characteristics, including biodegradability and surface modifiability, polymeric NP are preferred. Various NP types (nanophytopolyphenols) are illustrated in Fig. 7.1 (Kalušević et al. 2017; Rudrapal et al. 2022).

### *7.4.1 Lipid-Based Nanocarriers*

Solid lipid NPs (SLN) are similar to nanoemulsions, with the distinction that lipids serve as the solid phase. SLN are colloidal nanocarriers composed of lipids that are compatible with physiological conditions, sized between 50 and 1,000 nm. They are dispersed in an aqueous solution containing dissolved surfactants. Micro-fluidization (also known as high-pressure homogenization) and ultra-sonication are among the methods used to prepare SLN (Desai and Thakkar 2016). SLN have the advantages of being submicron in size, which offers a large surface area, an increased drug payload, and enhanced interfacial interactions. SLN face limitations, including the lack of effective targeting and controlled release from colloidal nanocarriers like polymeric NP, liposomes, nanoemulsions, and microemulsions. The drug payload is accomplished through two methods. First, the drug can be incorporated into the polymeric core; second, it can be affixed to the surface of the polymeric component. SLN are seen to have a significant advantage in that they can stabilize the lipophilic drug without resorting to toxic organic solvents. SLN have additional benefits, such as their distinct size-related characteristics and capability for drug incorporation, allowing for the loading of both hydrophilic and lipophilic drugs. It is easier to

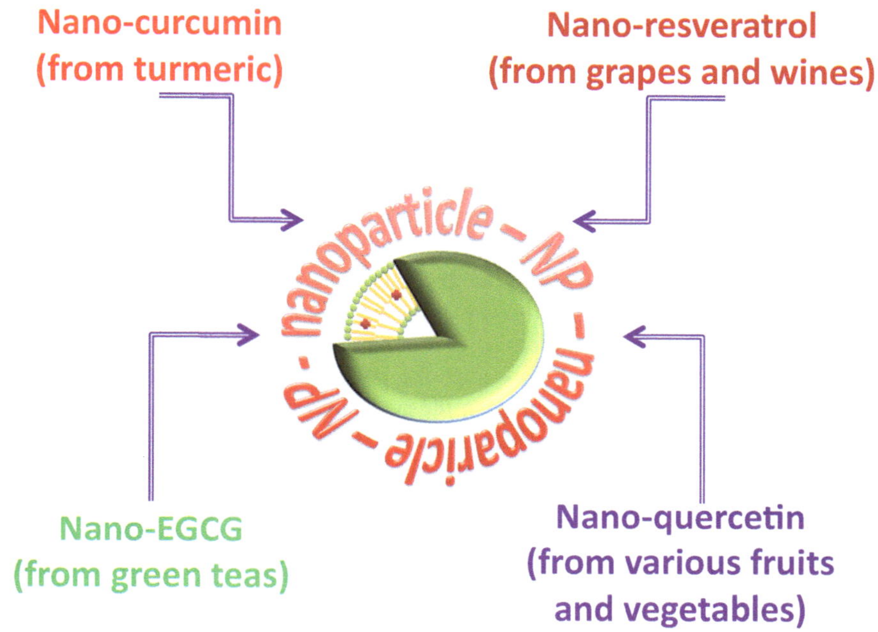


**Fig. 7.1** Simple graphical illustration of nanophytopolyphenols derived from key dietary polyphenols and their typical food sources (Rudrapal et al. 2022): nanocurcumin, nanoresveratrol, nano-EGCG, and nanoquercetin, each shown together with representative natural matrices such as turmeric, grapes/wine, green tea, and fruits/vegetables, respectively. This picture has been created by the authors

achieve large-scale production and higher bioavailability with minimal toxic effects (Carlotti et al. 2012; El Monfalouti and Eddine Kartah 2024).

For the purpose of targeting the brain in Alzheimer's disease, Loureiro and coworkers created SLN with attached OX26 antibodies containing resveratrol and grape extracts. Therapeutic applications of resveratrol are constrained due to its low solubility and limited permeability across the blood–brain barrier (BBB). In order to enable transferrin receptor-mediated BBB transport, OX26-functionalized SLN were created using cetyl palmitate and polysorbate 80. The optimized formulation exhibited a particle size of approximately 180 nm, an encapsulation efficiency ranging from 75 to 100%, and good stability. In vitro studies demonstrated that BBB transport was four times greater and cellular uptake was two times greater compared to non-targeted SLN. The combination of resveratrol and grape extract notably reduced the amyloid-beta aggregation (by as much as 97%). In summary, SLN that are functionalized with OX26 represent a promising nanocarrier aimed at the brain for improving neuroprotective treatment in Alzheimer's disease. They merit further in vivo evaluation (Loureiro et al. 2017).

Liposomes make for outstanding drug delivery candidates because they are biocompatible and have adjustable physicochemical and surface features, all while being composed of natural or synthetic phospholipids that encase an aqueous core. Since 2008, liposomes, which were first suggested by Gregoriadis and Florence (1993), have been the only available NP on the market. The successful commercialization of various drugs such as doxorubicin, amphotericin B, verteporfin, cytarabine, vincristine, and inactivated hepatitis B virus has provided them with an advantage through accumulated technical and efficacy knowledge (Feng 2006; Yeh et al. 2011). Numerous firms currently manufacture good manufacturing practice (GMP)-grade liposomes (Wagner and Vorauer-Uhl 2011).

Liposomes can enclose hydrophilic drugs in their aqueous core and lipophilic drugs within their lipid bilayer; however, the encapsulation of lipophilic drugs is often inefficient and can compromise membrane integrity. Even so, research has been conducted on the use of liposomes for polyphenol encapsulation; however, their stability is highly contingent upon the structure of the polyphenol, especially the quantity of rings and OH groups that influence solubility in surfactants. It is worth mentioning that the majority of liposomal polyphenol formulations documented in literature aim at food applications due to their antioxidant properties, rather than at pharmaceutical uses. Nevertheless, these formulations are crucial for comprehending interactions between polyphenols and membranes (Fahr et al. 2005; Maurer et al. 2001).

Jovanović and coworkers created liposome-based carriers able to improve the stability and delivery of polyphenols derived from wild thyme (*Thymus serpyllum* L.) tea residues, thus converting agricultural waste into a value-added product. Polyphenols are unstable bioactive compounds and have low absorption rates, making an effective encapsulation system essential.

Liposomes, created from phospholipids via the proliposome method, were supplemented with either cholesterol or β-sitosterol (10–30 mol %). It was anticipated that β-sitosterol, a plant sterol, would enhance encapsulation and stability. Extracts abundant in polyphenols were produced using microwave-assisted ethanol extraction, and the resulting liposomes were assessed for encapsulation efficiency (EE), particle size, zeta potential, stability, antioxidant capacity, and release behavior in simulated gastric and intestinal fluids. Formulations with 10 mol % sterol demonstrated the highest EE (approximately 92–94%), and liposomes based on β-sitosterol displayed improved stability and smaller vesicle sizes. All formulations remained stable during refrigeration, ultraviolet light-exposure, and lyophilization, showing no significant leakage. Encapsulation contributed to a reduction in polyphenol release and provided protection against oxidation. At the same time, β-sitosterol enhanced retention and regulated release, particularly under intestinal conditions. To sum up, liposomes made of β-sitosterol demonstrated superior performance compared to those based on cholesterol, providing high stability, robust antioxidant protection, and sustained release. This method successfully transforms thyme waste into functional nutraceutical components with possible industrial applications (Jovanović et al. 2025).

### 7.4.2 Polymeric NP and Nanogel

Polymeric NP, including polymeric nanospheres and nanocapsules, are colloidal systems typically ranging in size from 5 to 1,000 nm (with a common range of 100–500 nm). Bioactive molecules such as drugs, genes, and fluorescent agents can be either adsorbed on the surface or encapsulated within solid matrix particles known as nanospheres. Unlike them, nanocapsules consist of bioactive substances encased in an aqueous core that is enveloped by a polymeric shell (Banik et al. 2016; Lu et al. 2011). Natural or synthetic polymers can be used to create polymeric NP. Natural polymers (e.g., starch, gelatin, chitin, cellulose, alginate, albumin, and olyhydroxyalkanoates) are favored for their biodegradability and compatibility with living organisms, while synthetic options like polyethylene glycol (PEG), poly(lactic-co-glycolic acid (PLGA), polyvinyl alcohol, polyvinylpyrrolidone, polyethylene, polyanhydrides, and polyorthoesters are also utilized. To enhance targeting, ligand binding, and immune evasion, their surfaces can be modified with metal ions, small molecules, surfactants, or polymers (Quintanar-Guerrero et al. 1998). When constructed from biodegradable polymers, polymeric NP provide a high drug encapsulation efficiency, improved cellular uptake, increased stability, and biocompatibility. They can be designed for targeted delivery, enhancing therapeutic outcomes and minimizing side effects. Numerous polymeric NP are already utilized commercially in biomedical applications as versatile and multifunctional drug carriers, with ongoing developments aimed at new materials, enhanced functionalities, and better drug performance (Hoffman 2008; Kreuter 2007). Polymeric NP may have drawbacks such as potential non-biodegradability, susceptibility to breakage, high manufacturing costs, and the presence of toxic solvent residues. To design polymeric NP that are safe and effective, it is essential to grasp their underlying principles and to optimize factors including biocompatibility, pharmacokinetics, targeting efficiency, and cost-effectiveness. It is particularly important to reduce toxicity, which is why in vitro cytotoxicity studies are vital for confirming the therapeutic potential of these agents (Singh and Nalwa 2007). Polymeric NP consist of polymeric micelles, dendrimers, nanospheres, nanocomplexes, nanocapsules, nanogels, and polymersomes (Chaudhari et al. 2024).

Quercetin, a powerful antioxidant flavonoid, has limitations for therapeutic use due to its poor solubility, instability, and low bioavailability. To tackle this issue, quercetin was encapsulated within biodegradable poly(D,L-lactide) (PLA) NP created using the solvent evaporation technique. Produced NP were smooth and spherical, measuring approximately 130 nm. They displayed a high EE of 96.7% and retained their antioxidant activity. In vitro investigations demonstrated a biphasic release pattern, characterized by an initial burst of 45% followed by a sustained release that reached 87.6% over the course of 96 h. The results show that PLA NP significantly improve the stability, solubility, and controlled release of quercetin. Thus, they represent a promising method for delivering hydrophobic bioactive compounds (Kumari et al. 2010).

Nanogels serve as efficient nanocarriers for transporting food-grade active compounds to the gastrointestinal tract. This effect makes it easier for them to be transformed into particles that are essential for digestion and absorption. Nanogels made from carboxymethyl starch and chitosan hydrochloride, created via chemical cross-linking, serve as effective carriers for bioactive substances like curcumin (El Monfalouti and Eddine Kartah 2024; Sharma et al. 2024).

Resveratrol, a powerful natural antioxidant with neuroprotective and cardioprotective effects, encounters significant limitations because of its poor solubility, instability, and low bioavailability. To address these challenges, Radeva and colleagues created a biodegradable nanogel made from natural citric acid and pentane-1,2,5-triol to encapsulate resveratrol. With an EE of 94.5%, the nanogel (approximately 220 nm) allowed for complete drug release within 24 h, in contrast to free resveratrol, which had a release rate of only 11%. Antioxidant assays verified that encapsulation maintained resveratrol-related activity, and cellular studies showed a marked enhancement in the protection of neuroblastoma and fibroblast cells from oxidative stress. Encapsulated resveratrol offered protection of up to 84% in brain microsomes within tissue models, exceeding that of free resveratrol. This natural nanogel system significantly improved resveratrol solubility, stability, and therapeutic efficacy, underscoring its promise as a multipurpose delivery platform for hydrophobic bioactive compounds in conditions linked to oxidative stress (Radeva et al. 2023).

### *7.4.3 Inorganic NP*

Inorganic NP have attracted interest for improving the bioavailability of polyphenolic compounds because of their controllable size and shape, along with their large specific surface area. The most commonly utilized nanocarriers for improving the bioavailability of polyphenols are gold and silver NP (Meena et al. 2020). Moreover, NP of silica, titanium dioxide, and magnetic iron oxide have been utilized in drug delivery systems (Bruckmann et al. 2022; Vasić et al. 2024; Ziental et al. 2020). To improve wound healing, a novel nanoformulation was created by Martinez-Cuazitl and coworkers: it combines gold NP with polyphenolic compounds derived from *Bacopa procumbens* (AuNP–BP). In a rat excision wound model, the use of AuNP–BP markedly sped up tissue repair, enabling a reduction of the plant extract dosage by 100 times while still achieving better healing results than with the extract alone. The characterization confirmed that the conjugation was successful and that NP stability was enhanced. Compared to controls, rats treated with AuNP–BP exhibited a faster wound closure (approximately 85%), reduced inflammation, enhanced collagen organization, and improved tissue regeneration. Higher collagen maturity and enhanced tissue strength were indicated by histological and Fourier transform infrared spectroscopy analyses. Gold NP served as effective carriers, improving bioavailability, stability, and therapeutic efficacy. Overall, AuNP–BP showed an impressive ability to promote wound healing at lower doses, underscoring its potential as a cutting-edge and affordable nanotherapeutic for both chronic and acute wound management (Martínez-Cuazitl et al. 2023).

### 7.4.4 Nanoemulsions and Phytosomes

Nanoemulsions consist of combinations of at least two liquids that do not mix, one being water and the other an oil component utilized in the formulation. These kinds of nanoformulations are produced through chemical or mechanical methods. In the case of a chemical process, nanoemulsion droplets result from the interaction of hydrophobic compounds with emulsifiers. In contrast, a mechanical process involves high shear homogenization of large emulsion droplets to create nanodroplets. The size and shape of emulsion droplets produced (ranging from 20 to 200 nm) is a fundamental concept that distinguishes nanoemulsion from conventional emulsion (Ishkeh et al. 2021).

Phytosomes are vesicles containing complex phospholipids and herbal drug extracts, where phytoconstituents are encapsulated by a phospholipid layer. Due to their hydrophilic head and lipophilic tail, phospholipids possess the characteristic of an active emulsifier. The phytosome emulsifier property enhances the bioavailability of phytopolyphenols by aiding their transition from an aqueous to a lipophilic environment at the cell membrane (Kumar et al. 2020). Rutin-loaded phytosomes were created by Das and colleagues to improve the transdermal administration of rutin, a naturally occurring anti-inflammatory flavonoid that has low bioavailability and poor solubility. Researchers developed a lipophilic–hydrophilic hybrid system that enhanced solubility, permeability, and skin retention by complexing rutin with phosphatidylcholine. Ex vivo skin penetration increased by 2.5 times, drug retention in skin tissue increased by 44%, and solubility increased from 0.426 to 2.025 mg/mL in the improved formulation (F3, 1:1 ratio). Amorphous, spherical vesicles with robust physical connections between phospholipids and rutin were validated by characterization. Rutin's transport was enhanced via the phytosome, allowing for prolonged local anti-inflammatory effect with reduced systemic exposure. A promising framework for creating potent topical treatments for inflammatory diseases like arthritis is provided by this technology (Das and Kalita 2014).

## 7.5 Mechanisms of Enhanced Bioavailability Via Nanocarriers

Reduced therapeutic efficacy is frequently the result of conventional drug delivery's poor solubility, fast metabolism, and non-specific dispersion. By shielding medications from enzymatic breakdown, nanocarriers enable site-specific, regulated release via ligand-receptor or stimuli-responsive processes. Dendrimers offer accurate drug loading and targeting, polymeric NP such as PLGA guarantee regulated release, and liposomes effectively transport both hydrophilic and hydrophobic medications. Notwithstanding their potential, problems like cost, toxicity, regulatory concerns, and formulation complexity still exist (Ponnuswami 2024).

By utilizing the material's advantageous qualities, including pH-responsiveness, bioadhesion, biocompatibility, biodegradability, modifiability, and processability, nanodrug delivery systems can increase drug bioavailability (Bhaskar et al. 2010). Lipid components in nanodrug delivery systems can be broken down by bile and pancreatic enzymes, enhancing drug absorption and transmembrane transport. SLN use their biodegradability and biocompatibility to shield medications from chemical deterioration, which improves cellular uptake (Lu et al. 2005; Chai et al. 2017). By using channels including betaine transporters, bile acids, and folate, nanodrug delivery devices can enhance active transport (Fang et al. 2013). The enhanced permeability and retention action of nanocarriers increases the possibility of tumor cell uptake by passively targeting tumor tissues and enriching in tumor locations (Li et al. 2012a, b). Additionally, by altering their surface ligand, nanocarriers can actively target particular cells or tissues via receptor-mediated endocytosis, enhancing the likelihood that target cells will absorb them (Akhter et al. 2021). To firmly encapsulate medications under ideal pH values and prevent contact with unfavorable environments, several delivery methods use pH-responsive technologies. They can break down or swell to release the medication at neutral pH (Farheen Akhter et al. 2023; Ricordel et al. 2015). Enzyme-sensitive formulations, such as PEGylated liposomes modified with the cell-penetrating peptide PepFects (PF), can be used to create nanodrug delivery systems. When matrix metalloproteinase (MMP)-2 acts on these liposomes, PEG-2 is shed, exposing PF. This change causes the liposome's surface charge to reverse from negative to positive, which makes it easier for tumor tissues to absorb it (Wang et al. 2020).

Nanodrug delivery systems can use the size difference between pharmaceuticals and digestive enzymes to build porous inorganic nanoparticle carriers, shielding the drug against enzymatic breakdown (Cikrikci et al. 2018). These devices can also protect the medicine from degradation by taking advantage of the hydrophobic interactions between the nanocarriers and digestive enzymes (Qi et al. 2018). Hepatic macrophages primarily phagocytize unmodified conventional nanodrugs once they enter the bloodstream, resulting in a brief half-life and poor bioavailability. In contrast, NP treated with PEG exhibit greater solubility and considerably enhanced capacity to elude macrophage identification. This considerable reduction in phagocytosis by hepatic macrophages prolongs blood circulation time, so significantly improving the drug's bioavailability (Deodhar et al. 2020).

## 7.6 Therapeutic Applications of Nanocarrier-Delivered Polyphenols

### 7.6.1 *Cancer Therapy*

Natural polyphenols, particularly epigallocatechin gallate (EGCG), resveratrol, and curcumin exert multi-target anticancer effects but are limited by poor solubility, instability, rapid metabolism, and low bioavailability. Engineered nanocarriers (liposomes, SLN, polymeric micelles/NP, dendrimers, hybrid/lipid–polymer systems, and mesoporous silica) enhance delivery via improved stability, controlled/triggered release, and passive/active tumor targeting (Chimento et al. 2023; Kim et al. 2023). Beyond improving exposure, polyphenol nanomedicines modulate pathways central to multidrug resistance, such as P-glycoprotein efflux, mitochondrial metabolism, redox balance, and nuclear factor kappa-light-chain-enhancer of activated B cells (NF-κB) signaling, thereby resensitizing tumors and enabling synergy with chemo-, radio-, immuno-, and photo-therapies (Yang et al. 2024).

EGCG's anticancer actions include suppression of NF-κB and MMP-9, apoptosis induction, and chemosensitization; yet free EGCG is poorly stable and rapidly cleared. Nanoencapsulation—liposomes, SLN, nanostructured lipid carriers (NLC), and polymeric NP—improves physicochemical stability, cellular uptake, and in vivo antitumor activity. Folic acid-functionalized EGCG-loaded NLC enhanced intestinal permeability in Caco-2 models (supporting oral delivery potential), and nanoEGCG slowed lung cancer cell proliferation via AMP-activated protein kinase (AMPK) activation in preclinical systems (Sun et al. n.d.). Hyaluronic acid-modified EGCG nanoliposomes are being developed to target CD44-high tumors, improving stability and in vitro anticancer effects. EGCG also synergizes with agents such as doxorubicin by augmenting apoptosis and suppressing migration (Kciuk et al. 2023).

Curcumin demonstrates broad antineoplastic activity (apoptosis/autophagy induction, NF-κB and PI3K/AKT inhibition) but suffers from extreme bioavailability limits when free. Nanodelivery (liposomes, polymeric micelles/NP, SLN/NLC, nanocrystals) markedly improves solubility and systemic exposure, enabling combination therapy. A single-center, open-label phase I study of liposomal curcumin (Lipocurc™) in patients with metastatic tumors (weekly IV infusions for 8 weeks) established safety across dose escalation (100–300 mg/m$^2$), with pharmacokinetic characterization and preliminary antitumor signals reported. Building on this, trials are exploring liposomal curcumin with radiation and temozolomide in high-grade gliomas. Reviews in 2024 synthesize preclinical and clinical data, underscoring nanocarriers' role in overcoming solubility and stability constraints and enabling synergistic regimens (e.g., with paclitaxel) (Greil et al. 2018).

### 7.6.2 Cardiovascular Disease Management

Polyphenols exert cardioprotective effects through multifactorial mechanisms: antioxidant scavenging of reactive oxygen species (ROS), downregulation of vascular inflammation (NF-κB, cytokines), improvement of endothelial nitric oxide availability, and attenuation of lipid peroxidation and platelet activation (Caiati and Jirillo 2025). Nanocarrier-based delivery significantly enhances solubility, stability, and circulation time, allowing targeted myocardial accumulation. For instance, curcumin-loaded NP reduced ROS and improved endothelial function in hypertensive rat models. Similarly, resveratrol nanoformulations attenuated vascular inflammation via NF-κB inhibition and improved lipid metabolism. EGCG NP demonstrated reduced atherosclerotic lesion formation by modulating oxidative stress markers. These findings highlight that nanocarrier-mediated delivery amplifies the therapeutic potential of polyphenols in cardiovascular disease management by overcoming pharmacokinetic barriers (Jacob et al. 2024).

In hypertension models, nanoparticulate resveratrol and curcumin preparations have shown improved vascular reactivity and blood-pressure modulation compared with free compounds, attributable to sustained plasma levels and enhanced endothelial targeting. For atherosclerosis, nanoencapsulation of polyphenols increases arterial wall deposition and reduces plaque-associated oxidative stress and inflammation in preclinical studies, resulting in smaller necrotic cores and reduced macrophage infiltration (Ghaeini Hesarooeyeh et al. 2024). Nanocarriers can also be engineered for myocardial protection: intravenous or cardiac-targeted NP delivering antioxidant polyphenols reduce ischemia–reperfusion injury, limit infarct size, and preserve post-ischemic contractile function in animal models (Li et al. 2024).

### 7.6.3 Neuroprotective Effects

Polyphenols such as curcumin, resveratrol, and EGCG exhibit potent neuroprotective effects through antioxidant, anti-inflammatory, and anti-amyloidogenic mechanisms. They reduce ROS, modulate mitochondrial function, and inhibit neuroinflammatory pathways, including NF-κB and mitogen-activated protein kinase. Despite strong therapeutic potential, their translation is hindered by poor solubility, rapid metabolism, and limited BBB permeability.

Nanocarrier-based delivery offers a solution by enhancing BBB penetration and sustained release. Polymeric NP, SLN, and liposomes have been shown to increase brain accumulation of polyphenols. For instance, curcumin-loaded PLGA NP improved memory retention and significantly reduced amyloid plaque deposition in Alzheimer's disease (AD) mouse models (Mourtas et al. 2014). Similarly, resveratrol nanoemulsions enhanced hippocampal bioavailability and attenuated dopaminergic neuronal loss in Parkinson's disease (PD) models. EGCG-loaded nanoliposomes demonstrated superior antioxidant capacity and reduced α-synuclein aggregation, a hallmark of PD. Collectively, nanocarrier-delivered polyphenols show promise as disease-modifying therapies in AD and PD, offering sustained neuroprotection, improved pharmacokinetics, and targeted central nervous system delivery. However, long-term safety evaluation and clinical validation remain essential for translation into neurotherapeutics (Xiong et al. 2020).

### *7.6.4 Metabolic and Inflammatory Disorders*

Polyphenols are increasingly recognized as modulators of metabolic and inflammatory disorders, including diabetes, obesity, and autoimmune diseases. Their pleiotropic actions involve antioxidant activity, suppression of pro-inflammatory cytokines (TNF-α, IL-6, IL-1β), activation of AMPK, and improved insulin sensitivity. However, poor oral bioavailability and rapid systemic clearance limit their therapeutic applications. Nanocarrier-based systems such as lipid NP, polymeric micelles, and chitosan-based formulations improve solubility, protect against enzymatic degradation, and enhance controlled release (Zhang et al. 2021).

In diabetes, nanocurcumin demonstrated superior efficacy over free curcumin by lowering fasting glucose, HbA1c, and oxidative stress in both preclinical and clinical studies (Rahimi et al. 2016). Resveratrol NP enhanced AMPK activation, reduced hepatic steatosis, and improved insulin signaling in obese animal models. Similarly, EGCG-loaded NP protected pancreatic β-cells from oxidative damage and improved glucose tolerance. In autoimmune diseases such as rheumatoid arthritis, nanocurcumin and resveratrol nanocarriers reduced joint inflammation, suppressed NF-κB signaling, and improved histopathological outcomes in rodent models (Payne et al. 2022).

Overall, nanocarrier-delivered polyphenols address critical pharmacokinetic barriers, enabling improved delivery to target tissues, reduced dosing frequency, and enhanced disease-modifying effects in metabolic and inflammatory pathologies. These advances highlight their translational potential as adjuncts or alternatives to conventional therapies.

## 7.7 Recent Advances and Case Studies in Polyphenol Nanoformulations

Over the last decade, polyphenol nanoformulations have moved from proof-of-concept studies to increasingly sophisticated, translationally minded platforms that demonstrably improve pharmacokinetics, biodistribution, and therapeutic index. A landmark clinical signal is the phase I dose-escalation study of liposomal curcumin (Lipocurc™) in patients with metastatic cancer, which established intravenous tolerability and provided the first human pharmacokinetic (PK) characterization of a curcumin nanomedicine, validating IV nanodelivery as a feasible route for otherwise poorly bioavailable polyphenols (Greil et al. 2018). Preclinical case studies illustrate clear pharmacodynamic gains. Co-delivery NP carrying doxorubicin and resveratrol achieved higher intratumoral accumulation, stronger tumor suppression, and reduced cardiotoxicity relative to free drug or separate dosing, demonstrating how fixed-ratio co-encapsulation can blunt multidrug resistance (via efflux inhibition and survival-pathway modulation) while lowering systemic toxicity (Zhao et al. 2016). Lipid-based platforms have been particularly successful for catechins: EGCG loaded in liposomes or SLN shows markedly improved stability, transmucosal absorption, and in vivo efficacy in cancer and metabolic models (Farabegoli and Pinheiro 2022).

Innovative formulation strategies are accelerating pharmacodynamic improvements. Hybrid lipid–polymer NP, pH/redox-responsive polymeric micelles, and albumin/chitosan composite nanogels enable (i) controlled release profiles able to maintain therapeutic tissue exposures, (ii) active targeting via ligands (folate, arginylglycylaspartic acid, and hyaluronic acid), and (iii) co-delivery of polyphenols with small-molecule chemotherapeutics, small interfering RNA, or immunomodulators to produce synergistic effects in tumor, neurodegenerative and inflammatory models. Examples now report multi-fold increases in tumor area under the curve and significant enhancements in efficacy metrics (tumor growth inhibition, survival) compared with free compounds (Radeva et al. 2024).

Clinical development: beyond lipocurcumin, multiple randomized pilot trials using nanocurcumin formulations have reported measurable clinical signals, add-on anti-inflammatory effects in hospitalized coronavirus disease patients, reductions in HbA1c, and fasting glucose in Type 2 diabetes mellitus-cohorts, and symptomatic improvements in diabetic peripheral neuropathy supporting the translational potential of clinically acceptable oral or IV nanoforms when well-characterized. These human studies, while small, provide crucial pharmacokinetic-pharmacodynamic (PK/PD) and safety benchmarks for larger efficacy trials (Ahmadi et al. 2023; Mansour et al. 2025).

However, moving from bench to clinic exposes several persistent challenges. Scale-up and reproducibility remain nontrivial: many laboratory-scale nanoparticle methods (sonication, solvent evaporation) do not translate to industrial capacity without process redesign (microfluidic mixing, continuous manufacturing) to ensure tight control of size, polydispersity index (PDI), drug loading, and sterility. Reviews and practical reports emphasize that physicochemical heterogeneity during scale-up leads to inconsistent PK/PD and complicates comparability across batches (Paliwal et al. 2014).

Regulatory pathways are evolving but demanding: Agencies now expect thorough characterization of nanomaterial attributes that affect product performance (size, shape, surface chemistry, and release kinetics), as highlighted in the Food and Drug Administration (FDA) guidance on drug products containing nanomaterials. Sponsors must integrate advanced analytics (orthogonal particle sizing, stability under ICH conditions,[1] endotoxin/sterility, and complement activation testing) and build in PK/PD biomarkers early to support dose selection and safety monitoring. Regulatory filings increasingly require bridging data that link formulation attributes to clinical exposure and effects (FDA 2022).

Finally, formulation complexity and multi-component systems raise cost and chemistry, manufacturing, and controls hurdles: complex co-loaded or ligand-decorated NP require more elaborate quality controls and supply-chain oversight. To overcome these barriers, the field is converging on modular, scalable platforms (lipid NP with standardized excipients, polymer libraries amenable to microfluidic manufacture) and on early regulatory engagement coupled with robust PK/PD and manufacturability studies to de-risk translation. Continued progress will depend on standardized in vivo benchmarks, head-to-head comparisons with best-available formulations, and adequately powered clinical trials testing clinically meaningful endpoints rather than only biomarker changes.

## 7.8 Safety, Toxicity, and Regulatory Considerations of Nanocarriers

Since some nanomaterials may have distinct toxicological profiles that call for thorough safety evaluations, nanomaterial toxicity is a crucial factor to take into account. Understanding how nanomaterials bioaccumulate and distribute inside the body is essential to reducing any possible consequences; therefore, bioaccumulation and biodistribution are also important. Furthermore, immunogenicity and allergic reactions are issues since nanomaterials may cause allergic reactions or immunogenic responses, which calls for assessment of their immunogenic potential. Environmental impact is another important consideration since ecological safety is called into question by the disposal of nano-based materials and their possible effects on the environment. Nanocarrier-based formulations must be carefully evaluated, risk-managed, and developed responsibly in order to mitigate potential dangers and guarantee their safety (Kardani 2024).

Understanding the possible toxicological profiles of nanocarrier systems can be facilitated by using natural nanosized particles as models. Natural nanomaterial exposure data from the workplace and environment have served as the foundation for the

[1] The ICH acronym is for: International Council for Harmonisation of Technical Requirements for Pharmaceuticals for Human Use (https://www.ich.org).

majority of current toxicological research. Since both types of NP adhere to identical toxicological principles, these findings can be applied to artificial NP. Notably, nano-sized particles can be produced at quantities significantly higher than those normally seen in the environment under specific work-related conditions. Nevertheless, little study has been done expressly to date on the toxicity of nanocarriers (Vega-Villa et al. 2008). Curtis and coworkers suggested a number of potential explanations for the toxicity of NP. First, bulk compounds like heavy metals have a well-established toxicity that is reasonably simple to measure. Second, unlike bulk materials, NP have unique electrical characteristics that allow them to produce or scavenge free radicals and ROS. Third, research on ultrafine particles in the respiratory system suggests that the toxicity of NP may be directly correlated with their size. The size, propensity to aggregate, and capacity to penetrate biological barriers like the epidermis, vascular endothelium, and blood–brain barrier all affect how harmful ultrafine particles are absorbed, distributed, and eliminated. Fourth, toxicity can also be influenced by particle shape; for example, carbon nanotubes have shown hazardous effects that are depending on their shape. Lastly, there is increasing worry about the possibility that NP could result in allergic or hypersensitive reactions, even if the exact mechanisms by which they trigger immune responses are still unknown (Curtis et al. 2006).

The lack of defined preclinical research and characterization methodologies has hindered the clinical translation of nanomedicines, even though their number is increasing. The FDA, the European Medicines Agency, and Japanese PDMA and MHLW Institutions[2] have all made strides toward harmonization in recent years, despite the fact that worldwide regulatory alignment is still developing. At present, regulatory review frequently modifies procedures created for traditional pharmaceutical goods, focusing on the active pharmaceutical ingredient and its safety, physicochemical, and biological properties. Strict quality control procedures are required for evaluating parameters like size, morphology, charge, drug release, metabolism, and cellular uptake because of the special nature of nanomedicines, where even minor changes in raw materials or manufacturing processes can drastically change structure, biodistribution, and efficacy. Biocompatibility and immunotoxicity must also be taken into account during preclinical examination because of interactions with plasma proteins and immune cells. While many nanomedicines improve safety and therapeutic accuracy, others, such as carbon nanotubes, quantum dots, and dendrimers, cause toxicity issues. Process analytical technologies and ICH Q8–Q10 recommendations are used in quality-by-design techniques to address manufacturing difficulties, including scale-up and process variability, in order to guarantee consistent product quality. The goal of regulatory initiatives, such as the EMA's working group on nanomedicines, is to make quality, safety, and efficacy criteria more clear. The clinical and social value of nanotechnology-based therapies will be determined in the future by pharmacoeconomic analyses that measure quality-adjusted life years and healthcare cost reductions (Sainz et al. 2015).

[2] The PMDA acronym is for: the Pharmaceuticals and Medical Devices.
Agency, Japan; the MHLW acronym means:the Ministry of Health, Labour and Welfare, Japan.

## 7.9 Future Perspectives and Emerging Trends

Advancements in nanotechnology offer considerable promise for enhancing polyphenol delivery through the development of smart, stimuli-responsive nanocarriers (Liu et al. 2024). These systems can finely regulate the release of polyphenols in response to specific internal triggers such as pH changes, redox conditions, or enzymatic activity, as well as external cues like light or temperature (Costa et al. 2021). Such targeted delivery not only improves bioavailability but also reduces off-target effects, enabling more effective and personalized treatments. Emerging multi-responsive nanoplatforms that respond to several stimuli simultaneously are poised to further refine therapeutic precision tailored to individual patient needs (Kurul et al. 2025).

Personalized medicine represents a critical frontier where polyphenol nanoformulations can be customized based on genetic, metabolic, and microbiome factors unique to each patient, thus maximizing efficacy and minimizing variability in response. The integration of artificial intelligence and computational modeling is accelerating this progress by allowing rapid design optimization and predictive simulation of nanocarrier behavior, reducing experimental burden and development time (Kashyap et al. 2017).

Additionally, combination therapies utilizing nanoformulations that co-deliver polyphenols alongside other therapeutic agents hold great potential, especially in multifaceted diseases like cancer and neurodegeneration, by enabling synergistic effects and dose reduction. Expanding applications into immunomodulation, tissue regeneration, and microbiome-targeted interventions offer additional scope for innovation (Yang et al. 2020).

Despite these advances, challenges including scalable manufacturing, thorough safety evaluation, and navigating regulatory pathways must be addressed. Ongoing interdisciplinary collaboration and rigorous clinical investigations are essential to translate these breakthroughs into safe, effective nanomedicines that improve patient outcomes and broaden the therapeutic impact of polyphenols.

## 7.10 Conclusions

Nanotechnology offers a promising solution to the challenges limiting polyphenol delivery, including poor solubility, rapid metabolism, and low bioavailability. Employing various nanocarriers—lipid-based, polymeric, inorganic, and biomimetic—has enhanced polyphenol stability, targeted delivery, and controlled release, resulting in improved therapeutic outcomes across diseases such as cancer, cardiovascular and neurodegenerative disorders, metabolic syndromes, and infections.

However, challenges persist, including the need for standardized formulations, thorough safety evaluations, scalable manufacturing, and clear regulatory pathways.

Variability in patient responses further underscores the importance of personalized delivery approaches.

Future directions point toward integrating nanotechnology with advanced computational methods, including artificial intelligence, for designing smarter, stimuli-responsive systems and combinational therapies. Continued collaborative research is vital to translate preclinical advancements into clinical practice, thereby establishing polyphenol-based nanomedicine as an effective and versatile tool in contemporary disease management.

## References


Aatif A (2023) Current understanding of polyphenols to enhance bioavailability for better therapies. Biomed 11(7):2078. https://doi.org/10.3390/biomedicines11072078

Ahmadi S, Mehrabi Z, Zare M, Ghadir S, Masoumi SJ (2023) Efficacy of nanocurcumin as an add-on treatment for patients hospitalized with COVID-19: a double-blind, randomized clinical trial. Int J Clin Pract 1:5734675. https://doi.org/10.1155/2023/5734675

Ajayi IO, Otemuyiwa IO, Adeyanju AA, Falade OS (2021) Vegetable polyphenols inhibit starch digestibility and phenolic availability from composite carbohydrate foods in-vitro. J Agric Food Res 3:100116. https://doi.org/10.1016/j.jafr.2021.100116

Akhter MH, Beg S, Tarique M, Malik A, Afaq S, Choudhry H, Hosawi S (2021) Receptor-based targeting of engineered nanocarrier against solid tumors: recent progress and challenges ahead. Biochim Biophys Acta-General Subjects 1865(2):129777

Akhter MH, Beg S, Tarique M et al (2021) Receptor-based targeting of engineered nanocarrier against solid tumors: recent progress and challenges ahead. Biochimica et Biophysica Acta (BBA) Gen Subj 1865, 2:129777. https://doi.org/10.1016/j.bbagen.2020.129777

Amidon GL, Lennernäs H, Shah VP, Crison JR (1995) A theoretical basis for a biopharmaceutic drug classification: the correlation of in vitro drug product dissolution and in vivo bioavailability. Pharm Res 12(3):413–420. https://doi.org/10.1023/A:1016212804288

Anand P, Kunnumakkara AB, Newman RA, Aggarwal BB (2007) Bioavailability of curcumin: problems and promises. Mol Pharm 4(6):807–818. https://doi.org/10.1021/mp700113r

Anand S, Sowbhagya R, Ansari MA, Alzohairy MA, Alomary MN, Almalik AI, Ahmad W, Tripathi T, Elderdery AY (2022) Polyphenols and their nanoformulations: protective effects against human diseases. Life 12(10):1639. https://doi.org/10.3390/life12101639

Aravind SM, Wichienchot S, Tsao R, Ramakrishnan S, Chakkaravarthi S (2021) Role of dietary polyphenols on gut microbiota, their metabolites and health benefits. Food Res Int 142:110189. https://doi.org/10.1016/j.foodres.2021.110189

Balata GF, Essa EA, Shamardl HA, Zaidan SH, Abourehab MA (2016) Self-emulsifying drug delivery systems as a tool to improve solubility and bioavailability of resveratrol. Drug des, Develop Ther 10:117–128. https://doi.org/10.2147/DDDT.S95905

Banik BL, Fattahi P, Brown JL (2016) Polymeric nanoparticles: the future of nanomedicine. Wiley Interdiscipl Rev Nanomed Nanobiotechnol 8(2):271–299. https://doi.org/10.1002/wnan.1364

Bertelli A, Biagi M, Corsini M, Baini G, Cappellucci G, Miraldi E (2021) Polyphenols: from theory to practice. Foods 10(11):2595. https://doi.org/10.3390/foods10112595

Bhaskar S, Tian F, Stoeger T, Kreyling W, de la Fuente JM, Grazú V, Borm P, Estrada G, Ntziachristos V, Razansky D (2010) Multifunctional Nanocarriers for diagnostics, drug delivery and targeted treatment across blood-brain barrier: perspectives on tracking and neuroimaging. Part Fibre Toxicol 7(1):3. https://doi.org/10.1186/1743-8977-7-3

Brglez Mojzer E, Knez Hrnčič M, Škerget M, Knez Ž, Bren U (2016) Polyphenols: extraction methods, antioxidative action, bioavailability and anticarcinogenic effects. Mol 21(7):901. https://doi.org/10.3390/molecules21070901

Bruckmann FDS, Nunes FB, Salles TDR, Franco C, Cadoná FC, Bohn Rhoden CR (2022) Biological applications of silica-based nanoparticles. Magnetochem 8, 10:131. https://doi.org/10.3390/magnetochemistry8100131

Caiati C, Jirillo E (2025) Cellular and molecular bases for the application of polyphenols in the prevention and treatment of cardiovascular disease. Dis 13(7):221. https://doi.org/10.3390/diseases13070221

Carlotti ME, Sapino S, Ugazio E, Gallarate M, Morel S (2012) Resveratrol in solid lipid nanoparticles. J Disp Sci Technol 33(4):465–471. https://doi.org/10.1080/01932691.2010.548274

Chai Z, Hu X, Wei X, Zhan C, Lu L, Jiang K, Su B, Ruan H, Ran D, Fang RH, Zhang L, Lu W (2017) A facile approach to functionalizing cell membrane-coated nanoparticles with neurotoxin-derived peptide for brain-targeted drug delivery. J Control Release 264:102–111. https://doi.org/10.1016/j.jconrel.2017.08.027

Chaudhari A, Vadodariya P, Vyas A, Patel D, Teli D (2024) Composition of nanoparticles. Nanocarrier vaccines: biopharmaceutics-based fast track development. In: Chavda VP, Apostolopoulos V (eds) Nanocarrier vaccines, 1st edn. Scrivener Publishing LLC, Beverly, MA, pp 25–62. https://doi.org/10.1002/9781394175482.ch2

Chimento A, D'Amico M, De Luca A, Conforti FL, Pezzi V, De Amicis F (2023) Resveratrol, epigallocatechin gallate and curcumin for cancer therapy: challenges from their pro-apoptotic properties. Life 13(2):261. https://doi.org/10.3390/life13020261

Cikrikci S, Mert B, Oztop MH (2018) Development of pH sensitive Alginate/Gum tragacanth based hydrogels for oral insulin delivery. J Agric Food Chem 66(44):11784–11796. https://doi.org/10.1021/acs.jafc.8b02525

Costa GT, Vasconcelos QDJS, Abreu GC, Albuquerque AO, Vilar JL, Aragão GF (2021) Systematic review of the ingestion of fructooligosaccharides on the absorption of minerals and trace elements versus control groups. Clin Nutr ESPEN 41:68–76. https://doi.org/10.1016/j.clnesp.2020.11.007.

Curtis J, Greenberg M, Kester J, Phillips S, Krieger G (2006) Nanotechnology and nanotoxicology. Toxicol Rev 25(4):245–260. https://doi.org/10.2165/00139709-200625040-00005

Das MK, Kalita B (2014) Design and evaluation of phyto-phospholipid complexes (Phytosomes) of rutin for transdermal application. J Appl Pharm Sci 4(10):51–57. https://doi.org/10.7324/JAPS.2014.401010

Deodhar S, Dash AK, North EJ, Hulce M (2020) Development and in vitro evaluation of long circulating liposomes for targeted delivery of gemcitabine and irinotecan in pancreatic ductal adenocarcinoma. AAPS Pharm Sci Tech 21(6):231. https://doi.org/10.1208/s12249-020-01745-6

Desai J, Thakkar H (2016) Effect of particle size on oral bioavailability of darunavir-loaded solid lipid nanoparticles. J Microencapsul 33(7):669–678. https://doi.org/10.1080/02652048.2016.1245363

Di Lorenzo C, Colombo F, Biella S, Stockley C, Restani P (2021) Polyphenols and Human health: the role of bioavailability. Nutr 13(1):273. https://doi.org/10.3390/nu13010273

Egbuna C, Găman MA, Jeevanandam J (eds) (2022) Applications of nanotechnology in drug discovery and delivery. Elsevier, Amsterdam, Oxford, and Cambridge

El Monfalouti H, Kartah BE (2024) Enhancing polyphenol bioavailability through nanotechnology: current trends and challenges. In: Gouvinhas I, Novo Barros A (eds) Biochemistry, vol 66. IntechOpen, London. https://doi.org/10.5772/intechopen.1005764

Fahr A, Van Hoogevest P, May S, Bergstrand N, Leigh MLS (2005) Transfer of lipophilic drugs between liposomal membranes and biological interfaces: consequences for drug delivery. Eur J Pharm Sci 26(3–4):251–265. https://doi.org/10.1016/j.ejps.2005.05.012

Fang RH, Hu CMJ, Chen KN, Luk BT, Carpenter CW, Gao W, Lid S, Zhang DE, Lue W, Zhang L (2013) Lipid-insertion enables targeting functionalization of erythrocyte membrane-cloaked nanoparticles. Nanoscale 5(19):8884–8888. https://doi.org/10.1039/c3nr03064d

Farabegoli F, Pinheiro M (2022) Epigallocatechin-3-Gallate delivery in lipid-based nanoparticles: potentiality and perspectives for future applications in cancer chemoprevention and therapy. Front Pharmacol 13:809706. https://doi.org/10.3389/fphar.2022.809706

Farhan M, Rizvi A, Aatif M, Ahmad A (2023) Current understanding of flavonoids in cancer therapy and prevention. Metabol 13(4):481. https://doi.org/10.3390/metabo13040481

Farheen Akhter MH, Chitme H, Suliman M, Jaremko M, Emwas A (2023) Surface-modified biobased polymeric nanoparticles for dual delivery of doxorubicin and gefitinib in glioma cell lines. ACS Omega 8(31):28165–28184. https://doi.org/10.1021/acsomega.3c01375

FDA (2022) Drug Products, Including Biological Products, that Contain Nanomaterials—Guidance for Industry, April 21. Food and Drug Administration (FDA), Center for Drug Evaluation and Research, Washington, D.C. https://www.fda.gov/regulatory-information/search-fda-guidance-documents/drug-products-including-biological-products-contain-nanomaterials-guidance-industry. Accessed 05 Jan 2026

Feng SS (2006) New-concept chemotherapy by nanoparticles of biodegradable polymers: where are we now? Nanomed 1(3):297–309. https://doi.org/10.2217/17435889.1.3.297

Ghaeini Hesarooeyeh Z, Basham A, Sheybani-Arani M, Abbaszadeh M, Salimi Asl A, Moghbeli M, Saburi E (2024) Effect of resveratrol and curcumin and the potential synergism on hypertension: a mini-review of human and animal model studies. Phytother Res 38(1):42–58. https://doi.org/10.1002/ptr.8023

Gregoriadis G, Florence AT (1993) Liposomes in drug delivery: clinical, diagnostic and ophthalmic potential. Drugs 45(1):15–28. https://doi.org/10.2165/00003495-199345010-00003

Greil R, Greil-Ressler S, Weiss L, Schönlieb C, Magnes T, Radl B, Bolger GT, Vcelar B, Sordillo PP (2018) A phase 1 dose-escalation study on the safety, tolerability and activity of liposomal curcumin (Lipocurc™) in patients with locally advanced or metastatic cancer. Cancer Chemother Pharmacol 82(4):695–706. https://doi.org/10.1007/s00280-018-3654-0

Guo Y, Sun Q, Wu F, Dai Y, Chen X (2021) Polyphenol-Containing nanoparticles: synthesis, properties, and therapeutic delivery. Adv Mat 33(22):e2007356. https://doi.org/10.1002/adma.202007356

Hoffman AS (2008) The origins and evolution of 'Controlled' drug delivery systems. J Control Release 132(3):153–163. https://doi.org/10.1016/j.jconrel.2008.08.012

Imam HT, Marr PC, Marr AC (2021) Enzyme entrapment, biocatalyst immobilization without covalent attachment. Green Chem 23(14):4980–5005. https://doi.org/10.1039/D1GC01852C

Ishkeh SR, Shirzad H, Asghari M, Alirezalu A, Pateiro M, Lorenzo JM (2021) Effect of chitosan nanoemulsion on enhancing the phytochemical contents, health-promoting components, and shelf life of raspberry (Rubus sanctus Schreber). Appl Sci 11(5):2224. https://doi.org/10.3390/app11052224

Jacob S, Kather FS, Morsy MA, Boddu SH, Attimarad M, Shah J, Shinu P, Nair AB (2024) Advances in nanocarrier systems for overcoming formulation challenges of curcumin: current insights. Nanomater 14(8):672. https://doi.org/10.3390/nano14080672

Jovanović AA, Balanč B, Petrović PM, Čutović N, Marković SB, Djordjević VB, Bugarski BM (2025) Liposome-based encapsulation of extract from wild thyme (Thymus serpyllum L.) tea processing residues for delivery of polyphenols. Foods 14, 15:2626. https://doi.org/10.3390/foods14152626

Kalušević AM, Lević SM, Čalija BR, Milić JR, Pavlović VB, Bugarski BM, Nedović VA (2017) Effects of different carrier materials on physicochemical properties of microencapsulated grape skin extract. J Food Sci Technol 54(11):3411–3420. https://doi.org/10.1007/s13197-017-2790-6

Kardani SL (2024) Nanocarrier-based formulations: regulatory challenges, ethical and safety considerations in pharmaceuticals. As J Pharm 18(2):346–351. https://doi.org/10.22377/ajp.v18i02.5444

Karnwal A, Sharma V, Kumar G, Jassim AY, Dohroo A, Sivanesan I (2024) Transforming medicine with nanobiotechnology: nanocarriers and their biomedical applications. Pharm 16(9):1114. https://doi.org/10.3390/pharmaceutics16091114

Kashyap PC, Chia N, Nelson H, Segal E, Elinav E (2017) Microbiome at the frontier of personalized medicine. Mayo Clin Proc 92(12):1855–1864. https://doi.org/10.1016/j.mayocp.2017.10.004

Kaur H, Kaur G (2014) A critical appraisal of solubility enhancement techniques of polyphenols. J Pharm 2014:1–14. https://doi.org/10.1155/2014/180845

Kciuk M, Alam M, Ali N, Rashid S, Głowacka P, Sundaraj R, Celik I, Yahya EB, Dubey A, Zerroug E, Kontek R (2023) Epigallocatechin-3-gallate therapeutic potential in cancer: mechanism of action and clinical implications. Mol 28(13):5246. https://doi.org/10.3390/molecules28135246

Khan J, Rudrapal M, Bhat EA, Ali A, Alaidarous M, Alshehri B, Banwas S, Ismail R, Egbuna C (2021) Perspective insights to bio-nanomaterials for the treatment of neurological disorders. Front Bioeng Biotechnol 9:724158. https://doi.org/10.3389/fbioe.2021.724158

Kim KH, Ki M, Min KH, Pack SP (2023) Advanced delivery system of polyphenols for effective cancer prevention and therapy. Antioxid 12(5):1048. https://doi.org/10.3390/antiox12051048

Kipp J (2004) The role of solid nanoparticle technology in the parenteral delivery of poorly water-soluble drugs. Int J Pharm 284(1–2):109–122. https://doi.org/10.1016/j.ijpharm.2004.07.019

Kreuter J (2007) Nanoparticles—a historical perspective. Int J Pharm 331(1):1–10. https://doi.org/10.1016/j.ijpharm.2006.10.021

Kumar D, Vats N, Saroha K, Rana AC (2020) Phytosomes as emerging nanotechnology for herbal drug delivery. In: Saneja A, Panda AK, Lichtfouse E (eds) Sustainable agriculture reviews, vol. 43. Springer International Publishing, Cham, pp 217–237. https://doi.org/10.1007/978-3-030-41838-0_7

Kumari A, Yadav SK, Pakade YB, Singh B, Yadav SC (2010) Development of biodegradable nanoparticles for delivery of quercetin. Coll Surf B Biointerfac 80(2):184–192. https://doi.org/10.1016/j.colsurfb.2010.06.002

Kurul F, Turkmen H, Cetin AE, Topkaya SN (2025) Nanomedicine: how nanomaterials are transforming drug delivery, bio-imaging, and diagnosis. Next Nanotechnol 7:100129. https://doi.org/10.1016/j.nxnano.2024.100129

Lee Y, Chiang Y, Chen M, Chiang P (2025) Micronization combined ultrasound-assisted extraction enhances the sustainability of polyphenols from pineapple and lemon peels utilizing acidified ethanol. Foods 14(16):2872. https://doi.org/10.3390/foods14162872

Li B, Harich K, Wegiel L, Taylor LS, Edgar KJ (2012a) Stability and solubility enhancement of ellagic acid in cellulose ester solid dispersions. Carbohydr Polym 92(2):1443–1450. https://doi.org/10.1016/j.carbpol.2012.10.051

Li Y, He H, Jia X, Lu W, Lou J, Wei Y (2012b) A dual-targeting nanocarrier based on poly(amidoamine) dendrimers conjugated with transferrin and tamoxifen for treating brain gliomas. Biomater 33(15):3899–3908. https://doi.org/10.1016/j.biomaterials.2012.02.004

Li C, Feng J, Huang WY, An XT (2013) Composition of polyphenols and antioxidant activity of rabbiteye blueberry (*Vaccinium ashei*) in Nanjing. J Agric Food Chem 61(3):523–531. https://doi.org/10.1021/jf3046158

Li S, Li F, Wang Y, Li W, Wu J, Hu X, Tang T, Liu X (2024) Multiple delivery strategies of nanocarriers for myocardial ischemia-reperfusion injury: current strategies and future prospective. Drug Deliv 31(1):2298514. https://doi.org/10.1080/10717544.2023.2298514

Lin D, Xiao M, Zhao J, Li Z, Xing B, Li X, Kong M, Li L, Zhang Q, Liu Y, Chen H, Qin W, Wu H, Chen S (2016) An overview of plant phenolic compounds and their importance in human nutrition and management of type 2 diabetes. Mol 21(10):1374. https://doi.org/10.3390/molecules21101374

Liu X, He F, Liu M (2024) New opportunities of stimulus-responsive smart nanocarriers in cancer therapy. Nano Mater Sci. in press. https://doi.org/10.1016/j.nanoms.2024.10.013

Loureiro JA, Andrade S, Duarte A, Neves AR, Queiroz JF, Nunes C, Sevin E, Fenart L, Gosselet F, Coelho MAN, Pereira MC (2017) Resveratrol and grape extract-loaded solid lipid nanoparticles

for the treatment of Alzheimer's disease. Mol 22(2):277. https://doi.org/10.3390/molecules22020277
Lu W, Tan Y, Hu K, Jiang X (2005) Cationic albumin conjugated pegylated nanoparticle with its transcytosis ability and little toxicity against blood–brain barrier. Int J Pharm 295(1–2):247–260. https://doi.org/10.1016/j.ijpharm.2005.01.043
Lu X, Wu D, Li Z, Chen G (2011) Polymer nanoparticles. Prog Mol Biol Transl Sci 104:299–323. https://doi.org/10.1016/B978-0-12-416020-0.00007-3
Mansour A, Rahimi HR, Gerami H, Khorasanian AS, Esmaeili AH, Amrollahi Bioky A, Mohajeri-Tehrani MR, Fateh HR, Sajjadi-Jazi SM (2025) The effectiveness and safety of nanocurcumin supplementation for diabetic peripheral neuropathy in patients with type 2 diabetes: a randomized double-blind clinical trial. Nutr J 24(1):115. https://doi.org/10.1186/s12937-025-01184-8
Martínez-Cuazitl A, Gómez-García MDC, Pérez-Mora S, Rojas-López M, Delgado-Macuil RJ, Ocampo-López J, Vázquez-Zapién GJ, Mata-Miranda MM, Pérez-Ishiwara DG (2023) Polyphenolic compounds nanostructurated with gold nanoparticles enhance wound repair. Int J Mol Sci 24(24):17138. https://doi.org/10.3390/ijms242417138
Maurer N, Fenske DB, Cullis PR (2001) Developments in liposomal drug delivery systems. Expert Opin Biol Ther 1(6):923–947. https://doi.org/10.1517/14712598.1.6.923
Meena J, Gupta A, Ahuja R, Singh M, Bhaskar S, Panda AK (2020) Inorganic nanoparticles for natural product delivery: a review. Environ Chem Lett 18(6):2107–2118. https://doi.org/10.1007/s10311-020-01061-2
Mignet N, Seguin J, Chabot G (2013) Bioavailability of polyphenol liposomes: a challenge ahead. Pharm 5(3):457–471. https://doi.org/10.3390/pharmaceutics5030457
Milinčić DD, Popović DA, Lević SM, Kostić AŽ, Tešić ŽL, Nedović VA, Pešić MB (2019) Application of polyphenol-loaded nanoparticles in food industry. Nanomat 9(11):1629. https://doi.org/10.3390/nano9111629
Morzel M, Canon F, Guyot S (2022) Interactions between salivary proteins and dietary polyphenols: potential consequences on gastrointestinal digestive events. J Agric Food Chem 70(21):6317–6327. https://doi.org/10.1021/acs.jafc.2c01183
Mourtas S, Lazar AN, Markoutsa E, Duyckaerts C, Antimisiaris SG (2014) Multifunctional nanoliposomes with curcumin–lipid derivative and brain targeting functionality with potential applications for Alzheimer disease. Eur J Med Chem 80:175–183. https://doi.org/10.1016/j.ejmech.2014.04.050
Nedovic V, Kalusevic A, Manojlovic V, Levic S, Bugarski B (2011) An overview of encapsulation technologies for food applications. Procedia Food Sci 1:1806–1815. https://doi.org/10.1016/j.profoo.2011.09.265
Ntuli S, Leuschner M, Bester MJ, Serem JC (2022) Stability, morphology, and effects of in vitro digestion on the antioxidant properties of polyphenol inclusion complexes with β-Cyclodextrin. Mol 27(12):3808. https://doi.org/10.3390/molecules27123808
Oliver S, Vittorio O, Cirillo G, Boyer C (2016) Enhancing the therapeutic effects of polyphenols with macromolecules. Polym Chem 7(8):1529–1544. https://doi.org/10.1039/C5PY01912E
Paliwal R, Babu RJ, Palakurthi S (2014) Nanomedicine scale-up technologies: feasibilities and challenges. AAPS Pharm Sci Tech 15(6):1527–1534. https://doi.org/10.1208/s12249-014-0177-9
Payne A, Nahashon S, Taka E, Adinew GM, Soliman KFA (2022) Epigallocatechin-3-Gallate (EGCG): new therapeutic perspectives for neuroprotection, aging, and neuroinflammation for the modern age. Biomol 12(3):371. https://doi.org/10.3390/biom12030371
Ponnuswami A (2024) Exploring the role of nanocarriers in enhancing drug bioavailability at the nanoscale. J Nanomed Biotherap Discov 14(5):272. https://doi.org/10.4172/2155-983X.24.14.272
Qi X, Yuan Y, Zhang J, Bulte JWM, Dong W (2018) Oral administration of Salecan-based hydrogels for controlled insulin delivery. J Agric Food Chem 66(40):10479–10489. https://doi.org/10.1021/acs.jafc.8b02879

Quintanar-Guerrero D, Allémann E, Fessi H, Doelker E (1998) Preparation techniques and mechanisms of formation of biodegradable nanoparticles from preformed polymers. Drug Develop Ind Pharm 24(12):1113–1128. https://doi.org/10.3109/03639049809108571

Radeva L, Stefanova D, Yordanov Y, Kamenova K, Petrov PD, Marinova MK, Simeonov SP, Kondeva-Burdina M, Tzankova V, Yoncheva K (2023) Incorporation of resveratrol in polymeric nanogel for improvement of its protective effects on cellular and microsomal oxidative stress models. Gels 9(6):450. https://doi.org/10.3390/gels9060450

Radeva L, Yordanov Y, Spassova I, Kovacheva D, Tzankova V, Yoncheva K (2024) Double encapsulation of resveratrol and doxorubicin in composite Nanogel—an opportunity to reduce cardio- and neurotoxicity of doxorubicin. Gels 10(11):699. https://doi.org/10.3390/gels10110699

Rahimi HR, Mohammadpour AH, Dastani M, Jaafari MR, Abnous K, Mobarhan MG, Oskuee RK (2016) The effect of nano-curcumin on HbA1c, fasting blood glucose, and lipid profile in diabetic subjects: a randomized clinical trial. Avicenna J Phytomed 6(5):567–577

Rauta PR, Mohanta YK, Nayak D (eds) (2019) Nanotechnology in biology and medicine: research advancements and future perspectives. CRC Press, Boca Raton

Ricordel C, Labalette-Tiercin M, Lespagnol A, Kerjouan M, Dugast C, Mosser J, Desrues B, Léna H (2015) EFGR-mutant lung adenocarcinoma and Li-Fraumeni syndrome: report of two cases and review of the literature. Lung Cancer 87(1):80–84. https://doi.org/10.1016/j.lungcan.2014.11.005

Rostami M, Yousefi M, Khezerlou A, Mohammadi MA, Jafari SM (2019) Application of different biopolymers for nanoencapsulation of antioxidants via electrohydrodynamic processes. Food Hydrocoll 97:105170. https://doi.org/10.1016/j.foodhyd.2019.06.015

Rudrapal M, Mishra AK, Rani L, Sarwa KK, Zothantluanga JH, Khan J, Kamal M, Palai S, Bendale AR, Talele SG, Pathan VT9, Borse LB, Neharkar VS, Sahoo PK (2022) Nanodelivery of dietary polyphenols for therapeutic applications. Mol 27, 24:8706. https://doi.org/10.3390/molecules27248706

Sainz V, Conniot J, Matos AI, Peres C, Zupanǒiǒ E, Moura L, Silva LC, Florindo HF, Gaspar RS (2015) Regulatory aspects on nanomedicines. Biochem Biophys Res Commun 468(3):504–510. https://doi.org/10.1016/j.bbrc.2015.08.023

Sharma Y, Bhardwaj R, Kaur A, Singh G, Kulkarni S, Bhatia A, Bala K (2024) Nano-gel formulation of polyphenolic fraction of tobacco stem for wound healing and its inhibitory efficacies against the receptors of chronic wound development. J Biomol Struct Dyn 43(15):8280–8294. https://doi.org/10.1080/07391102.2024.2317986

Singh S, Nalwa HS (2007) Nanotechnology and Health safety—toxicity and risk assessments of nanostructured materials on human health. J Nanosci Nanotechnol 7(9):3048–3070. https://doi.org/10.1166/jnn.2007.922

Squillaro T, Cimini A, Peluso G, Giordano A, Melone M (2018) Nano-delivery systems for encapsulation of dietary polyphenols: an experimental approach for neurodegenerative diseases and brain tumors. Biochem Pharmacol 154:303–317. https://doi.org/10.1016/j.bcp.2018.05.016

Tabrez S, Jabir NR, Adhami VM, Khan MI, Moulay M, Kamal MA, Mukhtar H (2020) Nanoencapsulated dietary polyphenols for cancer prevention and treatment: successes and challenges. Nanomed 15(11):1147–1162. https://doi.org/10.2217/nnm-2019-0398

Vasić K, Knez Ž, Leitgeb M (2024) Multifunctional iron oxide nanoparticles as promising magnetic biomaterials in drug delivery: a review. J Funct Biomater 15(8):227. https://doi.org/10.3390/jfb15080227

Vega-Villa KR, Takemoto JK, Yáñez JA, Remsberg CM, Forrest ML, Davies NM (2008) Clinical toxicities of nanocarrier systems. Adv Drug Deliv Rev 60(8):929–938. https://doi.org/10.1016/j.addr.2007.11.007

Wagner A, Vorauer-Uhl K (2011) Liposome technology for industrial purposes. J Drug Deliv 2011(1):591325. https://doi.org/10.1155/2011/591325

Wang Z, Liu J, Zhao N, Li C, Lv S, Hu Y, Lv H, Wang D, Wang S (2020) Cancer cell macrophage membrane camouflaged persistent luminescent nanoparticles for Imaging-Guided photothermal therapy of colorectal cancer. ACS Appl Nano Mat 3(7):7105–7118. https://doi.org/10.1021/acsanm.0c01433

Yang B, Dong Y, Wang F, Zhang Y (2020) Nanoformulations to enhance the bioavailability and physiological functions of polyphenols. Mol 25(20):4613. https://doi.org/10.3390/molecules25204613

Xiong J, Lipsitz O, Nasri F, Lui LMW, Gill H, Phan L, Chen-Li D, Iacobucci M, Ho R, Majeed A, McIntyre RS (2020) Impact of COVID-19 pandemic on mental health in the general population: a systematic review. J Affect Disord 277:55–64. https://doi.org/10.1016/j.jad.2020.08.001

Yang M, He Y, Ni Q, Zhou M, Chen H, Li G, Yu J, Wu X, Zhang X (2024) Polyphenolic nanomedicine regulating mitochondria REDOX for innovative cancer treatment. Pharm 16(8):972. https://doi.org/10.3390/pharmaceutics16080972

Yeh M, Chang NH, Cheng NM (2011) Clinical development of liposome based drugs: formulation, characterization, and therapeutic efficacy. Int J Nanomed 7:49. https://doi.org/10.2147/IJN.S26766

Zhang S, Xu M, Zhang W, Liu C, Chen S (2021) Natural polyphenols in metabolic Syndrome: protective mechanisms and clinical applications. Int J Mol Sci 22(11):6110. https://doi.org/10.3390/ijms22116110

Zhao Y, Huan M, Liu M, Cheng Y, Sun Y, Cui H, Liu D, Mei Q, Zhou S (2016) Doxorubicin and resveratrol co-delivery nanoparticle to overcome doxorubicin resistance. Sci Rep 6(1):35267. https://doi.org/10.1038/srep35267

Ziental D, Czarczynska-Goslinska B, Mlynarczyk DT, Glowacka-Sobotta A, Stanisz B, Goslinski T, Sobotta L (2020) Titanium dioxide nanoparticles: prospects and applications in medicine. Nanomat 10(2):387. https://doi.org/10.3390/nano10020387

# Chapter 8
# Emerging Research on Marine Polyphenols. Unique Properties, Health Benefits, and Future Research Direction

**Viswapriya Viswalingam and Dileep Kumar**

**Abstract** Marine polyphenols refer to a distinct category of bioactive molecules that have drawn on to them growing interest as a source of therapy in human health. Opposite to the terrestrial polyphenols, marine polyphenols are produced by algae, fish, crustaceans, and other organisms that inhabit the sea and have their own chemical structures due to the harsh conditions of the sea. This review initially discusses the importance of marine polyphenols and key natural sources of polyphenols with special reference to microalgae, macroalgae, fish, and marine invertebrates. Health-promoting properties of marine polyphenols are critically evaluated with reference to their antioxidant, anti-inflammatory, antimicrobial, antidiabetic, cardioprotective, neuroprotective, and anticancer properties, mechanisms of action, and reasons of that activity including redox regulation, enzyme-inhibitory, along with cell-signaling and modulatory activity. With reference to the future of marine polyphenols applications, the use of these compounds is already present in the industry (nutraceuticals, pharmaceuticals, cosmetic, and functional foods). It is necessary to explore the existing applications of polyphenols in the future. The issues of extraction, purification, stability, bioavailability, safety, and large-scale production are mentioned in order to present the picture of related challenges in a realistic way regarding their translational prospects. Future directions underline the concept of sustainable sourcing, improved technologies in extractions and delivery, and the significance of stringent preclinical and clinical validation.

**Keyword** Algae · Marine fungus · Marine polyphenol · Safety · Sea urchin · Shellfish · Stability

V. Viswalingam
Department of Pharmaceutical Chemistry, Poona College of Pharmacy, Bharati Vidyapeeth (Deemed to Be University), Pune, Maharashtra, India

D. Kumar (✉)
Department of Pharm Chemistry, Manipal College of Pharmaceutical Sciences, Manipal Academy of Higher Education, Manipal, Karnataka, India
e-mail: dileep.kumar@manipal.edu



R. K. Singla et al. (eds.), *Polyphenols in Health*,
https://doi.org/10.1007/978-3-032-27324-6_8

## 8.1 Introduction

Polyphenols are a large group of naturally occurring bioactive compounds that have gained significant attention due to their protective roles against a wide range of chronic diseases, including hyperglycemia, hyperlipidemia, cardiovascular disorders, neurodegenerative diseases, and cancer (Pandey and Rizvi 2009; Yu et al. 2025). Traditionally, polyphenols have been extensively studied in terrestrial plants, particularly fruits and vegetables, and are commonly classified based on their chemical structure into flavonoids, phenolic acids, lignans, and related subclasses (Ciupei et al. 2024; Zhang et al. 2022a, b). These compounds are widely recognized for their antioxidant, anti-inflammatory, and anticancer properties. For these reasons, they have been incorporated into functional foods and nutraceutical formulations aimed at disease prevention and health promotion.

Despite their proven biological activities, the practical application of plant-derived polyphenols faces important limitations, particularly related to poor water solubility, low oral bioavailability, and rapid metabolism (Aatif 2023; Rumman and Khan 2026). These challenges significantly restrict their effectiveness when administered through conventional dietary or pharmaceutical routes. To overcome these barriers, advanced strategies such as food-grade and bio-based nanocarriers have been explored to enhance polyphenol stability, protection, and delivery, thereby improving their absorption and biological efficacy (Ghosh et al. 2025; Yang et al. 2020a, b).

With developing technology, the marine environment has emerged as a promising and largely untapped reservoir of bioactive compounds. Covering more than 70% of the Earth's surface, marine ecosystems represent the largest and most diverse biological system on the planet (Karthikeyan et al. 2022; Simon et al. 2025). These environments are characterized by extreme physicochemical conditions, including high salinity, low temperatures, limited light availability, and elevated pressure. Such harsh conditions drive marine organisms to produce structurally unique secondary metabolites, including polyphenols, as adaptive survival strategies (Donato et al. 2018).

Marine polyphenols have recently attracted growing scientific interest as natural alternatives for disease prevention and treatment. These compounds exhibit a broad spectrum of biological activities, including antioxidant, anti-inflammatory, antidiabetic, neuroprotective, and antitumor effects (Hossain et al. 2025; Salehi and Rashidinejad 2025). Compared with terrestrial polyphenols, marine polyphenols often possess unique chemical structures able to offer enhanced bioactivity or novel mechanisms of action. Their multifunctional nature positions them as promising candidates for applications in health, nutrition, and medicine (Sanjeewa et al. 2016).

Algae represent the primary and most extensively studied source of marine polyphenols. Both microalgae and macroalgae synthesize a wide range of phenolic compounds, including flavonoids, phenolics, organic acids, and phlorotannins (Menaa et al. 2021). In addition to algae, marine animals such as fish and crustaceans contribute to dietary polyphenol intake through compounds such as catechins and phenolic acids, either synthesized endogenously or accumulated through the marine food chain (de Oliveira et al. 2025; Gaglione et al. 2019). These diverse sources expand the availability and applicability of marine polyphenols across different sectors.

Many researchers have shown that marine polyphenols play an important role in the prevention of several major health conditions (Vauzour et al. 2010). These compounds reduce oxidative stress and chronic inflammation, thereby lowering the risk of cardiovascular diseases and improving vascular function. Marine polyphenols have also demonstrated antidiabetic properties by regulating glucose metabolism and enhancing insulin sensitivity. Furthermore, their protecting neural effects, mediated through antioxidant and anti-inflammatory pathways, suggest potential benefits in preventing or slowing the progression of neurodegenerative disorders (Aryaeian et al. 2017). In oncology research, marine polyphenols have shown promise by inhibiting tumor growth, inducing apoptosis, and modulating cancer-related signaling pathways.

The growing recognition of these health benefits has fueled interest in the utilization of marine polyphenols in nutraceutical, pharmaceutical, and functional food products (Rosales and Fabi 2023; Zhang et al. 2022a, b). However, despite encouraging preclinical evidence, significant gaps remain in understanding their safety, optimal dosage, long-term efficacy, and pharmacokinetic behavior in humans (Davies et al. 2020; Ettlin et al. 2010). Challenges related to extraction, purification, stability, bioavailability, and large-scale production must also be addressed to support industrial and clinical translation (Nair et al. 2026; Petrovic et al. 2024).

This review aims to provide a comprehensive overview of marine polyphenols, focusing on their sources, health benefits, and applications across existing industries. It further explores their mechanisms of action, safety considerations, and the challenges associated with their development, including extraction and purification strategies. By adopting a circular and integrative approach, this review highlights the future prospects of marine polyphenols as novel therapeutic and nutraceutical agents for the prevention and management of cardiovascular diseases, diabetes, neurodegenerative disorders, and cancer.

## 8.2 Natural Reservoirs of Polyphenols in Marine Organisms

Marine polyphenols are naturally found in a broad variety of marine life, including algae, fish, crustaceans, and mollusks (Wijethunga et al. 2025). Of all these sources, marine algae are the most significant and widely investigated contributors of polyphenolic compounds. This section is based on the key polyphenols marine sources and the most important bioactive compounds related to them.

### 8.2.1 Sea Urchins

Sea urchin is becoming a resource of significance in the marine bioresource as it contains a large number of bioactive compounds with potential therapeutic benefits in aging and age-related conditions. Carotenoids, polyhydroxynaphthoquinones (PHNQ), and flavonoids are some of the metabolites with high antioxidant and anti-inflammatory properties and evidence of positive effects on cell viability and protection of cells as evidenced in Fig. 8.1 (Escobar-Doncel et al. 2026). Laboratory-based and emerging clinical research has produced experimental evidence that these compounds can exhibit protective effects against age-related associated conditions such as cardiovascular diseases, neurodegenerative, ocular, dermatological, and post-menopausal health disorders (Bjørklund et al. 2022; Morkovin et al. 2024). Besides containing bioactive metabolites, sea urchins are also highly nutritional sources of proteins, essential minerals, and trace elements that make their further usefulness as a functional marine food and therapeutic resource (Suleria et al. 2015; Zhang et al. 2022a, b). PHNQ is unique bioactive pigments that are found in sea urchins and they have received a lot of attention due to their great antioxidant effects (Hou et al. 2020). They are mainly found and are mostly concentrated in shells and gonads; in addition, these molecules are functionally related to polyphenolic antioxidants in plants found on the land (Aneklaphakij et al. 2021; Lourenço et al. 2019). PHNQ has been proven to be very effective as lipid peroxidation inhibitors; moreover, they can preserve biological membranes and food systems against oxidative degradation (Melotti et al. 2023). Their polyhydroxylated quinone structures are thought to be one of the factors that make them effective antioxidant agents (Gulcin 2025). Nevertheless, due to low extraction yield and its brown or orange color, PHNQ has restricted practical applications due to the lack of ability to use them in some biomedical and nutraceutical preparations (Apak et al. 2016; Morkovin et al. 2024). Despite these difficulties, the special chemical structure of PHNQ promises to be of great biomedical use. The quinone frameworks enable them to undergo reversible oxidation–reduction reactions, thus stability is of particular concern when speaking of future developments (Cores et al. 2023; Li et al. 2025). The limitations of instability and bioavailability could be overcome with structural changes or formulation plans. In addition, the structural variability of PHNQ and other compounds in the sea urchin group may open up new bioactivities, including antioxidant, anti-inflammatory, and cytoprotective effects as well as signaling-modulating activities (Ghelani et al. 2022; Zhao et al. 2023). Their molecular mechanisms of action need to be clarified. Moreover, differences in bioactivity species-specific need to be evaluated, the extraction and stabilization have to be effectively optimized, and well-designed preclinical and clinical research should be done (Adelusi et al. 2022). Such initiatives will be critical in the complete realization of the translational capability of the sea urchin-derived compounds in promoting health and prevention of aging-related diseases.

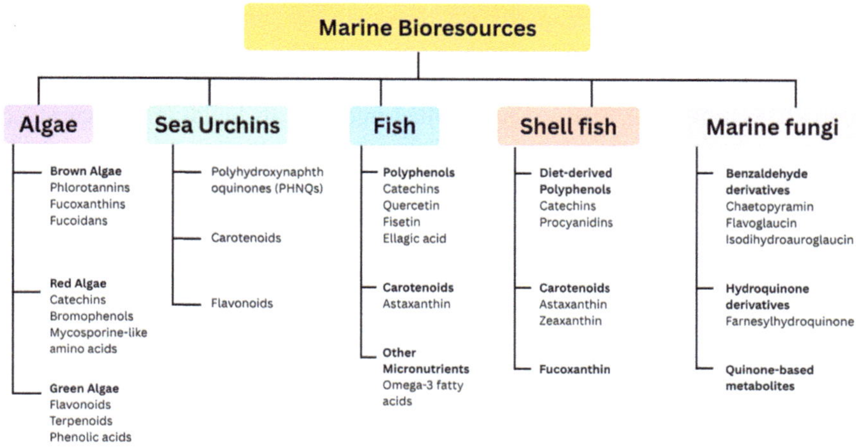


**Fig. 8.1** Schematic representation of major marine sources of polyphenols and associated bioactive compounds. This picture has been created by the authors

## *8.2.2 Fish*

Marine polyphenols and other insignificant bioactive nutrients in marine life forms are found in large quantities in fish, especially fatty fish, including tuna (*Thunnus orientalis*), salmon (*Salmo salar*), and sardines (*Sardina pilchardus*). Catechins, phenolic acids, and carotenoids are polyphenols present also in fish (Durazzo et al. 2022; Noreen et al. 2025).

Catechins are flavonoids that are well known for their antioxidant and anti-inflammatory properties, and phenolic acids are ubiquitous bioactive compounds with similar biological actions (Fan et al. 2017; Mutha et al. 2021).

Carotenoids, including astaxanthin, are natural pigments found in certain types of fish, and these molecules have been reported to exhibit intense antioxidant and anti-inflammatory properties.

Curcumin is a natural polyphenol which is found in certain fish species including tambaqui (*Colossoma macropomum*) (de Carvalho and Caramujo 2017; Nakano and Wiegertjes 2020). The most common yellow color in turmeric is ascribed to curcumin, which has been extensively researched due to its great antioxidant and anti-inflammatory effects. Scientific evidence depicts that curcumin can prevent or treat several inflammatory disorders, such as arthritis, inflammatory bowel disease, and cardiovascular disorders. It has also been shown to decrease the level of cholesterol in blood hence decreasing the chances of heart diseases (Yadav et al. 2025). Generally, curcumin is thought to be safe in moderate doses but its bioavailability is not very high and this fact may restrict its use in some form of application.

Catechins are a category of polyphenols that are heavily documented to have antioxidant and anti-inflammatory effects and are present in a variety of foodstuffs such as fish like tuna and salmon (Cione et al. 2019; Sharifi-Rad et al. 2020). These substances can be used to contrast inflammatory substances, known as free radicals, which are unstable molecules formed in the organism under the influence of stress factors, environmental influence and during metabolic activities (Chandimali et al. 2025; Lobo et al. 2010). Uncontrolled free radicals build-up may cause cell damage and promote the occurrence of chronic illnesses such as cancer, cardiovascular malady, and neurodegenerative diseases. Besides antioxidant effect, catechins also have anti-inflammatory effects preventing long-term inflammation and regulating immune response (Jówko 2015; Kim and Heo 2022; Truong and Jeong 2022). Catechins have been shown to have anticancer effects by inhibiting the proliferation of tumor cell growth and angiogenesis which is a process vital in tumor growth. However, many studies concerning catechins have been conducted in reference to plant sources and related products like green tea. Dietary supplementation with catechin-containing fish may be the supplementary and complementary source of these bioactive components (Botten et al. 2015; Ferrari et al. 2025). Quercetin is a flavonol that is also included in the flavonoid category and it is highly abundant in plant foods (fruits, vegetables, and herbs). It has also been found in some fish species such as salmon and trout (Vollmannová et al. 2024). Antioxidant and anti-inflammatory effects of quercetin are considered strong and have led to the highest study of this flavonoid. This molecule acts by eliminating the free radicals and hence safeguarding the cells against oxidative damages that can result in the onset of chronic disease (Rajesh and Dhanaraj 2023; Yang et al. 2020a, b). Quercetin also promotes cardiovascular health by decreasing low-density lipoprotein cholesterol, elevating high-density lipoprotein cholesterol, decreasing blood pressure, and defending vascular tissues against oxidative stress. In addition, in vitro and in vivo research indicates that quercetin can also prevent cancer cell proliferation, angiogenesis, and improve the action of standard cancer treatment (Alharbi et al. 2025).

Another naturally occurring phenolic acid belonging under the fruit and vegetable category—though it has also been detected in some fish species—is ellagic acid. Reported studies indicate that ellagic acid can have anti-canceric effect and it could be used in the prevention and treatment of different cancer types. Moreover, ellagic acid is also an anti-inflammatory and antimicrobial agent (Sharifi-Rad et al. 2020; Wang et al. 2024). Ellagic acid has been found in the muscles of fish like salmon and rainbow trout (*Oncorhynchus mykiss*), but in low levels, and usually not as high as in plants. Consequently, fish consumption does not play a pivotal role in the totality of ellagic acid intake, while a varied diet can certainly play a vital role in achieving the right intake on the amount of this substance.

Fisetin is a flavonoid which is present in fruits and vegetables, such as strawberries, grapes, apples, onions, and cucumbers. It has also been reported in some fish species including salmon (Khan et al. 2013; Mirza et al. 2025). Fisetin has been shown to possess excellent antioxidant, anti-inflammatory, neuroprotective, and cardioprotective effects. As a result of mitigating oxidative stress, fisetin helps in prevention of chronic illnesses such as cancer, cardiovascular diseases, and neuro-degenerative diseases. Research indicates that fisetin can be used to enhance the health of the cardiovascular system by increasing blood circulation, decreasing inflammation of arteries, and lowering blood pressure (Hassan et al. 2022). Moreover, fisetin has been demonstrated to possess anticancer effects through growth and invasion of different types of cancer cells such as breast, prostate, and colon cancer cells. Experimental literature also shows that fisetin has the potential to improve brain health because it can improve cognitive ability and memory.

### *8.2.3 Shellfish*

Shellfish—such as shrimps, clams, oysters, other bivalves, and crustaceans—also constitutes another source of marine bioactive compounds with possible health-benefit potential. These organisms are also rich in polyphenols and other small nutrients, and carotenoids are some of the most noticeable bioactive substances in them (Nag et al. 2022; Ngandjui et al. 2024). Carotenoids (astaxanthin and zeaxanthin) that are found in shellfish tissues are highly potent antioxidants and anti-inflammatory agents. These substances facilitate a protective effect against oxidative stress neutralizing reactive oxygen species (ROS) and regulating the inflammatory cascade, which in the end has an effect on cellular protection and health as a whole (Galasso et al. 2017; Kabir et al. 2022). These polyphenols are mainly determined by available feed for shellfish since most of them are stored as a result of feeding on algae and other marine organisms that are gleaned on during filter feeding (Cory et al. 2018; Lomartire and Gonçalves 2023). Catechins and procyanidins, which are present in fruits, tea, vegetables, and brown seaweeds on land, are among the polyphenolic compounds that are found in shellfish. These substances have been linked to diverse biological functions such as antioxidant, anti-inflammatory, and cardioprotective effects (Patanè et al. 2023; Rathod et al. 2023). By way of scavenging free radicals and controlling inflammation processes, shellfish-derived polyphenols can be associated with prevention of chronic diseases, especially those associated with oxidative stress and low-grade inflammation (Kulkarni et al. 2025; Singh et al. 2025). Besides these polyphenols, shellfish can also be a source of fucoxanthin which is a carotenoid mostly produced by brown seaweeds (Miyashita et al. 2020). The use of fucoxanthin as an antioxidant, anti-inflammatory, and anti-obesity agent has also cast a significant level of interest owing to its possible effects on metabolism (Gammone et al. 2015; Zhang et al. 2015). Fucoxanthin levels in shellfish are usually not as high as fucoxanthin levels in macroalgae, but they also increase the nutritional and functional value of shellfish as a food additive. Shellfish offer a complex combination of bioactive compounds that have the potential of synergistic health benefits. Future studies ought to be aimed at defining species-specific bioactive profiles, enhancement of extraction and bioavailability, and long-term health effects of shellfish-derived polyphenols by conducting sound preclinical and clinical studies (Ebrahimi et al. 2023; Ingallina et al. 2025).

### 8.2.4 Marine Fungi

Marine fungi have emerged as an important and relatively underexplored source of polyphenolic and phenolic-derived bioactive compounds with strong antioxidant potential. Among these, benzaldehyde derivatives produced by marine-derived fungi have attracted particular interest due to their effective free radical scavenging properties (Azhari et al. 2025; Jin et al. 2016).

Several benzaldehyde-type metabolites have been isolated from marine fungal species and evaluated for their antioxidant activity (Chen et al. 2022). Chaetopyramin is one such scavenging compound that was identified and characterized from the marine fungus *Chaetomium globosum* (phylum *Ascomycota*), as well as from the red alga *Polysiphonia stricta* (formerly *Polysiphonia urceolata*) (Wang et al. 2006). In addition to chaetopyramin, structurally related benzaldehyde derivatives such as isotetrahydroauroglaucin and 2-(2′,3′-epoxy-1′,3′-heptadienyl)-6-hydroxy-5-(3-methyl-2-butenyl)benzaldehyde have also been synthesized and evaluated. These compounds exhibited notable free radical scavenging activity in DPPH assays, indicating their potential as natural antioxidants.

Further investigations concerning marine fungi have led to the isolation of benzaldehyde derivatives, including flavoglaucin and isodihydroauroglaucin, from a marine-derived *Microsporum* species (Kim et al. 2014). These metabolites demonstrated strong antioxidant activity, which is largely attributed to the presence of multiple phenolic hydroxyl groups within their chemical structures. Their scavenging effectiveness was shown to surpass that of commonly used antioxidant standards, highlighting their promise as potent natural antioxidants (Kostić et al. 2023; Lang et al. 2024).

In addition to benzaldehyde derivatives, hydroquinone-based compounds have also been identified in marine fungi. Farnesyl hydroquinone and its oxidized form, a sesquiterpene quinone, were isolated from a marine-derived *Penicillium* species. Farnesyl hydroquinone displayed strong free radical scavenging activity, exceeding that of standard antioxidants such as ascorbic acid in comparative assays. The presence of both hydroquinone and quinone moieties contributes significantly to the redox activity of these compounds (Son et al. 2002).

The antioxidant properties of marine fungal metabolites are closely related to their unique chemical structures, particularly the number and position of hydroxyl groups, conjugated double bonds, and aromatic rings. Such structural characteristics improve their ability to donate hydrogen atoms or electrons, thereby facilitating ROS neutralization and reducing oxidative damage (Vitale et al. 2020).

Beyond antioxidant activity, many marine fungal phenolic compounds exhibit additional biological effects, including anti-inflammatory, antimicrobial, cytoprotective, and anticancer activities. These multifunctional properties increase their potential value in pharmaceutical, nutraceutical, and cosmeceutical applications. Moreover, marine fungi thrive under extreme environmental conditions such as high salinity, pressure, and limited nutrients, which often drive the biosynthesis of structurally novel and biologically potent secondary metabolites.

### *8.2.5 Algae*

Marine polyphenols are widely distributed among different groups of macroalgae, including green, brown, and red algae. Each algal group possesses a distinct chemical composition, resulting in variations in both the type and concentration of polyphenolic compounds (Lomartire and Gonçalves 2023). Algae are particularly rich in compounds such as fucoxanthins, phlorotannins, and fucoidans.

Fucoxanthins are carotenoid pigments primarily found in brown algae and are known for their antioxidant, anti-inflammatory, and antitumor properties (Méresse et al. 2020). Phlorotannins are unique phenolic compounds exclusive to brown seaweeds and exhibit strong antioxidant, anti-inflammatory, and antitumor activities. Fucoidans are sulfated polysaccharides present in brown algae: they have demonstrated anticoagulant, antitumor, anti-inflammatory, and immunomodulatory effects (Erpel et al. 2020).

Algal polyphenols mainly consist of phenolic compounds, including catechins, phlorotannins, fucoidans, and fucoxanthins (Pereira and Cotas 2023). Catechins are known for their antioxidant and anti-inflammatory properties, with red algae serving as their main source. Brown algae, on the other hand, are the primary producers of phlorotannins and fucoidans, whereas green and red algae mainly synthesize flavonoids, bromophenols, terpenoids, and mycosporine-like amino acids, particularly under environmental stress conditions.

Seaweed-derived polyphenols exhibit a wide range of biological activities that are beneficial to health. In addition to antioxidant and anti-inflammatory effects, these compounds demonstrate antibacterial, antihypertensive, antidiabetic, and anticancer properties (Sadeghi et al. 2024). Due to these bioactivities, macroalgal polyphenols are increasingly incorporated into cosmetic and cosmeceutical formulations, particularly in skincare products, where they contribute to anti-aging, moisturizing, photoprotective, and ultraviolet-protective effects (Jayawardhana et al. 2023).

Numerous studies have reported the isolation of individual phenolic compounds and polyphenol groups from seaweeds, including flavonoids, phlorotannins, bromophenols, mycosporine-like amino acids, and terpenoids (Cotas et al. 2020). Biological effects of phenolic compounds depend largely on their molecular structure, including the number of phenyl rings and hydroxyl group positioning (Kumar and Goel 2019).

In cosmetic applications, phlorotannins are known to inhibit enzymes such as hyaluronidase, tyrosinase, elastase, collagenase, and matrix metalloproteinases, thereby contributing to anti-allergic, anti-wrinkle, skin-whitening, anti-aging, and photoprotective effects (Manandhar et al. 2019). Additionally, seaweed-derived phenolic compounds exhibit inhibitory activity against angiotensin-converting enzyme-1, cyclooxygenases, lipoxygenase, dipeptidyl peptidase-4, and hydroxymethylglutaryl-CoA reductase, highlighting their potential in both therapeutic and cosmeceutical applications (Purcell et al. 2022).

It is important to note that the concentration of polyphenols in seaweeds varies significantly depending on species, habitat, environmental conditions, developmental stage, and extraction methods. Therefore, systematic studies are required to identify optimal algal species and extraction conditions to maximize the yield and efficacy of bioactive compounds.

Among the most promising sources, red macroalgae such as *Neorhodomela larix*, *Callophycus serratus*, *Rhodomela confervoides*, *Tichocarpus crinitus*, Chondrus crispus, *Kappaphycus* species, *Porphyra* species, and *Symphyocladia latiuscula* show high potential (Ismail et al. 2020). Brown macroalgae including *Ecklonia cava*, *Eisenia bicyclis*, *Ishige okamurae*, *Fucus vesiculosus*, *Fucus spiralis*, *Gongolaria nodicaulis*, *Laminaria digitata*, *Sargassum* species, *Lessonia spicata*, *Padina gymnospora*, and *Macrocystis pyrifera* are also notable sources. Green macroalgae such as *Caulerpa racemosa*, *Cladophora socialis*, *Monostroma grevillei*, and *Ulva* species further contribute to the diversity of marine polyphenol sources (Rubiño et al. 2022).

## 8.3 Health Benefits of Polyphenols

### *8.3.1 Neurodegenerative Diseases*

Seaweed polyphenols have a strong potential with regard to acting on neurodegenerative-related molecular pathways. These substances serve as potent reactive oxygen and nitrogen radicals scavengers, which lessen oxidative stress and curt the fatalities of neuronal cells. Besides the antioxidant activity, polyphenols present in seaweed regulate inflammatory signal transduction, and cellular survival processes, which include apoptotic, mitochondrial, and neuronal regenerative pathways (Jomova et al. 2023; Rananaware et al. 2025). New data indicate that neuroprotection is not the direct effect of antioxidant activity, and polyphenols also control intracellular signaling networks that favor neuronal survival and plasticity. Neuroprotection can be described as a process able to maintain the neuronal structure, also functioning and inhibiting neuronal damage, dysfunction, and cell death in the central nervous system (Michaels 2024). Neurodegenerative diseases are defined as the progressive depletion of a particular population of neurons and include diseases like Alzheimer, Parkinson, frontotemporal dementia, and amyotrophic lateral sclerosis (Dugger and Dickson 2017; Gadhave et al. 2024). Even though the conditions have different clinical manifestations and anatomical localization, they are associated with multiple similar pathological processes such as oxidative stress, chronic neuroinflammation, mitochondrial dysfunction, protein misfolding, and cellular repair processes (Brezic et al. 2025; Guo et al. 2013). Existing treatment options are still focused on control of symptoms and fail to prevent the course of the disease, which explains the necessity of developing new neuroprotective interventions. A number of polyphenolic compounds like flavonoid quercetin, luteolin, myricetin, naringenin, and tannic acid have been shown to have neuroprotective properties in in vitro and in vivo models

(Minocha et al. 2022; Rudrapal et al. 2022). These compounds have been demonstrated to stimulate neuronal differentiation, neurite outgrowth, enhanced synaptic functioning, and neurons against oxidative and inflammatory injury. In a similar vein, experimental research suggests that the polyphenols can also facilitate neurogenesis and neuronal repair to aid functional restoration in neurodegeneration models (Arias-Sánchez et al. 2023; Chelliah et al. 2025). It is due to its renewable and sustainable nature and low costs that seaweeds are considered to be attractive candidates as sources of these bioactive polyphenols to pharmaceutical and nutraceutical applications. Phlorotannins, carotenoids, sulfated polysaccharides and sterols in seaweeds are a part of polyphenols that play a role in their neuroprotective properties by alleviating oxidative stress, inhibiting neuroinflammation, and modulating neuronal survival-related signaling pathways (Duan et al. 2025). These features put seaweed-derived polyphenols in good prospects as potential natural compounds in the prevention and management of neurodegenerative diseases.

### *8.3.2 Cancer and Oncology Intervention*

The development of marine polyphenols has become a topic of significant interest in cancer studies because of their structural diversity and remarkable diversity of biological activities. The sea environment and organisms like algae, sponges, tunicates, mollusks, and marine microorganisms are also abundant sources of polyphenolic compounds which have extensive anticancer properties (Alves et al. 2018; Tamzi et al. 2024). These compounds are classified into various chemical subcategories. With their antitumor effects, they are mediated in a variety of ways including the following actions: stimulation of apoptosis; angiogenesis inhibition; immune response modulation; interruption of cell division in tumor progression; and regulation of vital tumor development and metastasis signaling pathways (Ahmed et al. 2025; Nan et al. 2023). The translational value of marine bioactive compounds is emphasized by clinical development of a number of marine-based therapies, which indicates a further interest in the investigation of marine polyphenols as potential anticancer agents. The therapeutic potential of marine natural products and marine polyphenol-related pathways is exemplified by a number of clinically relevant anticancer agents (Khalifa et al. 2019). Early ones such as cytarabine, a derivative of marine sponge metabolites, have stayed a mainstay in the therapy of hematological malignancies. Trabectedin, a marine tunicates isolated compound, is used in the treatment of the soft tissue sarcoma and ovarian cancer (Dyshlovoy and Honecker 2019; Pereira et al. 2019). Eribulin mesylate, a synthetic analog based on compounds derived by synthesis of sponges, is found in metastatic breast cancer (Oey et al. 2024). Moreover, antibody–drug conjugates with marine-based cytotoxic groups have shown high clinical efficacy in lymphomas. Such achievements highlight the significance of marine bioresources as sources of clinically useful anticancer agents such as polyphenol-rich organisms (Barreca et al. 2020). Marine polyphenols use a combination of multiple and complementary molecular mechanisms to exert their anticancer effects. A number of these compounds cause programmed cell death through activation of apoptotic pathway and interference with protein synthesis leading to arrest of cell cycle (Chimento et al. 2023; Yoshioka et al. 2022). Some of them prevent tumor angiogenesis by restraining the pro-angiogenic signaling and the growth of endothelial cells. Marine polyphenols can also control the immune responses as it can regulate the inflammatory signal pathways and increase the immune cells tumor cell recognition (Liu et al. 2023; Wang et al. 2022). Other polyphenols have epigenetic effects, including inhibition of histone-modifying enzymes, and this fact means that the tumor suppressor genes are reactivated and the cells become more susceptible to immune-mediated cytotoxicity (Rajendran et al. 2022). Moreover, marine polyphenols engage significant oncogenic signaling pathways, such as pathways in microtubule dynamics, mitochondrial activity, and transcriptional regulation, eventually inhibiting tumor progression and development (Cháirez-Ramírez et al. 2021; Guan et al. 2025). Although marine polyphenols have therapeutic potential, there are some factors that limit their clinical use (low natural concentration, complex structure, low aqueous solubility, and low bioavailability). With marine biotechnology, the sustainable production of marine polyphenols has been achieved using the methods of fermentation and genome mining, as well as synthetic biology (Veluchamy et al. 2025). Moreover, new drug delivery techniques, including nanoparticle-based, liposomes, and targeted conjugate, provide good means to improve the stability, bioavailability, and treatment effect. Further studies in the marine biodiversity and molecular target as well as strategic combination with known chemotherapeutic agents will be necessary to achieve full potential of marine polyphenols in cancer treatment.

### *8.3.3 Diabetes*

Insulin treatment and synthetic oral antidiabetic medications are usually employed in the treatment of diabetes mellitus. Though these treatments are good strategies when speaking of control of blood glucose levels, their long-term usage is coupled with slow recuperation and various negative effects (Corathers et al. 2013). In this regard, marine polyphenols have become more of an interesting topic of exploration as potential natural alternatives to control diabetes as phenols have high potential of antioxidant as well as broad biological applications.

Marine organisms create diverse polyphenolic compounds that have the potential to regulate major metabolic routes of glucose homeostasis and insulin sensitivity and have been of interest in the prevention and treatment of diabetes (Rampengan et al. 2025; Tong et al. 2024). Marine polyphenols obtained from algae and other marine resources have antidiabetic effects that occur in several ways. The dysregulation of enzymes like alpha-glucosidase, alpha-amylose, protein tyrosine phosphatase, and aldose reductase is critical in the progression of type two diabetes and its complications (Ansari et al. 2024; Praise et al. 2025). Marine polyphenols are reported to prevent these enzymes, thus decreasing the intestinal uptake of glucose and enhancing the postprandial glycemia (Coe and Ryan 2016; Golovinskaia and

Wang 2023). Moreover, some polyphenols stimulate both glucose transporter type four and adenosine monophosphate-activated protein kinase signaling pathways that promote the peripheral glucose uptake and insulin sensitivity. Phlorotannins and fucoxanthin and other polyphenolic compounds found in brown and red algae have exhibited the capacity to control adipokines and inflammatory mediators, which lead to the enhancement of insulin resistance and metabolic homeostasis (Goya and Mateos 2024). Brown algae are especially high in polyphenols, dietary fibers, sulfated polysaccharides, pigments, omega-three fatty acids, and essential amino acids which are some of the attributes that make it antidiabetic (Tong et al. 2024). Extraction of phlorotannins contained in seaweeds has exhibited great inhibitory features against alpha-glucosidase and alpha-amylase which are direct enzymes in the digestion of carbohydrates (Kumar et al. 2022). By these means, marine polyphenols contribute to the regulation of blood glucose levels, insulin sensitivity, glucose uptake by peripheral tissues, so decreasing the possibility of developing diabetes-related complications. Along with algae, marine microorganisms, fungi, and sponge-associated microbes also generate polyphenol-related metabolites, which increase the variety of marine antidiabetic agents (Wang et al. 2025). Marine polyphenols also have a protective effect by the redox effect and stimulation of cellular antioxidant defense mechanisms. Oxidative stress is the key factor in the pathogenesis of pancreatic beta-cell malfunction and insulin resistance (Yesupatham and Saraswathy 2025). Marine polyphenols stimulate nuclear factor erythroid two-related factor two and heme oxygenase-one pathway, an important cytoprotective mechanism of oxidative damage (Zhang and Zhang 2025). The stimulation of this pathway leads to an increase in the expression of antioxidant enzymes, a decrease in intracellular ROS, and maintenance of cellular redox homeostasis (Jena et al. 2023; Marengo et al. 2016). Some of the marine pigments, like PHNQ derivatives, exhibit high levels of free radical scavenging and metal-chelating effect, which also contributes to the prevention of oxidative damage of beta cells (Kikuchi et al. 2019). Besides being antioxidants, marine polyphenols reduce inflammation by inhibiting enzymes engaged in the synthesis of the pro-inflammatory mediators like prostaglandins and leukotrienes (Fernando et al. 2016; Yahfoufi et al. 2018). They also suppress inflammatory cytokines and oxidative stress and chronic inflammation, which are contributors to insulin resistance and disease progression, through down regulation of inflammatory cytokines and transcription factors (Berbudi et al. 2025). Marine polyphenols provide a multi-targeted approach to the control of diabetes by addressing bio-markers of oxidative stress, inflammation, enzyme abnormalities, and insulin signaling.

### *8.3.4 Hypertension*

Flavonoids and flavanols as well as anthocyanins, phenolic acids, tannins, and other phenolic compounds are marine polyphenols that are effective vasodilators (Mutha et al. 2021). These compounds are able to relax smooth muscle of the vessels by increasing endothelial responsiveness and by regulating signaling pathways that control vascular tone. The research has shown that polyphenols have the ability to modulate the activity of calcium channels, increase nitric oxide production, and decrease the inflammation of the endothelium, all of which lead to reduced blood pressure and better vascular performance (Serreli and Deiana 2023). Curcumin, a natural polyphenol and also found in some marine-derived food chains as well as food formulations that are of marine origin, has shown to have a vasodilatory effect and antihypertensive effect (Moldoveanu et al. 2024). Experimental results show that curcumin has the ability to cause the relaxation of vascular tissues without influencing the functioning of other antihypertensives that are regularly used like calcium channel blockers (Bai et al. 2025; Li et al. 2022). This fact implies that polyphenols can be used safely as part of dietary or therapeutic interventions in addition to standard antihypertensive therapies with complementary effects that does not reduce the effects of drugs.

### *8.3.5 Application of Polyphenols in Major Industries*

Marine polyphenols are bioactive compounds that have received a lot of interest in various industries due to their variety of bioactivities, natural source, and sustainability (Bolat et al. 2024; Karthikeyan et al. 2022). These compounds have great antioxidant, anti-inflammatory, antimicrobial, and protection effects because they are derived out of marine ecosystems that contain vast biological diversity (Elbandy 2022; Vasarri et al. 2022). Due to the change in consumer demand toward natural, environmental friendly, and health-promoting items, marine polyphenols are coming under the light as functional ingredients in nutraceutical, pharmaceutical, cosmetic, and food industries. Their multisensory capability enables them to help in health, preventive diseases, and value addition to products in all industries. Marine polyphenols are used in nutraceutical and functional food market largely derived as algae, both microalgae and macroalgae (Zambelli et al. 2026). These organisms offer polyphenols including phlorotannins, phenolics, and pigments that aid in metabolic well-being, cardiovascular operations, and glycemic regulation. Due to their antioxidant effects, enzyme modifying effects and gut and immune health advantages, marine polyphenols are commonly found in supplements, fortified foods and beverages (Iqbal et al. 2023; Pereira and Cotas 2023). The fact that they are associated with dietary fibers, minerals, and essential fatty acids also lends more support to their application as holistic nutraceutical ingredients. Marine polyphenols have also been found to have therapeutic properties in the pharmaceutical industry (El-Gammal et al. 2024). These compounds are used in disease prevention and management because they target molecular pathways which are implicated in inflammation, oxidative damage, and cell dysfunction. Marine polyphenols have been promising in supplementing the treatment of diabetes, high blood pressure, neurodegenerative diseases, as well as cancer (De Brito Alves et al. 2023; Saad et al. 2025). Their capacity to regulate signaling pathways, shield cells against oxidative injury, and to enhance endogenous defense systems makes them useful drug development and adjunct therapeutic candidates. Marine polyphenols find extensive application in the cosmetic and cosmeceuticals industry in skin health and anti-aging and numerous others shown in Fig. 8.2. Polyphenols that are extracted in seaweeds and other marine species are added to skin-care products because of their antioxidant, photoprotective, anti-inflammatory, and collagen-preserving effects (Chinjoo and Golzary 2025; López-Hortas et al. 2021). These compounds serve to delay the aging process of skin, enhance the protective capabilities of the skin, foster repair processes, and guard against environmental stresses. Natural and sustainable image of the marine polyphenols corresponds with the increasing demand of clean-label and eco-conscious cosmetic items (Couceiro et al. 2025; Fonseca et al. 2023). Other than in the health and beauty industries, marine polyphenols are also useful in food, aquaculture, and biotechnology industries. Their antioxidant and antimicrobial properties are useful in the preservation of food to enhance shelf life and product stability in food preservation. Marine polyphenols are considered as functional feed additives to be exploited in aquaculture and animal nutrition to improve immunity and stress resistance (Petcu et al. 2023). Although they can be used across a wide range, there are still issues of sustainability sourcing, stability of compounds, bioavailability, and scalable production. Further development of marine biotechnology, regulated growth and formulation technologies is the key to wrapping all industrial potential of marine polyphenols in various industries.

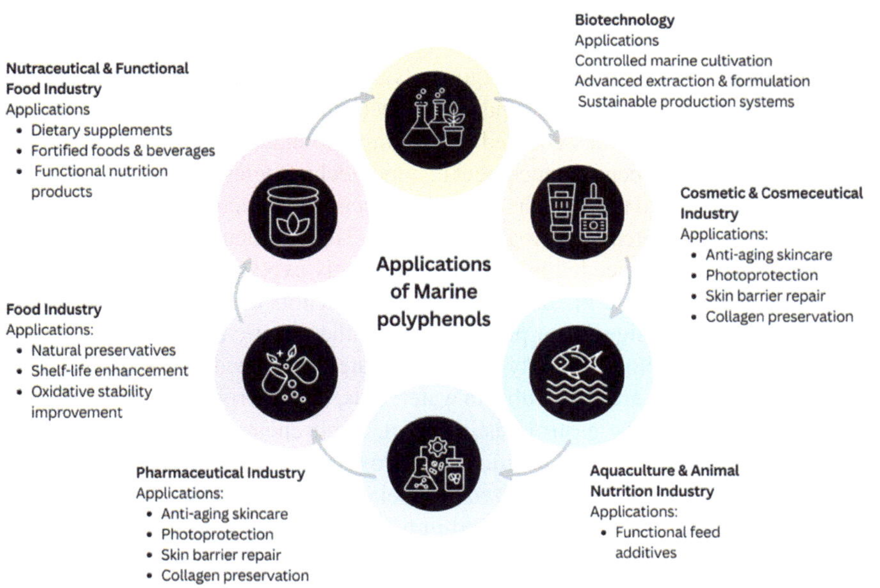


**Fig. 8.2** Industrial applications of marine polyphenols highlighting their multifunctional roles in nutraceutical, pharmaceutical, cosmetic, food, aquaculture, and biotechnology sectors. This picture has been created by the authors

### 8.3.6 Challenges and Prospects

Despite the fact that many studies have established the cardioprotective effect of marine polyphenolic compounds, a number of critical issues need to be overcome to facilitate their effective therapeutic transcription (Khurana et al. 2013; Ugariogu et al. 2025). A significant difficulty comes as a result of the complexity of marine ecosystems. Marine settings are very diverse in that they cover coral reefs, open oceans, deep-sea vents, and polar area among others to mention a few, which have different biological and chemical conditions (Costello and Chaudhary 2017). This biodiversity leads to a broad range of variation in the composition, concentration and bioactivity of polyphenolic compounds (Roik et al. 2022). The polyphenol biosynthesis is further affected by the environmental factors of temperature, salinity, pressure, and the availability of nutrients. Consequently, the technical separation and repetitive production of certain marine polyphenols that have cardiovascular effects have not yet been reached (Hollman et al. 2011). To solve this complexity, it is necessary to use a multidisciplinary approach involving marine biology, natural product chemistry, pharmacology, and biotechnology. The great variety of structure that marine ecosystems create also makes it difficult to develop structure–activity relations. Knowledge of structure–activity relationships is critical when speaking of the role of particular molecular features in the antioxidant, anti-inflammatory, and cardioprotective actions of marine polyphenols (Nagahawatta et al. 2024). Nevertheless, this process is difficult due to the broad spectrum of different types of molecular architecture, the extent of polymerization, and stereochemical diversity. The detailed structural characterization and activity prediction require advanced analytical methods like nuclear magnetic resonance spectroscopy and mass spectrometry in combination with high-throughput screening studies coupled with computational modeling and molecular docking studies. The recent advancement in omics technologies and integrative bioinformatics has enhanced the analysis of large datasets and detection of structural characteristics that are related to biological activity (Dai and Shen 2022). Further interdisciplinary studies are needed in order to completely explain the structure–activity relations and to maximize the use of marine polyphenols as cardiovascular agents. The second and third significant problems in the therapeutic application of marine polyphenols are low bioavailability and complicated pharmacokinetic profile. A lot of marine polyphenols are not soluble in water; thus, the absorption of these polyphenols is reduced inside the gastrointestinal tract. Their chemical structure and more specifically the availability and position of hydroxyl groups and aromatic rings make them relatively insoluble and stabilized (Ciupei et al. 2024; Kumar and Goel 2019). Consequently, organic solvents are used in most cases to extract such compounds, instead of aqueous systems. Glycosylation and other strategies have also reported to increase water solubility without changing the structure, but enzymatic methods of glycosylation might have cost and technical limitations (Andreu et al. 2023; Ding et al. 2024). Besides this, solubility and absorption will also be affected by interactions with dietary constituents, proteins, and polysaccharides (Fontes-Candia et al. 2025). The stability and bioavailability of marine polyphenols are also influenced by physiological factors including gastrointestinal pH, transit time, enzymatic degradation, biliary excretion, and intestinal fermentation (Kupikowska-Stobba et al. 2025; Scott et al. 2022). A large number of compounds are not stable in acidic gastric conditions or are quickly metabolized, leading to a decrease in systemic availability and a short half-life of circulation. Moreover, secondary metabolites can be produced through metabolism by gut microbiota with changes in biological activity or reduced biological activity (Zhang et al. 2023). All these pharmacokinetic constraints such as rapid metabolic rates, low tissue penetration, and low half-life greatly impair clinical efficacy despite excellent in vitro activity. As a way of addressing these hurdles, sophisticated formulation planning is being considered.

The use of nanotechnology in delivery systems such as nanoparticles, nanoliposomes, and other nanocarriers have demonstrated potential in enhancing solubility, stability, absorption and targeted delivery (Elumalai et al. 2024; Patra et al. 2018). Another viable approach is prodrug design, where active compounds can be enzymatically converted in vivo by an inactive (pro)-precursor. Inhibition of efflux transporters or metabolic enzymes by the use of bioenhancers as well as controlled-release formulations could be used to positively influence bioavailability and therapeutic efficacy (Jornada et al. 2015; Markovic et al. 2020). Another important obstacle in the process of developing marine polyphenol-based therapeutics is safety and toxicity assessment. Even though these compounds are mostly considered bioactive and helpful, their long-term safety needs to be thoroughly evaluated. The fact that standardized test cases are not developed on specific toxicity and that little information exists on chronic exposure creates concerns about possible adverse effects, e.g., reproductive toxicity, endocrine disruption and carcinogenicity (Aware et al. 2022; Dwyer 2023). Also, biotransformations result in the formation of metabolites which are potentially more toxic compared to the original compounds. Personal differences based on genetic, environmental, and lifestyle diversity also make the safety assessment more problematic, with possible drug-polyphenol interactions (Ayoub et al. 2023; Rathee et al. 2023). Lastly, clinical trials have significant challenges in the design and implementation. The biggest challenges include recruiting adequate and representative participants, negotiating regulatory demands, the prolonged follow-up, and obtaining adequate funds. Natural compounds that are used in clinical trials typically take long periods of time to determine efficacy and safety, compounding the logistical aspects and cost. Recent progress in recruitment methods such as partnership with patient advocacy organizations, digital recruitment platforms, and streamlined delivery systems (to maintain sufficient bioavailability) will be critical with concern to the future development of marine polyphenols in clinical practice.

## 8.4 Conclusion

Due to the broad spectrum of health-promoting benefits, scientific interest has increasingly been attracted to the therapeutic potential of polyphenols and other marine-derived micronutrients due to their diverse biomolecular behaviors and medical uses. Marine polyphenols are natural products of compounds generated by different marine organisms and have been widely known to be antioxidative and anti-inflammatory. Besides polyphenols, marine sources are an excellent source of essential micronutrients including omega-3 fatty acids, vitamins, and minerals which synergically result in better physiological functioning and disease prevention. Clinical studies indicate that sea-derived polyphenols have a protective effect in the treatment of cardiovascular diseases, diabetes, cancer, and neurodegenerative diseases. The most frequently recognized effects of omega-3 fatty acids are their cardioprotective properties (thus a reduction in inflammation, an improvement in lipid profiles, and a decreasing effect on blood pressure). Another potential that marine polyphenols possess is anticancer such that they inhibit proliferation of tumor cells and promote programmed cell death. In addition to therapeutic uses, marine polyphenols and micronutrients are now being considered as an ingredient in cosmetic and skincare products. These compounds are found to be beneficial in maintaining the health of skin, inflammation, boosting collagen production, as well as protection against ultraviolet radiation and environmental stressors. These natural compounds possess multifunctional bioactivity, which makes them good ingredients in cosmetic products. Although they have good prospective advantages, more studies need to be done to learn more about molecular pathways of the biological effects of marine-derived polyphenols and micronutrients, their long-term safety, and their possible side effects. Additional research in the form of effective preclinical and clinical trials is necessary in order to determine the best dosage, bioavailability and therapeutic activity. In short, marine-derived polyphenols and micronutrients are an unexploited resource that has high potential benefits to supplement human health. The future of marine bioresources in health and disease management practices will be ensured through continued research efforts to understand their mechanisms of action and safety profiles to ensure the creation of efficacious therapeutic and preventive interventions.

## Abbreviations

| | |
|---|---|
| Polyhydroxynaphthoquinone | PHNQ |
| Reactive oxygen species | ROS |

## References

Aatif A (2023) Current understanding of polyphenols to enhance bioavailability for better therapies. Biomedicines 11(7):2078. https://doi.org/10.3390/biomedicines11072078

Adelusi TI, Oyedele AQK, Boyenle ID, Ogunlana AT, Adeyemi RO, Ukachi CD, Idris MO, Olaoba OT, Adedotun IO, Kolawole OE, Xiaoxing Y, Abdul-Hammed M (2022) Molecular modeling in drug discovery. Inf Med Unlocked 29:100880. https://doi.org/10.1016/j.imu.2022.100880

Ahmed ZSO, Khan E, Elias N, Elshebiny A, Dou Q, Ahmed ZSO, Khan E, Elias N, Elshebiny A, Dou Q (2025) Updated review on natural polyphenols: molecular mechanisms, biological effects, and clinical applications for cancer management. Biomolecules 15(5):629. https://doi.org/10.3390/biom15050629

Alharbi HOA, Alshebremi M, Babiker AY, Rahmani AH, Alharbi HOA, Alshebremi M, Babiker AY, Rahmani AH (2025) The role of quercetin, a flavonoid in the management of pathogenesis through regulation of oxidative stress, inflammation, and biological activities. Biomolecules 15(1):151. https://doi.org/10.3390/biom15010151

Alves C, Silva J, Pinteus S, Gaspar H, Alpoim MC, Botana LM, Pedrosa R (2018) From marine origin to therapeutics: the antitumor potential of marine algae-derived compounds. Front Pharm 9:777. https://doi.org/10.3389/fphar.2018.00777

Andreu A, Ćorović M, Garcia-Sanz C, Santos AS, Milivojević A, Ortega-Nieto C, Mateo C, Bezbradica D, Palomo JM (2023) Enzymatic glycosylation strategies in the production of bioactive compounds. Catalysts 13(10):1359. https://doi.org/10.3390/catal13101359

Aneklaphakij C, Saigo T, Watanabe M, Naake T, Fernie AR, Bunsupa S, Satitpatipan V, Tohge T (2021) Diversity of chemical structures and biosynthesis of polyphenols in nut-bearing species. Front Plant Sci 12:642581. https://doi.org/10.3389/fpls.2021.642581

Ansari P, Khan JT, Chowdhury S, Reberio AD, Kumar S, Seidel V, Abdel-Wahab YHA, Flatt PR (2024) Plant-based diets and phytochemicals in the management of diabetes mellitus and prevention of its complications: a review. Nutrients 16(21):3709. https://doi.org/10.3390/nu16213709

Apak R, Özyürek M, Güçlü K, Çapanoğlu E (2016) Antioxidant activity/capacity measurement. 1. Classification, physicochemical principles, mechanisms, and electron transfer (ET)-based assays. J Agric Food Chem 64(5):997–1027. https://doi.org/10.1021/acs.jafc.5b04739

Arias-Sánchez RA, Torner L, Fenton Navarro B (2023) Polyphenols and neurodegenerative diseases: potential effects and mechanisms of neuroprotection. Molecules 28(14):5415. https://doi.org/10.3390/molecules28145415

Aryaeian N, Sedehi SK, Arablou T (2017) Polyphenols and their effects on diabetes management: a review. Med J Islam Repub Iran 31:134. https://doi.org/10.14196/mjiri.31.134

Aware CB, Patil DN, Suryawanshi SS, Mali PR, Rane MR, Gurav RG, Jadhav JP (2022) Natural bioactive products as promising therapeutics: a review of natural product-based drug development. S Afr J Bot 151:512–528. https://doi.org/10.1016/j.sajb.2022.05.028

Ayoub WS, Ritu Zahoor I, Dar AH, Farooq S, Mir TA, Ganaie TA, Srivastava S, Pandey VK, Altaf A (2023) Exploiting the polyphenolic potential of honey in the prevention of chronic diseases. Food Chem Adv 3:100373. https://doi.org/10.1016/j.focha.2023.100373

Azhari M, Merliani N, Singgih M, Arai M, Julianti E (2025) Insights into natural products from marine-derived fungi with antimycobacterial properties: opportunities and challenges. Mar Drugs 23(7):279. https://doi.org/10.3390/md23070279

Bai X, Hao J, Feng Q, Guo C, Pang M, Li J, Wang J, Xing J (2025) Curcumin alone not combined with piperine exerts cardioprotective effects in pressure-overload rats by increasing glucagon-like peptide-1 receptor signaling and additional properties. J Tradit Complement Med 15(5):522–535. https://doi.org/10.1016/j.jtcme.2024.07.007

Barreca M, Spanò V, Montalbano A, Cueto M, Díaz Marrero AR, Deniz I, Erdoğan A, Lukić Bilela L, Moulin C, Taffin-de-Givenchy E, Spriano F, Perale G, Mehiri M, Rotter AP, Thomas O, Barraja P, Gaudêncio SP, Bertoni F (2020) Marine anticancer agents: an overview with a particular focus on their chemical classes. Mar Drugs 18(12):619. https://doi.org/10.3390/md18120619

Berbudi A, Khairani S, Tjahjadi AI (2025) Interplay between insulin resistance and immune dysregulation in type 2 diabetes mellitus: implications for therapeutic interventions. ImmunoTarg Ther 14:359–382. https://doi.org/10.2147/ITT.S499605

Bjørklund G, Shanaida M, Lysiuk R, Butnariu M, Peana M, Sarac I, Strus O, Smetanina K, Chirumbolo S (2022) Natural compounds and products from an anti-aging perspective. Molecules 27(20):7084. https://doi.org/10.3390/molecules27207084

Bolat E, Sarıtaş S, Duman H, Eker F, Akdaşçi E, Karav S, Witkowska AM (2024) Polyphenols: secondary metabolites with a biological impression. Nutrients 16(15):2550. https://doi.org/10.3390/nu16152550
Botten D, Fugallo G, Fraternali F, Molteni C (2015) Structural properties of green tea catechins. J Phys Chem B 119(40):12860–12867. https://doi.org/10.1021/acs.jpcb.5b08737
Brezic N, Gligorevic S, Sic A, Knezevic NN (2025) Protein misfolding and aggregation as a mechanistic link between chronic pain and neurodegenerative diseases. Curr Issues Mol Biol 47(4):259. https://doi.org/10.3390/cimb47040259
Cháirez-Ramírez MH, de la Cruz-López KG, García-Carrancá A (2021) Polyphenols as antitumor agents targeting key players in cancer-driving signaling pathways. Front Pharmacol 12:710304. https://doi.org/10.3389/fphar.2021.710304
Chandimali N, Bak SG, Park EH, Lim HJ, Won YS, Kim EK, Park SI, Lee SJ (2025) Free radicals and their impact on health and antioxidant defenses: a review. Cell Death Discov 11(1):19. https://doi.org/10.1038/s41420-024-02278-8
Chelliah R, Ghimeray S, Mettu S, Barathikannan K, Wei S, Pabba M, Banat F, Oh DH (2025) Dietary polyphenols and gut–brain axis modulation: the neuroprotective role of quercetin, allicin, and ferulic acid in traumatic brain injury and neurodegeneration. J Funct Foods 135:107091. https://doi.org/10.1016/j.jff.2025.107091
Chen M, Liang J, Wang Y, Liu Y, Zhou C, Hong P, Zhang Y, Q ZJ (2022) A new benzaldehyde from the coral-derived fungus Aspergillus terreus C23-3 and its anti-inflammatory effects via suppression of MAPK signaling pathway in RAW264.7 cells. J Zhejiang Univ Sci B 23(3):230–240. https://doi.org/10.1631/jzus.B2100807
Chimento A, De Luca A, D'Amico M, De Amicis F, Pezzi V (2023) The involvement of natural polyphenols in molecular mechanisms inducing apoptosis in tumor cells: a promising adjuvant in cancer therapy. Int J Mol Sci 24(2):1680. https://doi.org/10.3390/ijms24021680
Chinjoo N, Golzary A (2025) Microalgae: revolutionizing skin repair and enhancement. Biotechnol Rep 47:e00911. https://doi.org/10.1016/j.btre.2025.e00911
Cione E, Torre CL, Cannataro R, Caroleo MC, Plastina P, Gallelli L (2019) Quercetin, epigallocatechin gallate, curcumin, and resveratrol: from dietary sources to human MicroRNA modulation. Molecules 25(1):63. https://doi.org/10.3390/molecules25010063
Ciupei D, Colişar A, Leopold L, Stănilă A, Diaconeasa ZM (2024) Polyphenols: from classification to therapeutic potential and bioavailability. Foods 13(24):4131. https://doi.org/10.3390/foods13244131
Coe S, Ryan L (2016) Impact of polyphenol-rich sources on acute postprandial glycaemia: a systematic review. J Nutr Sci 5:e24. https://doi.org/10.1017/jns.2016.11
Corathers SD, Peavie S, Salehi M (2013) Complications of diabetes therapy. Endocrinol Metab Clin N Am 42(4):947–970. https://doi.org/10.1016/j.ecl.2013.06.005
Cores Á, Carmona-Zafra N, Clerigué J, Villacampa M, Menéndez JC (2023) Quinones as neuroprotective agents. Antioxidants 12(7):1464. https://doi.org/10.3390/antiox12071464
Cory H, Passarelli S, Szeto J, Tamez M, Mattei J (2018) The role of polyphenols in human health and food systems: a mini-review. Front Nutr 5:87. https://doi.org/10.3389/fnut.2018.00087
Costello MJ, Chaudhary C (2017) Marine biodiversity, biogeography, deep-sea gradients, and conservation. Curr Biol 27(11):R511–R527. https://doi.org/10.1016/j.cub.2017.04.060
Cotas J, Leandro A, Monteiro P, Pacheco D, Figueirinha A, Gonçalves AMM, da Silva GJ, Pereira L (2020) Seaweed phenolics: from extraction to applications. Mar Drugs 18(8):384. https://doi.org/10.3390/md18080384
Couceiro B, Hameed H, Vieira ACF, Singh SK, Dua K, Veiga F, Pires PC, Ferreira L, Paiva-Santos AC (2025) Promoting health and sustainability: exploring safer alternatives in cosmetics and regulatory perspectives. Sustain Chem Pharm 43:101901. https://doi.org/10.1016/j.scp.2024.101901
Dai X, Shen L (2022) Advances and trends in omics technology development. Front Med 9:911861. https://doi.org/10.3389/fmed.2022.911861

Davies M, Jones RDO, Grime K, Jansson-Löfmark R, Fretland AJ, Winiwarter S, Morgan P, McGinnity DF (2020) Improving the accuracy of predicted human pharmacokinetics: lessons learned from the astrazeneca drug pipeline over two decades. Trends Pharmacol Sci 41(6):390–408. https://doi.org/10.1016/j.tips.2020.03.004

De Brito Alves JL, Brasil JMA, Maia LA, Da Costa LM, Sampaio KB, De Souza EL (2023) Phenolic compounds in hypertension: targeting gut-brain interactions and endothelial dysfunction. J Funct Foods 104:105531. https://doi.org/10.1016/j.jff.2023.105531

de Carvalho CCCR, Caramujo MJ (2017) Carotenoids in aquatic ecosystems and aquaculture: a colorful business with implications for human health. Front Mar Sci 4:93. https://doi.org/10.3389/fmars.2017.00093

de Oliveira I, Santos-Buelga C, Aquino Y, Barros L, Heleno SA (2025) New frontiers in the exploration of phenolic compounds and other bioactives as natural preservatives. Food Biosci 68:106571. https://doi.org/10.1016/j.fbio.2025.106571

Ding J, Li X, Jin Z, Hachem MA, Bai Y (2024) Efficient glycosylation of polyphenols via dynamic complexation of cyclodextrin and synchronous coupling reaction with cyclodextrin glycosyltransferase in water. Int J Biol Macromol 280:136065. https://doi.org/10.1016/j.ijbiomac.2024.136065

Donato PD, Buono A, Poli A, Finore I, Abbamondi GR, Nicolaus B, Lama L (2018) Exploring marine environments for the identification of extremophiles and their enzymes for sustainable and green bioprocesses. Sustainability 11(1):149. https://doi.org/10.3390/su11010149

Duan X, Agar OT, Barrow CJ, Dunshea FR, Suleria HAR (2025) Improving potential strategies for biological activities of phlorotannins derived from seaweeds. Crit Rev Food Sci Nutr 65(5):833–855. https://doi.org/10.1080/10408398.2023.2282669

Dugger BN, Dickson DW (2017) Pathology of neurodegenerative diseases. Cold Spring Harb Perspect Biol 9(7):a028035. https://doi.org/10.1101/cshperspect.a028035

Durazzo A, Lena GD, Gabrielli P, Santini A, Lombardi-Boccia G, Lucarini M (2022) Nutrients and bioactive compounds in seafood: quantitative literature research analysis. Fishes 7(3):132. https://doi.org/10.3390/fishes7030132

Dwyer JT (2023) Have safety and efficacy assessments of bioactives come of age? Mol Asp Med 89:101103. https://doi.org/10.1016/j.mam.2022.101103

Dyshlovoy SA, Honecker F (2019) Marine compounds and cancer: the first two decades of XXI century. Mar Drugs 18(1):20. https://doi.org/10.3390/md18010020

Ebrahimi B, Baroutian S, Li J, Zhang B, Ying T, Lu J (2023) Combination of marine bioactive compounds and extracts for the prevention and treatment of chronic diseases. Front Nutr 9:1047026. https://doi.org/10.3389/fnut.2022.1047026

Elbandy M (2022) Anti-inflammatory effects of marine bioactive compounds and their potential as functional food ingredients in the prevention and treatment of neuroinflammatory disorders. Molecules 28(1):2. https://doi.org/10.3390/molecules28010002

El-Gammal MI, Abou-Dobara MI, Ibrahim HAH, Abdulhafith SA, Okbah MA (2024) Polyphenols in selected marine algae and aromatic herbs with antimicrobial properties: a comparative study. Egypt J Aquat Res 50(1):71–77. https://doi.org/10.1016/j.ejar.2023.12.006

Elumalai K, Srinivasan S, Shanmugam A (2024) Review of the efficacy of nanoparticle-based drug delivery systems for cancer treatment. Biomed Technol 5:109–122. https://doi.org/10.1016/j.bmt.2023.09.001

Erpel F, Mateos R, Pérez-Jiménez J, Pérez-Correa JR (2020) Phlorotannins: from isolation and structural characterization, to the evaluation of their antidiabetic and anticancer potential. Food Res Int 137:109589. https://doi.org/10.1016/j.foodres.2020.109589

Escobar-Doncel B, Papakosta A, Storm-Mathisen J, Fang EF (2026) Sea urchin bioactive compounds: emerging interventions for age-related diseases. Ageing Res Rev 114:102979. https://doi.org/10.1016/j.arr.2025.102979

Ettlin RA, Kuroda J, Plassmann S, Prentice DE (2010) Successful drug development despite adverse preclinical findings part1: processes to address issues and most important findings. J Toxicol Pathol 23(4):189–211. https://doi.org/10.1293/tox.23.189

Fan FY, Sang LX, Jiang M, Fan FY, Sang LX, Jiang M (2017) Catechins and their therapeutic benefits to inflammatory bowel disease. Molecules 22(3):484. https://doi.org/10.3390/molecules22030484

Fernando IPS, Nah JW, Jeon YJ (2016) Potential anti-inflammatory natural products from marine algae. Environ Toxicol Pharmacol 48:22–30. https://doi.org/10.1016/j.etap.2016.09.023

Ferrari E, Naponelli V, Ferrari E, Naponelli V (2025) Catechins and human health: breakthroughs from clinical trials. Molecules 30(15):3128. https://doi.org/10.3390/molecules30153128

Fonseca S, Amaral MN, Reis CP, Custódio L (2023) Marine natural products as innovative cosmetic ingredients. Mar Drugs 21(3):170. https://doi.org/10.3390/md21030170

Fontes-Candia C, Díaz-Piñero L, Vega-Gómez LM, Molina-Gilarranz I, Martínez-Sanz M (2025) Relevance of protein-polysaccharide interactions on nutritional quality and gastrointestinal digestion of protein-based foods. J Agric Food Chem 73(9):4998–5004. https://doi.org/10.1021/acs.jafc.4c11949

Gadhave DG, Sugandhi VV, Jha SK, Nangare SN, Gupta G, Singh SK, Dua K, Cho H, Hansbro PM, Paudel KR (2024) Neurodegenerative disorders: mechanisms of degeneration and therapeutic approaches with their clinical relevance. Ageing Res Rev 99:102357. https://doi.org/10.1016/j.arr.2024.102357

Gaglione R, Cesaro A, Dell'Olmo E, Della Ventura B, Casillo A, Di Girolamo R, Velotta R, Notomista E, Veldhuizen EJA, Corsaro MM, De Rosa C, Arciello A (2019) Effects of human antimicrobial cryptides identified in apolipoprotein B depend on specific features of bacterial strains. Sci Rep 9(1):6728. https://doi.org/10.1038/s41598-019-43063-3

Galasso C, Corinaldesi C, Sansone C (2017) Carotenoids from marine organisms: biological functions and industrial applications. Antioxidants 6(4):96. https://doi.org/10.3390/antiox6040096

Gammone MA, D'Orazio N, Gammone MA, D'Orazio N (2015) Anti-obesity activity of the marine carotenoid fucoxanthin. Mar Drugs 13(4):2196–2214. https://doi.org/10.3390/md13042196

Ghelani H, Khursheed M, Adrian TE, Jan RK (2022) Anti-inflammatory effects of compounds from echinoderms. Mar Drugs 20(11):693. https://doi.org/10.3390/md20110693

Ghosh S, Solanki R, Bhatia D, Sankaranarayanan S (2025) Nanomaterials for delivery of medicinal plant extracts and phytochemicals: potential applications and future perspectives. Plant Nano Biol 12:100161. https://doi.org/10.1016/j.plana.2025.100161

Golovinskaia O, Wang CK (2023) The hypoglycemic potential of phenolics from functional foods and their mechanisms. Food Sci Hum Wellness 12(4):986–1007. https://doi.org/10.1016/j.fshw.2022.10.020

Goya L, Mateos R (2024) Antioxidant and anti-inflammatory effects of marine phlorotannins and bromophenols supportive of their anticancer potential. Nutr Rev 83(3):e1225–e1242. https://doi.org/10.1093/nutrit/nuae066

Guan Y, Li L, Yang R, Lu Y, Tang J (2025) Targeting mitochondria with natural polyphenols for treating neurodegenerative diseases: a comprehensive scoping review from oxidative stress perspective. J Transl Med 23:572. https://doi.org/10.1186/s12967-025-06605-0

Gulcin İ (2025) Antioxidants: a comprehensive review. Arch Toxicol 99(5):1893–1997. https://doi.org/10.1007/s00204-025-03997-2

Guo C, Sun L, Chen X, Zhang D (2013) Oxidative stress, mitochondrial damage and neurodegenerative diseases. Neural Regen Res 8(21):2003–2014. https://doi.org/10.3969/j.issn.1673-5374.2013.21.009

Hassan SSU, Samanta S, Dash R, Karpiński TM, Habibi E, Sadiq A, Ahmadi A, Bungau S (2022) The neuroprotective effects of fisetin, a natural flavonoid in neurodegenerative diseases: focus on the role of oxidative stress. Front Pharmacol 13:1015835. https://doi.org/10.3389/fphar.2022.1015835

Hollman PCH, Cassidy A, Comte B, Heinonen M, Richelle M, Richling E, Serafini M, Scalbert A, Sies H, Vidry S (2011) The biological relevance of direct antioxidant effects of polyphenols for cardiovascular health in humans is not established1–4. J Nutr 141(5):989S-1009S. https://doi.org/10.3945/jn.110.131490

Hossain MS, Wazed MA, Asha S, Amin MR, Shimul IM (2025) Dietary phytochemicals in health and disease: mechanisms, clinical evidence, and applications—a comprehensive review. Food Sci Nutr 13(3):e70101. https://doi.org/10.1002/fsn3.70101

Hou Y, Carne A, McConnell M, Mros S, Vasileva EA, Mishchenko NP, Burrow K, Wang K, Bekhit AA, Bekhit AEDA (2020) PHNQ from evechinus chloroticus sea urchin supplemented with calcium promotes mineralization in saos-2 human bone cell line. Mar Drugs 18(7):373. https://doi.org/10.3390/md18070373

Ingallina C, Spano M, Prencipe SA, Vinci G, Di Sotto A, Ambroselli D, Vergine V, Crestoni ME, Di Meo C, Zoratto N, Izzo L, Navarré A, Adiletta G, Russo P, Di Matteo G, Mannina L, Giusti AM (2025) Enhancing human health through nutrient and bioactive compound recovery from agri-food by-products: a decade of progress. Nutrients 17(15):2528. https://doi.org/10.3390/nu17152528

Iqbal I, Wilairatana P, Saqib F, Nasir B, Wahid M, Latif MF, Iqbal A, Naz R, Mubarak MS (2023) Plant polyphenols and their potential benefits on cardiovascular health: a review. Molecules 28(17):6403. https://doi.org/10.3390/molecules28176403

Ismail MM, Alotaibi BS, El-Sheekh MM (2020) Therapeutic uses of red macroalgae. Molecules 25(19):4411. https://doi.org/10.3390/molecules25194411

Jayawardhana HHACK, Jayawardena TU, Sanjeewa KKA, Liyanage NM, Nagahawatta DP, Lee HG, Kim JI, Jeon YJ (2023) Marine algal polyphenols as skin protective agents: current status and future prospectives. Mar Drugs 21(5):285. https://doi.org/10.3390/md21050285

Jena AB, Samal RR, Bhol NK, Duttaroy AK (2023) Cellular red-ox system in health and disease: the latest update. Biomed Pharmacother 162:114606. https://doi.org/10.1016/j.biopha.2023.114606

Jin L, Quan C, Hou X, Fan S (2016) Potential pharmacological resources: natural bioactive compounds from marine-derived fungi. Mar Drugs 14(4):76. https://doi.org/10.3390/md14040076

Jomova K, Raptova R, Alomar SY, Alwasel SH, Nepovimova E, Kuca K, Valko M (2023) Reactive oxygen species, toxicity, oxidative stress, and antioxidants: chronic diseases and aging. Arch Toxicol 97(10):2499–2574. https://doi.org/10.1007/s00204-023-03562-9

Jornada DH, dos Santos Fernandes GF, Chiba DE, de Melo TRF, dos Santos JL, Chung MC (2015) The prodrug approach: a successful tool for improving drug solubility. Molecules 21(1):42. https://doi.org/10.3390/molecules21010042

Jówko E (2015) Green tea catechins and sport performance. In: Lamprecht M (ed) Antioxidants in sport nutrition. CRC Press/Taylor and Francis, Boca Raton

Kabir MT, Rahman MH, Shah M, Jamiruddin Mohd R, Basak D, Al-Harrasi A, Bhatia S, Ashraf GM, Najda A, El-kott AF, Mohamed HRH, Al-malky HS, Germoush MO, Altyar AE, Alwafai EB, Ghaboura N, Abdel-Daim MM (2022) Therapeutic promise of carotenoids as antioxidants and anti-inflammatory agents in neurodegenerative disorders. Biomed Pharmacother 146:112610. https://doi.org/10.1016/j.biopha.2021.112610

Karthikeyan A, Joseph A, Nair BG (2022) Promising bioactive compounds from the marine environment and their potential effects on various diseases. J Genet Eng Biotechnol 20:14. https://doi.org/10.1186/s43141-021-00290-4

Khalifa SAM, Elias N, Farag MA, Chen L, Saeed A, Hegazy MEF, Moustafa MS, Abd El-Wahed A, Al-Mousawi SM, Musharraf SG, Chang FR, Iwasaki A, Suenaga K, Alajlani M, Göransson U, El-Seedi HR (2019) Marine natural products: a source of novel anticancer drugs. Mar Drugs 17(9):491. https://doi.org/10.3390/md17090491

Khan N, Syed DN, Ahmad N, Mukhtar H (2013) Fisetin: a dietary antioxidant for health promotion. Antioxid Redox Signal 19(2):151–162. https://doi.org/10.1089/ars.2012.4901

Khurana S, Venkataraman K, Hollingsworth A, Piche M, Tai TC (2013) Polyphenols: benefits to the cardiovascular system in health and in aging. Nutrients 5(10):3779–3827. https://doi.org/10.3390/nu5103779

Kikuchi M, Hara N, Hasegawa M, Miyashita A, Kuwano R, Ikeuchi T, Nakaya A (2019) Enhancer variants associated with Alzheimer's disease affect gene expression via chromatin looping. BMC Med Gen 12:128. https://doi.org/10.1186/s12920-019-0574-8

Kim KS, Cui X, Lee DS, Ko W, Sohn JH, Yim JH, An RB, Kim YC, Oh H (2014) Inhibitory effects of benzaldehyde derivatives from the marine fungus Eurotium sp. SF-5989 on inflammatory mediators via the induction of heme oxygenase-1 in lipopolysaccharide-stimulated RAW 264.7 macrophages. Int J Mol Sci 15(12):23749–23765. https://doi.org/10.3390/ijms151223749

Kim JM, Heo HJ (2022) The roles of catechins in regulation of systemic inflammation. Food Sci Biotechnol 31(8):957–970. https://doi.org/10.1007/s10068-022-01069-0

Kostić K, Brborić J, Delogu G, Simić MR, Samardžić S, Maksimović Z, Dettori MA, Fabbri D, Kotur-Stevuljević J, Saso L (2023) Antioxidant activity of natural phenols and derived hydroxylated biphenyls. Molecules 28(6):2646. https://doi.org/10.3390/molecules28062646

Kulkarni A, Kulkarni CC, Pradeep SR, Poyya J, Kudva AK, Radhakrishnan V, Khandagale AS (2025) Role of anti-inflammatory and antioxidant properties of natural products in curing cardiovascular diseases. Curr Issues Mol Biol 47(11):955. https://doi.org/10.3390/cimb47110955

Kumar N, Goel N (2019) Phenolic acids: natural versatile molecules with promising therapeutic applications. Biotechnol Rep 24:e00370. https://doi.org/10.1016/j.btre.2019.e00370

Kumar LRG, Paul PT, Anas KK, Tejpal CS, Chatterjee NS, Anupama TK, Mathew S, Ravishankar CN (2022) Phlorotannins–bioactivity and extraction perspectives. J Appl Phycol 34(4):2173–2185. https://doi.org/10.1007/s10811-022-02749-4

Kupikowska-Stobba B, Niu H, Klojdová I, Agregán R, Lorenzo JM, Kasprzak M (2025) Controlled lipid digestion in the development of functional and personalized foods for a tailored delivery of dietary fats. Food Chem 466:142151. https://doi.org/10.1016/j.foodchem.2024.142151

Lang Y, Gao N, Zang Z, Meng X, Lin Y, Yang S, Yang Y, Jin Z, Li B (2024) Classification and antioxidant assays of polyphenols: a review. J Future Foods 4(3):193–204. https://doi.org/10.1016/j.jfutfo.2023.07.002

Li KX, Wang ZC, Machuki JO, Li MZ, Wu YJ, Niu MK, Yu KY, Lu QB, Sun HJ (2022) Benefits of curcumin in the vasculature: a therapeutic candidate for vascular remodeling in arterial hypertension and pulmonary arterial hypertension? Front Physiol 13:848867. https://doi.org/10.3389/fphys.2022.848867

Li B, Ming H, Qin S, Nice EC, Dong J, Du Z, Huang C (2025) Redox regulation: mechanisms, biology and therapeutic targets in diseases. Signal Transduct Target Ther 10(1):72. https://doi.org/10.1038/s41392-024-02095-6

Liu W, Cui X, Zhong Y, Ma R, Liu B, Xia Y (2023) Phenolic metabolites as therapeutic in inflammation and neoplasms: molecular pathways explaining their efficacy. Pharmacol Res 193:106812. https://doi.org/10.1016/j.phrs.2023.106812

Lobo V, Patil A, Phatak A, Chandra N (2010) Free radicals, antioxidants and functional foods: impact on human health. Pharmacogn Rev 4(8):118–126. https://doi.org/10.4103/0973-7847.70902

Lomartire S, Gonçalves AMM (2023) Marine macroalgae polyphenols as potential neuroprotective antioxidants in neurodegenerative diseases. Mar Drugs 21(5):261. https://doi.org/10.3390/md21050261

López-Hortas L, Flórez-Fernández N, Torres MD, Ferreira-Anta T, Casas MP, Balboa EM, Falqué E, Domínguez H (2021) Applying seaweed compounds in cosmetics, cosmeceuticals and nutricosmetics. Mar Drugs 19(10):552. https://doi.org/10.3390/md19100552

Lourenço SC, Moldão-Martins M, Alves VD (2019) Antioxidants of natural plant origins: from sources to food industry applications. Molecules 24(22):4132. https://doi.org/10.3390/molecules24224132

Manandhar B, Wagle A, Seong SH, Paudel P, Kim HR, Jung HA, Choi JS (2019) Phlorotannins with potential anti-tyrosinase and antioxidant activity isolated from the marine seaweed ecklonia stolonifera. Antioxidants 8(8):240. https://doi.org/10.3390/antiox8080240

Marengo B, Nitti M, Furfaro AL, Colla R, Ciucis CD, Marinari UM, Pronzato MA, Traverso N, Domenicotti C (2016) Redox homeostasis and cellular antioxidant systems: crucial players in cancer growth and therapy. Oxid Med Cell Longev 2016:6235641. https://doi.org/10.1155/2016/6235641

Markovic M, Ben-Shabat S, Dahan A (2020) Prodrugs for improved drug delivery: lessons learned from recently developed and marketed products. Pharmaceutics 12(11):1031. https://doi.org/10.3390/pharmaceutics12111031

Melotti L, Venerando A, Zivelonghi G, Carolo A, Marzorati S, Martinelli G, Sugni M, Maccatrozzo L, Patruno M (2023) A second life for seafood waste: therapeutical promises of polyhydroxynapthoquinones extracted from sea urchin by-products. Antioxidants 12(9):1730. https://doi.org/10.3390/antiox12091730

Menaa F, Wijesinghe U, Thiripuranathar G, Althobaiti NA, Albalawi AE, Khan BA, Menaa B (2021) Marine algae-derived bioactive compounds: a new wave of nanodrugs? Mar Drugs 19(9):484. https://doi.org/10.3390/md19090484

Méresse S, Fodil M, Fleury F, Chénais B (2020) Fucoxanthin, a marine-derived carotenoid from brown seaweeds and microalgae: a promising bioactive compound for cancer therapy. Int J Mol Sci 21(23):9273. https://doi.org/10.3390/ijms21239273

Michaels T (2024) Safeguarding the brain: understanding neuroprotection and its implications. Neurosci Psychiatry Open Access 7(2):188–190. https://doi.org/10.47532/npoa.2024.7(2).188-190

Minocha T, Birla H, Obaid AA, Rai V, Sushma P, Shivamallu C, Moustafa M, Al-Shehri M, Al-Emam A, Tikhonova MA, Yadav SK, Poeggeler B, Singh D, Singh SK (2022) Flavonoids as promising neuroprotectants and their therapeutic potential against alzheimer's disease. Oxid Med Cell Longev 2022:6038996. https://doi.org/10.1155/2022/6038996

Mirza MA, Padhi S, Mohapatra S, Mahmood S, Iqbal Z (2025) Fisetin from dietary supplement to a drug candidate: an assessment of potential. Curr Pharm Biotechnol 26(8):1143–1158. https://doi.org/10.2174/0113892010304479240415074928

Miyashita K, Beppu F, Hosokawa M, Liu X, Wang S (2020) Nutraceutical characteristics of the brown seaweed carotenoid fucoxanthin. Arch Biochem Biophys 686:108364. https://doi.org/10.1016/j.abb.2020.108364

Moldoveanu CA, Tomoaia-Cotisel M, Sevastre-Berghian A, Tomoaia G, Mocanu A, Pal-Racz C, Toma VA, Roman I, Ujica MA, Pop LC (2024) A review on current aspects of curcumin-based effects in relation to neurodegenerative, neuroinflammatory and cerebrovascular diseases. Molecules 30(1):43. https://doi.org/10.3390/molecules30010043

Morkovin E, Litvinov R, Koushner A, Babkov D, Morkovin E, Litvinov R, Koushner A, Babkov D (2024) Resveratrol and extra virgin olive oil: protective agents against age-related disease. Nutrients 16(24):4258. https://doi.org/10.3390/nu16244258

Mutha RE, Tatiya AU, Surana SJ (2021) Flavonoids as natural phenolic compounds and their role in therapeutics: an overview. Futur J Pharm Sci 7(1):25. https://doi.org/10.1186/s43094-020-00161-8

Nag M, Lahiri D, Dey A, Sarkar T, Pati S, Joshi S, Bunawan H, Mohammed A, Edinur HA, Ghosh S, Ray RR (2022) Seafood discards: a potent source of enzymes and biomacromolecules with nutritional and nutraceutical significance. Front Nutr 9:879929. https://doi.org/10.3389/fnut.2022.879929

Nagahawatta DP, Liyanage NM, Jayawardena TU, Jeon YJ (2024) Marine polyphenols in cardiovascular health: unraveling structure-activity relationships, mechanisms, and therapeutic implications. Int J Mol Sci 25(15):8419. https://doi.org/10.3390/ijms25158419

Nair A, Maity S, Pai V (2026) Sustainable extraction: a comprehensive review of advancements beyond conventional methods. Microchem J 220:116360. https://doi.org/10.1016/j.microc.2025.116360

Nakano T, Wiegertjes G (2020) Properties of carotenoids in fish fitness: a review. Mar Drugs 18(11):568. https://doi.org/10.3390/md18110568

Nan Y, Su H, Zhou B, Liu S (2023) The function of natural compounds in important anticancer mechanisms. Front Oncol 12:1049888. https://doi.org/10.3389/fonc.2022.1049888

NaS D, Alayudin ASM, Sofian-Seng N, Rahman HA, Razali NSM, Lim SJ, WaW M (2022) Brown algae as functional food source of fucoxanthin: a review. Foods 11(15):2235. https://doi.org/10.3390/foods11152235

Ngandjui YAT, Kereeditse TT, Kamika I, Madikizela LM, Msagati TAM (2024) Nutraceutical and medicinal importance of marine molluscs. Mar Drugs 22(5):201. https://doi.org/10.3390/md22050201

Noreen S, Hashmi B, Aja PM, Atoki AV (2025) Health benefits of fish and fish by-products—a nutritional and functional perspective. Front Nutr 12:1564315. https://doi.org/10.3389/fnut.2025.1564315

Oey O, Wijaya W, Redfern A (2024) Eribulin in breast cancer: current insights and therapeutic perspectives. World J Exp Med 14(2):92558. https://doi.org/10.5493/wjem.v14.i2.92558

Pandey KB, Rizvi SI (2009) Plant polyphenols as dietary antioxidants in human health and disease. Oxid Med Cell Longev 2(5):270–278. https://doi.org/10.4161/oxim.2.5.9498

Patanè GT, Putaggio S, Tellone E, Barreca D, Ficarra S, Maffei C, Calderaro A, Laganà G (2023) Catechins and proanthocyanidins involvement in metabolic syndrome. Int J Mol Sci 24(11):9228. https://doi.org/10.3390/ijms24119228

Patra JK, Das G, Fraceto LF, Campos EVR, Del Pilar R-T, Acosta-Torres LS, Diaz-Torres LA, Grillo R, Swamy MK, Sharma S, Habtemariam S, Shin H (2018) Nano based drug delivery systems: recent developments and future prospects. J Nanobiotechnol 16(1):71. https://doi.org/10.1186/s12951-018-0392-8

Pereira L, Cotas J (2023) Therapeutic potential of polyphenols and other micronutrients of marine origin. Mar Drugs 21(6):323. https://doi.org/10.3390/md21060323

Pereira RB, Evdokimov NM, Lefranc F, Valentão P, Kornienko A, Pereira DM, Andrade PB, Gomes NGM (2019) Marine-derived anticancer agents: clinical benefits, innovative mechanisms, and new targets. Mar Drugs 17(6):329. https://doi.org/10.3390/md17060329

Pereira CSGP, Carpena M, Barreira JCM, Silva CFC, Prieto MA, Oliveira MBPP (2025) Bioactive metabolites from algae: occurrence, extraction techniques, functional properties, food applications and therapeutic prospects. Futur Foods 13:100866. https://doi.org/10.1016/j.fufo.2025.100866

Petcu CD, Tăpăloagă D, Mihai OD, Gheorghe-Irimia RA, Negoiță C, Georgescu IM, Tăpăloagă PR, Borda C, Ghimpețeanu OM (2023) Harnessing natural antioxidants for enhancing food shelf life: exploring sources and applications in the food industry. Foods 12(17):3176. https://doi.org/10.3390/foods12173176

Petrovic S, Bita B, Barbinta-Patrascu ME (2024) Nanoformulations in pharmaceutical and biomedical applications: green perspectives. Int J Mol Sci 25(11):5842. https://doi.org/10.3390/ijms25115842

Praise SA, Kolawole AT, Olusanya MM, Bakare OS, Ezeuchenne IC (2025) Investigation of musa paradisiaca compounds as dipeptidyl peptidase-4, alpha-amylase and protein tyrosine phosphatase 1B inhibitors in diabetes via in-silico approach. Next Res 2(1):100108. https://doi.org/10.1016/j.nexres.2024.100108

Purcell D, Packer MA, Hayes M (2022) Angiotensin-I-converting enzyme inhibitory activity of protein hydrolysates generated from the macroalga laminaria digitata (hudson) JV lamouroux 1813. Foods 11(12):1792. https://doi.org/10.3390/foods11121792

Rajendran P, Abdelsalam SA, Renu K, Veeraraghavan V, Ben Ammar R, Ahmed EA (2022) Polyphenols as potent epigenetics agents for cancer. Int J Mol Sci 23(19):11712. https://doi.org/10.3390/ijms231911712

Rajesh RU, Dhanaraj S (2023) A critical review on quercetin bioflavonoid and its derivatives: scope, synthesis, and biological applications with future prospects. Arab J Chem 16(8):104881. https://doi.org/10.1016/j.arabjc.2023.104881

Rampengan DDCH, Lele JAJMN, Romano R, Taslim NA, Tjandrawinata RR, Santini A, Nurkolis F (2025) Harnessing the power of marine terpenoids against diabetes-associated oxidative stress. Front Nutr 12:1651804. https://doi.org/10.3389/fnut.2025.1651804

Rananaware P, Singh S, Brahmkhatri VP (2025) Scavenging of reactive oxygen and nitrogen species using nanoparticles and their applications in disease management. RSC Adv 15(56):47955–47980. https://doi.org/10.1039/d5ra05248c

Rathee P, Sehrawat R, Rathee P, Khatkar A, Akkol EK, Khatkar S, Redhu N, Türkcanoğlu G, Sobarzo-Sánchez E (2023) Polyphenols: natural preservatives with promising applications in food, cosmetics and pharma industries; problems and toxicity associated with synthetic preservatives; impact of misleading advertisements; recent trends in preservation and legislation. Materials 16(13):4793. https://doi.org/10.3390/ma16134793

Rathod NB, Elabed N, Punia S, Ozogul F, Kim SK, Rocha JM (2023) Recent developments in polyphenol applications on human health: a review with current knowledge. Plants 12(6):1217. https://doi.org/10.3390/plants12061217

Roik A, Reverter M, Pogoreutz C (2022) A roadmap to understanding diversity and function of coral reef-associated fungi. FEMS Microbiol Rev 46(6):fuac028. https://doi.org/10.1093/femsre/fuac028

Rosales TKO, Fabi JP (2023) Valorization of polyphenolic compounds from food industry by-products for application in polysaccharide-based nanoparticles. Front Nutr 10:1144677. https://doi.org/10.3389/fnut.2023.1144677

Rubiño S, Peteiro C, Aymerich T, Hortós M (2022) Brown macroalgae (phaeophyceae): a valuable reservoir of antimicrobial compounds on northern coast of Spain. Mar Drugs 20(12):775. https://doi.org/10.3390/md20120775

Rudrapal M, Khairnar SJ, Khan J, Dukhyil AB, Ansari MA, Alomary MN, Alshabrmi FM, Palai S, Deb PK, Devi R (2022) Dietary polyphenols and their role in oxidative stress-induced human diseases: insights into protective effects, antioxidant potentials and mechanism(s) of action. Front Pharmacol 13:806470. https://doi.org/10.3389/fphar.2022.806470

Rumman M, Khan MV (2026) Drug delivery systems for enhanced neuroprotective effects of polyphenols: current status and challenges. J Drug Deliv Sci Technol 115:107767. https://doi.org/10.1016/j.jddst.2025.107767

Saad AM, Mohammed DM, Alkafaas SS, Ghosh S, Negm SH, Salem HM, Fahmy MA, Semary HE, Ibrahim EH, Abu Qamar SF, El-Tarabily KA, El-Saadony MT (2025) Dietary polyphenols and human health: sources, biological activities, nutritional and immunological aspects, and bioavailability–a comprehensive review. Front Immunol 16:1653378. https://doi.org/10.3389/fimmu.2025.1653378

Sadeghi A, Rajabiyan A, Nabizade N, Meygoli Nezhad N, Zarei-Ahmady A (2024) Seaweed-derived phenolic compounds as diverse bioactive molecules: a review on identification, application, extraction and purification strategies. Int J Biol Macromol 266:131147. https://doi.org/10.1016/j.ijbiomac.2024.131147

Salehi M, Rashidinejad A (2025) Multifaceted roles of plant-derived bioactive polysaccharides: a review of their biological functions, delivery, bioavailability, and applications within the food and pharmaceutical sectors. Int J Biol Macromol 290:138855. https://doi.org/10.1016/j.ijbiomac.2024.138855

Sanjeewa KKA, Kim EA, Son KT, Jeon YJ (2016) Bioactive properties and potentials cosmeceutical applications of phlorotannins isolated from brown seaweeds: a review. J Photochem Photobiol B: Biol 162:100–105. https://doi.org/10.1016/j.jphotobiol.2016.06.027

Scott MB, Styring AK, McCullagh JSO (2022) Polyphenols: bioavailability, microbiome interactions and cellular effects on health in humans and animals. Pathogens 11(7):770. https://doi.org/10.3390/pathogens11070770

Serreli G, Deiana M (2023) Role of dietary polyphenols in the activity and expression of nitric oxide synthases: a review. Antioxidants 12(1):147. https://doi.org/10.3390/antiox12010147

Sharifi-Rad J, Rayess YE, Rizk AA, Sadaka C, Zgheib R, Zam W, Sestito S, Rapposelli S, Neffe-Skocińska K, Zielińska D, Salehi B, Setzer WN, Dosoky NS, Taheri Y, El Beyrouthy M, Martorell M, Ostrander EA, Suleria HAR, Cho WC, Maroyi A, Martins N (2020) Turmeric and its major compound curcumin on health: bioactive effects and safety profiles for food, pharmaceutical, biotechnological and medicinal applications. Front Pharmacol 11:01021. https://doi.org/10.3389/fphar.2020.01021

Simon SA, Aschmann V, Behrendt A, Hügler M, Engl LM, Pohlner M, Rolfes S, Brinkhoff T, Engelen B, Könneke M, Rodriguez-R LM, Bornemann TLV, Nuy JK, Rothe L, Stach TL,

Beblo-Vranesevic K, Leuko S, Runzheimer K, Möller R, Conrady M, Probst AJ (2025) Earth's most needed uncultivated aquatic prokaryotes. Water Res 273:122928. https://doi.org/10.1016/j.watres.2024.122928
Singh H, Kamal YT, Pandohee J, Mishra AK, Biswas A, Mohanto S, Kumar A, Nag S, Mishra A, Singh M, Gupta H, Chopra H (2025) Dietary phytochemicals alleviate the premature skin aging: a comprehensive review. Exp Gerontol 199:112660. https://doi.org/10.1016/j.exger.2024.112660
Son BW, Kim JC, Choi HD, Kang JS (2002) A radical scavenging farnesylhydroquinone from a marine-derived fungus Penicillium sp. Arch Pharm Res 25(1):77–79. https://doi.org/10.1007/BF02975266
Suleria HAR, Osborne S, Masci P, Gobe G (2015) Marine-based nutraceuticals: an innovative trend in the food and supplement industries. Mar Drugs 13(10):6336–6351. https://doi.org/10.3390/md13106336
Tamzi NN, Rahman MM, Das S (2024) Recent advances in marine-derived bioactives towards cancer therapy. Int J Translat Med 4(4):740–781. https://doi.org/10.3390/ijtm4040051
Tong A, Wang D, Jia N, Zheng Y, Qiu Y, Chen W, El-Seed HR, Zhao C (2024) Algal active ingredients and their involvement in managing diabetic mellitus. Biology 13(11):904. https://doi.org/10.3390/biology13110904
Truong VL, Jeong WS (2022) Antioxidant and anti-inflammatory roles of tea polyphenols in inflammatory bowel diseases. Food Sci Hum Wellness 11(3):502–511. https://doi.org/10.1016/j.fshw.2021.12.008
Ugariogu SN, Al-Tannak NF, Alenezi SS (2025) A review on recent insights on cardiovascular diseases drug design using natural products and in silico studies. Kuwait J Sci 52(4):100438. https://doi.org/10.1016/j.kjs.2025.100438
Vasarri M, Degl'Innocenti D (2022) Antioxidant and anti-inflammatory agents from the sea: a molecular treasure for new potential drugs. Mar Drugs 20(2):132. https://doi.org/10.3390/md20020132
Vauzour D, Rodriguez-Mateos A, Corona G, Oruna-Concha MJ, Spencer JPE (2010) Polyphenols and human health: prevention of disease and mechanisms of action. Nutrients 2(11):1106–1131. https://doi.org/10.3390/nu2111106
Veluchamy C, Nallakaruppan N, Nachiappan K, Sekaran M, Vartak RB, Zambare RB, Panneerselvam S, Chandrasekaran R, Sharma A, Thiagarajan K (2025) Marine microbial bioprospecting and bioactive bioproducts: seven decades of discovery and contemporary perspectives. Front Mar Sci 12:1698327. https://doi.org/10.3389/fmars.2025.1698327
Vitale GA, Coppola D, Palma Esposito F, Buonocore C, Ausuri J, Tortorella E, de Pascale D (2020) Antioxidant molecules from marine fungi: methodologies and perspectives. Antioxidants 9(12):1183. https://doi.org/10.3390/antiox9121183
Vollmannová A, Bojňanská T, Musilová J, Lidiková J, Cifrová M (2024) Quercetin as one of the most abundant represented biological valuable plant components with remarkable chemoprotective effects—a review. Heliyon 10(12):e33342. https://doi.org/10.1016/j.heliyon.2024.e33342
WaJP W, Jeon YJ (2012) Exploiting biological activities of brown seaweed Ecklonia cava for potential industrial applications: a review. Int J Food Sci Nutr 63(2):225–235. https://doi.org/10.3109/09637486.2011.619965
Wang Q, Yang B, Wang N, Gu J (2022) Tumor immunomodulatory effects of polyphenols. Front Immunol 13:1041138. https://doi.org/10.3389/fimmu.2022.1041138
Wang YZ, Zhao W, Moorjani P, Gross AL, Zhou X, Dey AB, Lee J, Smith JA, Kardia SLR (2024) Effect of apolipoprotein E ε4 and its modification by sociodemographic characteristics on cognitive measures in South Asians from LASI-DAD. Alzheimer's Dement 20(7):4854–4867. https://doi.org/10.1002/alz.14052
Wang Z, Zhao M, Yu Y, Kong F, Lin N, Wang Q (2025) Marine fungal metabolites as potential antidiabetic agents: a comprehensive review of their structures and enzyme inhibitory activities. Mar Drugs 23(4):142. https://doi.org/10.3390/md23040142

Wang S, Li XM Teuscher F, Li DL, Diesel A, Ebel R, Proksch P, Wang BG (2006) Chaetopyranin, a benzaldehyde derivative, and other related metabolites from Chaetomium globosum, an endophytic fungus derived from the marine red alga Polysiphonia urceolata. J Nat Prod 69(11):1622–1625. https://doi.org/10.1021/np060248n

Wijethunga T, Wijesekara T, Zhong S, Xu B (2025) Critical review on bioactive substances from sea food and their health promoting effects. J Futur Foods, in press. https://doi.org/10.1016/j.jfutfo.2025.04.022

Yadav R, Mishra S, Chaturvedi R, Pandey A (2025) Therapeutic potential of curcumin in cardiovascular disease: targeting atherosclerosis pathophysiology. Biomed Pharmacother 190:118412. https://doi.org/10.1016/j.biopha.2025.118412

Yahfoufi N, Alsadi N, Jambi M, Matar C (2018) The immunomodulatory and anti-inflammatory role of polyphenols. Nutrients 10(11):1618. https://doi.org/10.3390/nu10111618

Yang B, Dong Y, Wang F, Zhang Y (2020a) Nanoformulations to enhance the bioavailability and physiological functions of polyphenols. Molecules 25(20):4613. https://doi.org/10.3390/molecules25204613

Yang D, Wang T, Long M, Li P (2020b) Quercetin: its main pharmacological activity and potential application in clinical medicine. Oxid Med Cell Longev 2020:8825387. https://doi.org/10.1155/2020/8825387

Yesupatham A, Saraswathy R (2025) Role of oxidative stress in prediabetes development. Biochem Biophys Rep 43:102069. https://doi.org/10.1016/j.bbrep.2025.102069

Yoshioka Y, Ohishi T, Nakamura Y, Fukutomi R, Miyoshi N (2022) Anti-cancer effects of dietary polyphenols via ROS-mediated pathway with their modulation of MicroRNAs. Molecules 27(12):3816. https://doi.org/10.3390/molecules27123816

Yu W, Zhang Y, Lu Y, Ouyang Z, Peng J, Tu Y, He B (2025) Recent research on the bioactivity of polyphenols derived from edible fungi and their potential in chronic disease prevention. J Funct Foods 124:106627. https://doi.org/10.1016/j.jff.2024.106627

Zambelli RA, De Mendonça LG, Barroso AB (2026) Unconventional ingredients for flavors and textures. In: Sarkar T, Smaoui S (eds) Dietary, sensory and gastronomic applications, pp 241–269. Academic Press, Cambdrige, MA, Oxford, and Amsterdam. https://doi.org/10.1016/B978-0-443-33345-3.00009-2

Zhang YM, Zhang ZG (2025) Modulating nuclear factor erythroid 2-related factor 2 and heme oxygenase-1 in liver-brain axis disorders. World J Psychiatr 15(9):108382. https://doi.org/10.5498/wjp.v15.i9.108382

Zhang H, Tang Y, Zhang Y, Zhang S, Qu J, Wang X, Kong R, Han C, Liu Z (2015) Fucoxanthin: a promising medicinal and nutritional ingredient. Evid Based Compl Altern Med 2015:723515. https://doi.org/10.1155/2015/723515

Zhang Z, Li X, Sang S, McClements DJ, Chen L, Long J, Jiao A, Jin Z, Qiu C (2022b) Polyphenols as plant-based nutraceuticals: health effects, encapsulation, nano-delivery, and application. Foods 11(15):2189. https://doi.org/10.3390/foods11152189

Zhang Y, Chen R, Zhang D, Qi S, Liu Y (2023) Metabolite interactions between host and microbiota during health and disease: which feeds the other? Biomed Pharmacother 160:114295. https://doi.org/10.1016/j.biopha.2023.114295

Zhang Y, Cai P, Cheng G, Zhang Y (2022) A brief review of phenolic compounds identified from plants: their extraction, analysis, and biological activity. Nat Prod Commun 17(1):1934578X211069721. https://doi.org/10.1177/1934578X211069721

Zhao T, Zheng F, Liu Y, Khan A, Wang Z, Cheng G (2023) A comparative analysis of chemical constituents and antioxidant effects of dendrobium fimbriatum hook fractions with different polarities. Int J Mol Sci 24(16):12646. https://doi.org/10.3390/ijms241612646

# Chapter 9
# Polyphenols-Rich Functional Foods. Current Claims and Regulatory Perspectives

**Suni Mary Varghese, Carmelo Parisi, Vincenzo Barone, Vito Barone, Giorgia Punzo, Sara Petrelli, and Salvatore Parisi**

**Abstract** The problem of health and nutrition claims concerning polyphenols worldwide is strictly linked with the claimed function of these molecules and/or related classes when speaking of weight regulation, obesity and weight-related diseases in human medicine. These and other inflammatory diseases have to be considered as the most visible effect of biochemical strategies able to preserve functions and tissues of the subject from external and attacking agents of various nature. When speaking of health and nutrition claims, the use of dietary prescriptions should be considered as a possible co-treatment, and the management of weight-related disorders is surely connected with the consumption of favorable compounds. Polyphenols are extremely interesting in this ambit because of their natural origin and the absence of adverse effects, if compared with steroidal anti-inflammatory drugs. Dietary prescriptions concerning the proactive and predetermined use of polyphenols

S. M. Varghese · S. Parisi (✉)
Department of Nutrition and Food Processing Technology, Lourdes Matha Institute of Hotel Management and Catering Technology, Thiruvananthapuram, Kerala, India
e-mail: drparisi@inwind.it

S. M. Varghese
e-mail: sunimary@lmihmct.org

C. Parisi · V. Barone · V. Barone · G. Punzo
Palermo, Italy
e-mail: carmelo.parisi2012@libero.it

V. Barone
e-mail: vincenzo.barone2012@gmail.com

V. Barone
e-mail: vitobarone99@gmail.com

G. Punzo
e-mail: punzogiorgia01@gmail.com

S. Petrelli
Capaci (PA), Sicily, Italy
e-mail: sarettap25@gmail.com

 
R. K. Singla et al. (eds.), *Polyphenols in Health*,
https://doi.org/10.1007/978-3-032-27324-6_9

against metabolic disorders such as obesity are linked with two macro-categories: patient-related factors, and medical prescriptions-factors. The influence of specific polyphenols classes is a challenging argument so far. At the same time, health and nutrition claim should be absolutely demonstrable, reliable, and not be capable of raising doubts in the consumer about their dietary choices. These recommendations are considered worldwide, but National guidelines and associated regulations differ enough. More research—and improved cooperation between National regulations—is needed when speaking of the reliability of selected studies concerning the positive effects of polyphenols in selected medical ambits.

**Keywords** Chronic inflammation · Dietary prescription · Health claim · Metabolic disorder · Nutrition claim · Polyphenol · Weight management

## Abbreviations

| | |
|---|---|
| European Food Safety Authority | EFSA |
| Low-density lipoprotein | LDL |
| Food and Drug Administration | FDA |

## 9.1 Polyphenols and Regulatory Considerations for Health and Nutrition Claims. A Short Introduction

The problem of health and nutrition claims concerning polyphenols worldwide is strictly linked with the claimed function of these molecules and/or related classes when speaking of weight regulation, obesity and weight-related diseases in human medicine. As explained in Chap. 12, these and other inflammatory diseases have to be considered as the most visible effect of biochemical strategies able to preserve functions and tissues of the subject from external and attacking agents of various nature (Gasmi et al. 2022). The following disorders are included in this ambit (Black 2002; Dantzer et al. 2006; Hussain et al. 2016; Pasparakis et al. 2014; Rached et al. 2025; Santamarina et al. 2025; Zamani-Garmsiri et al. 2022).

(1) Metabolic disorders such as obesity and peculiar anti-dietary behaviors such as prolonged high-calorie consumption
(2) Chronic, autoimmune, and low-grade metabolic disorders including insulin resistance, type-2 diabetes, and cardiovascular diseases.

The rapidity and efficacy of inflammatory answers by the attacked human subject are discussed in Chap. 12. As a result, normal effects may not be immediate enough, and automated mechanisms should be not always expected. In these situations, health conditions may be not readily restored because of some important and endogenous factors:

(a) Subject-related effects (age, sex, etc.) can remarkably influence the result with the occurrence of severe symptoms
(b) Localization of attacked body organ(s) and tissues with notable effects on the efficacy and rapidity of counter-attacking answers (the occurrence of non-weight related symptoms associated to fever episodes, modification of blood composition, and well-known diseases including rheumatoid arthritis or bowel disease should be considered preventatively) (Gasmi et al. 2022; Homey et al. 2006; Yoon and Baek 2005).

Anyway, defense activation should not be always considered as a fully effective strategy (Chalmers et al. 2025; Guo et al. 2025; Harvanová and Duranková 2025; Pichard et al. 2025; Zhang et al. 2021). As an example, the deployment of adipokines with the aim of helping infiltration of the obese adipose tissue by white adipose tissue or inflammatory cells can be discussed (Engin 2017; Landechoet al. 2019; Nani et al. 2021). The problem is that such a deployment and associated/predictable effects could fail to define a good inflammatory answer. Should this situation occur, 'chronic inflammation' would follow the initial inflammatory answer, and physiological functions could not be fully restored. The possibility of concomitant autoimmune diseases (cardiovascular problems, metabolic syndrome, etc.) and patients' age should also be taken into account because of their roles as anti-inflammatory answers the general situation (Bucciantini et al. 2021).

## 9.2 Polyphenols and Regulatory Considerations for Health and Nutrition Claims. A Bird's Eye on Current Claims and Regulatory Considerations

When speaking of health and nutrition claims, the use of dietary prescriptions should be considered as a possible co-treatment, and the management of weight-related disorders is surely connected with the consumption of favorable compounds. Polyphenols are extremely interesting in this ambit because of their natural origin and the absence of adverse effects, if compared with steroidal anti-inflammatory drugs (Behl et al. 2022; Chen et al. 2025; Das et al. 2025; Kim et al. 2025; Li et al. 2025; Omar et al. 2025; Petersen et al. 2025; Ran et al. 2025; Tumbarski et al. 2025).

Dietary prescriptions concerning the proactive and predetermined use of polyphenols against metabolic disorders such as obesity are linked with two macro-categories of factors which should be considered carefully.

(a) Patient-related factors (Annuzzi et al. 2014; Bozzetto et al. 2015; Chong et al. 2013; Condezo-Hoyos et al. 2021; Macready et al. 2014; Malaveille et al. 2004; Vetrani et al. 2018–2020): age, classification of metabolic disorder (based on body mass index-measurement or other systems); physical measures/

features such as high waist circumference; chemical blood profiles; concomitant disorders and related classification such as grade I- or grade II-hypertension; concomitant abuse-factors such as smoking

(b) Medical prescriptions-factors: predominance of some polyphenols-rich foods or beverages with associated intake per day and/or per week (Annuzzi et al. 2014; Bozzetto et al. 2015; Chong et al. 2013; Condezo-Hoyos et al. 2021; Macready et al. 2014; Malaveille et al. 2004; Vetrani et al. 2018–2020): fruits (oranges, blueberries, strawberries and kiwis in descending order); vegetables (onion, rocket, spinach, cabbage, and tomatoes in descending order); decaffeinated green tea and coffee; cocoa-derived or containing foods, including dark chocolate; extra virgin olive oil; etc.

It has to be noted that the influence of specific polyphenols classes ('flavonoids': anthocyanidins, chalcones, flavanols, flavonols, flavanones, flavones, and isoflavones; and 'non-flavonoids': stilbenes, lignans, phenolic acids, tannins, curcominoids, and coumarins)—or specific molecules is a challenging argument so far (Chap. 12). In this ambit, well-known models such Mediterranean Diet or other 'sustainable and healthy' dietary habits can be studied with demonstrated results. However, there is not much information concerning the real importance of polyphenols amount in the ambit of health and nutrition dietary prescriptions.

At the same time, health and nutrition claim should be absolutely demonstrable. These advertising messages should not be capable of confusing the consumer in any way by implying that the food or supplement containing a polyphenol, or a polyphenol-related class of substances, has a certain effect on nutritional safety or adequacy (if compared to other foods or supplements). Other unacceptable behaviors include the suggestion of excessive consumption of the advertised foods or supplements, or the encouragement concerning the need to consume the advertised food or supplement to restore a diet. Moreover, these suggestions should not raise doubts in the consumer about their dietary choices (Patanè et al. 2019).

These recommendations are considered worldwide, but National guidelines and associated regulations differ enough. With specific reference to the European situation, many health and nutrition claims have been suggested in the last years, and repeatedly rejected by the European Food Safety Authority (EFSA), such as the most recent suggestion: 'Olive oil polyphenols contribute to normal lipid metabolism' (EFSA Panel on Dietetic Products, Nutrition and Allergies et al. 2025). On the other side, the following sentence has been allowed: 'Olive oil polyphenols contribute to the protection of blood lipids from oxidative stress' on the basis of human studies showing significantly reduced levels of oxidized low density lipoprotein (LDL) in plasma after virgin olive oil consumption (Dongo 2025; EFSA Panel on Dietetic Products, Nutrition and Allergies 2011a, b, 2012, 2018, 2025; European Commission 2008; Lockyer and Rowland 2014; Patanè et al. 2019). On the other side, the Food and Drug Administration (FDA) has a different approach when allowing food producers to state that daily consumption of about 1½ tablespoons (20 g) of oils containing high levels of oleic acid, when replacing fats and oils higher in saturated fat, and with the possible reduction of risks of coronary heart disease (Dongo 2025; Kavanaugh 2020).

Similar situations can be observed with reference to other products and claims, including pomegranate, several berries, grapes, and so on, at least in the European Union (EFSA Panel on Dietetic Products, Nutrition and Allergies 2010a, b, 2011; Patanè et al. 2019).

As a clear result, it appears that more research—and improved cooperation between National regulations—is needed when speaking of the reliability of selected studies concerning the positive effects of polyphenols in selected medical ambits. These studies can influence the regulatory situation both on the National and the International levels (Addepalli and Suryavanshi 2018; Akhtar and Haqqi 2012; Bhol et al. 2024; Cory et al. 2018; Darenskaya et al. 2021; Dubois-Deruy et al 2020; Huang et al. 2020; Izzo et al. 2021; Jiang et al. 2020; Jelic et al. 2021; Loke et al. 2008a, b, 2019a, b; Park et al. 2021; Robinson 2021; Ruskovska et al. 2020; Stevenson and Hurst 2007). Such a cooperative effort is surely needed with urgency. It has been recently demonstrated that food supplements containing polyphenols have one or more of the following claims, as displayed in Fig. 9.1 (Loke et al. 2024):

(a) Weight management (more than 30%)
(b) Antioxidant health properties (25%)
(c) Cardiovascular effects and immunity claims (both > 10%)
(d) Other claims (under 5% of commercial advertising messages) concerning anti-ageing, general antinflammatory properties, cognitive functions, mobility, urinary, and respiratory systems, digestive processes, hormonal functions, skin, sleep and/or stress-related disorders; and skin protection.

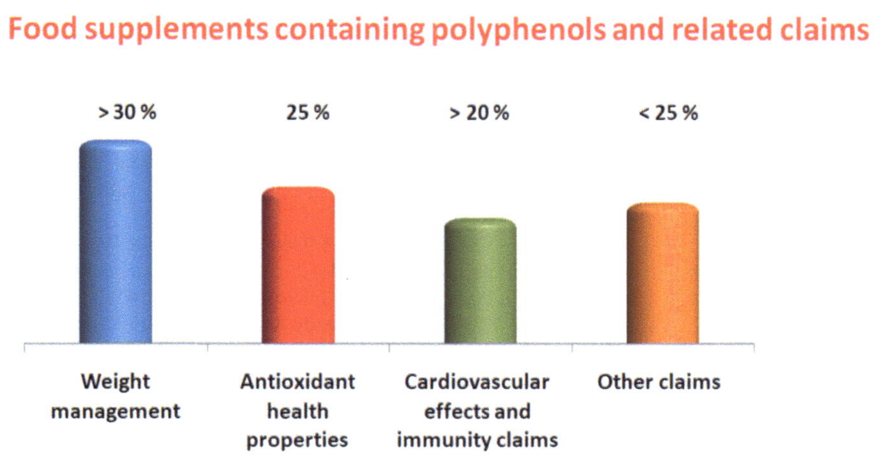


**Fig. 9.1** According to a recent research (Loke et al. 2024), food supplements containing polyphenols have one or more of the following claims as follows: weight management (more than 30%), antioxidant health properties (approximately 25%), cardiovascular effects and immunity claims (both > 10%), and other claims (all under 5%). This picture has been created by the authors

This and other studies demonstrate clearly the urgency of a good evaluation of the whole sector of supplements and also foods and beverages containing polyphenols, and with related health and nutrition claims.

## References


Abu Rached N, Dietrich JW, Ocker L, Stockfleth E, Haven Y, Myszkowski D, Bechara FG (2025) Endotyping insulin-glucose homeostasis in hidradenitis suppurativa: the impact of diabetes mellitus and inflammation. J Clin Med 14(7):2145. https://doi.org/10.3390/jcm14072145

Addepalli V, Suryavanshi SV (2018) Catechin attenuates diabetic auto-nomic neuropathy in streptozotocin induced diabetic rats. Biomed Pharmacother 108:1517–1523. https://doi.org/10.1016/j.biopha.2018.09.179

Akhtar N, Haqqi TM (2012) Current nutraceuticals in the management of osteoarthritis: a review. Therapeut Adv Musculoskeletal Dis 4:181–207. https://doi.org/10.1177/1759720X11436238

Annuzzi G, Bozzetto L, Costabile G, Giacco R, Mangione A, An-niballi G, Vitale M, Vetrani C, Cipriano P, Della Corte G, Pasanisi F, Riccardi G, Rivellese AA (2014) Diets naturally rich in polyphenols improve fasting and postprandial dyslipidemia and reduce oxidative stress: a randomized controlled trial. American J Clin Nutr 99(3):463–471. https://doi.org/10.3945/ajcn.113.073445

Behl T, Mehta K, Sehgal A, Singh S, Sharma N, Ahmadi A, Arora S, Bungau S (2022) Exploring the role of polyphenols in rheumatoid arthritis. Crit Rev Food Sci Nutr 62(19):5372–5393. https://doi.org/10.1080/10408398.2021.1924613

Bhol NK, Bhanjadeo MM, Singh AK, Dash UC, Ojha RR, Majhi S, Duttaray AK, Jena AB (2024) The interplay between cytokines, inflammation, and antioxidants: mechanistic insights and therapeutic potentials of various antioxidants and anti-cytokine compounds. Biomed Pharmacother 178:117177. https://doi.org/10.1016/j.biopha.2024.117177

Black PH (2002) Stress and the inflammatory response: a review of neurogenic inflammation. Brain Behav Immun 16(6):622–653. https://doi.org/10.1016/S0889-1591(02)00021-1

Bozzetto L, Annuzzi G, Pacini G, Costabile G, Vetrani C, Vitale M, Griffo E, Giacco A, De Natale C, Cocozza S, Della Pepa G, Tura A, Riccardi G, Rivellese AA (2015) Polyphenol-rich diets improve glucose metabolism in people at high cardio metabolic risk: a controlled random-ised intervention trial. Diabetolog 58(7):1551–1560. https://doi.org/10.1007/s00125-015-3592-x

Bucciantini M, Leri M, Nardiello P, Casamenti F, Stefani M (2021) Olive polyphenols: antioxidant and anti-inflammatory properties. Antiox 10(7):1044. https://doi.org/10.3390/antiox10071044

Chalmers JD, Metersky M, Aliberti S, Morgan L, Fucile S, Lauterio M, McDonald PP (2025) Neutrophilic inflammation in bronchiectasis. Eur Respir Rev 34(176):240179. https://doi.org/10.1183/16000617.0179-2024

Chen S, Xie Y, Guo D, Li T, Tan Z, Ran X, Wen X (2025) Effects of Salmonella Typhimurium infection on intestinal flora and intestinal tissue arachidonic acid metabolism in Wenchang chickens. Front Microbiol 16:1514115. https://doi.org/10.3389/fmicb.2025.1514115

Chong MF, George TW, Alimbetov D, Jin Y, Weech M, Macready AL, Spencer JPE, Kennedy OB, Minihane AM, Gordon MH, Lovegrove JA (2013) Impact of the quantity and fla-vonoid content of fruits and vegetables on markers of intake in adults with an increased risk of cardiovascular disease: the FLAVURS trial. Eur J Nutr 52(1):361–378. https://doi.org/10.1007/s00394-012-0343-3

Commission E (2008) Commission Regulation (EC) No 353/2008 of 18 April 2008 establishing implementing rules for applications for authorisation of health claims as provided for in Article 15 of Regulation (EC) No 1924/2006. Off J Eur Union L109:11–16

Condezo-Hoyos L, Gazi C, Pérez-Jiménez J (2021) Design of polyphenol-rich diets in clinical trials: a systematic review. Food Res Int 149:110655. https://doi.org/10.1016/j.foodres.2021.110655

Cory H, Passarelli S, Szeto J, Tamez M, Mattei J (2018) The role of polyphenols in human health and food systems: a mini-review. Front Nutr 5:87–93. https://doi.org/10.3389/fnut.2018.00087
Dantzer R, Castanon N, Lestage J, Moreau M, Capuron L (2006) Inflammation, sickness behaviour and depression. In: Steptoe A (ed) Depression and physical illness, pp 265–279. Cambridge University Press, Cambridge, UK. https://doi.org/10.1017/CBO9780511544293.013
Darenskaya MA, Kolesnikova LI, Kolesnikov SI (2021) Oxidative stress: pathogenetic role in diabetes mellitus and its complications and therapeutic approaches to correction. Bull Exp Biol Med 171:179–189. https://doi.org/10.1007/s10517-021-05191-7
Das UN, Hacimüftüoglu A, Akpinar E, Gul M, Abd El-Aty AM (2025) Crosstalk between renin and arachidonic acid (and its metabolites). Lipids Health Dis 24, 1:52.https://doi.org/10.1186/s12944-025-02463-3
Dongo D (2025) EFSA respinge health claim sui polifenoli dell'olio d'oliva. Food times, www.foodtimes. eu. https://www.foodtimes.eu/it/consumatori-e-salute/efsa-claim-polifenoli-olio-oliva/. Last accessed 20 Oct 2025
Dubois-Deruy E, Peugnet V, Turkieh A, Pinet F (2020) Oxidative stress in cardiovascular diseases. Antioxidants 9(9):864-870. https://doi.org/10.3390/antiox9090864
EFSA Panel on Dietetic Products, Nutrition and Allergies (2011a) Scientific Opinion on the substantiation of health claims related to polyphenols in olive and protection of LDL particles from oxidative damage (ID 1333, 1638, 1639, 1696, 2865), maintenance of normal blood HDL cholesterol concentrations (ID 1639), maintenance of normal blood pressure (ID 3781), 'anti-inflammatory properties' (ID 1882), 'contributes to the upper respiratory tract health' (ID 3468), 'can help to maintain a normal function of gastrointestinal tract' (3779), and 'contributes to body defences against external agents' (ID 3467) pursuant to article 13(1) of regulation (EC) No 1924/2006. EFSA J 9(4):2033. https://doi.org/10.2903/j.efsa.2011.2033
EFSA Panel on Dietetic Products, Nutrition and Allergies (2011b) Scientific Opinion on the substantiation of a health claim related to melatonin and reduction of sleep onset latency (ID 1698, 1780, 4080) pursuant to Article 13(1) of Regulation (EC) No 1924/2006. EFSA J 9(6):2241. https://doi.org/10.2903/j.efsa.2011.2241
EFSA Panel on Dietetic Products, Nutrition and Allergies (2012) Scientific Opinion on the substantiation of a health claim related to polyphenols in olive and maintenance of normal blood HDL cholesterol concentrations (ID 1639, further assessment) pursuant to article 13(1) of regulation (EC) No 1924/2006. EFSA J 10(8):2848. https://doi.org/10.2903/j.efsa.2012.2848
EFSA Panel on Dietetic Products, Nutrition and Allergies (2018) Guidance for the scientific requirements for health claims related to antioxidants, oxidative damage and cardiovas-cular health. EFSA J 16(1):5136. https://doi.org/10.2903/j.efsa.2018.5136
EFSA Panel on Dietetic Products, Nutrition and Allergies (2025) Olive oil polyphenols and the maintenance of normal HDL-cholesterol concentrations: evaluation of a health claim pursuant to Article 13(5) of Regulation (EC) No 1924/2006. EFSA J 23(5):e9372. https://doi.org/10.2903/j.efsa.2025.9372
EFSA Panel on Dietetic Products, Nutrition and Allergies (NDA) (2010a) Scientific Opinion on the substantiation of health claims related to pomegranate/pomegranate juice and maintenance of normal blood cholesterol concentrations (ID 1162, 1320, 2107, 2167), maintenance of normal erectile function (ID 1163), protection of lipids from oxidative damage (ID 1201, 1319, 2123), "antioxidant and anti-aging properties" (ID 1901), increase in appetite after unintentional weight loss leading to an increase in energy intake (ID 2122) and maintenance of normal blood glucose concentrations (ID 4471) pursuant to Article 13(1) of Regulation (EC) No 1924/2006. EFSA J 8(10):1750 [pp 22]. https://doi.org/10.2903/j.efsa.2010.1750
EFSA Panel on Dietetic Products, Nutrition and Allergies (NDA) (2010b) Scientific Opinion on the substantiation of health claims related to various food(s)/food constituent(s) and protection of cells from premature aging, antioxidant activity, antioxidant content and antioxidant properties, and protection of DNA, proteins and lipids from oxidative damage pursuant to Article 13(1) of Regulation (EC) No 1924/2006. EFSA J 8(2):1489. [pp 63]. https://doi.org/10.2903/j.efsa.2010.1489

EFSA Panel on Nutrition, Novel Foods and Food allergens (NDA), Turck D, Bohn T, Cámara M, Castenmiller J, de Henauw S, Hirsch-Ernst KI, Jos A, Maciuk A, Mangelsdorf I, McNulty B, Naska A, Pentieva K, Thies F, Craciun I, Fiolet T, Siani A (2025) Olive oil polyphenols and the maintenance of normal HDL-cholesterol concentrations: Evaluation of a health claim pursuant to Article 13 (5) of Regulation (EC) No 1924/2006. EFSA J 23(5):e9372. https://doi.org/10.2903/j.efsa.2025.9372

Engin A (2017) The pathogenesis of obesity-associated adipose tissue inflammation. In: Engin A (ed) Obesity and Lipotoxicity. Advances in experimental medicine and biology, vol 960. pp 221–245. Springer, Cham. https://doi.org/10.1007/978-3-319-48382-5_9

Gasmi A, Mujawdiya PK, Noor S, Lysiuk R, Darmohray R, Piscopo S,Lenchyk L, Antonyak H, Dehtiarova K, Shanaida M, Polishchuk A, Shanaida V, Peana M, Bjørklund G (2022) Polyphenols in metabolic diseases. Molecules 27(19):6280. https://doi.org/10.3390/molecules27196280

Guo D, Liu C, Zhu H, Cheng Y, Huo X, Guo Y, Qian H (2025) Food-induced adverse reactions: a review of physiological food quality control, mucosal defense mechanisms, and gastrointestinal physiology. Toxics 13(1):61. https://doi.org/10.3390/toxics13010061

Harvanová G, Duranková S (2025) Inflammatory process: Factors inducing inflammation, forms and manifestations of inflammation, immunological significance of the inflammatory reaction. Alergologia Polska-Pol J Allergol 12(1):54–61. https://doi.org/10.5114/pja.2025.147674

Homey B, Steinhoff M, Ruzicka T, Leung DY (2006) Cytokines and chemokines orchestrate atopic skin inflammation. J Allergy Clin Immunol 118(1):178–189. https://doi.org/10.1016/j.jaci.2006.03.047

Huang H, Liao D, Dong Y, Pu R (2020) Effect of quercetin supplementation on plasma lipid profiles, blood pressure, and glucose levels: a systematic review and meta-analysis. Nutr Rev 78(8):615–626. https://doi.org/10.1093/nutrit/nuz071

Hussain T, Tan B, Yin Y, Blachier F, Tossou MC (2016) Rahu N (2016) Oxidative stress and inflammation: what polyphenols can do for us? Oxidative Med Cell Longev 1:7432797. https://doi.org/10.1155/2016/7432797

Izzo C, Vitillo P, Di Pietro P, Visco V, Strianese A, Virtuoso N, Ciccarelli M, Galasso G, Carrizzo A, Vecchione C (2021) The role of oxidative stress in cardiovascular aging and cardiovascular diseases. Life 11(1):66. https://doi.org/10.3390/life11010060

Jelic MD, Mandic AD, Maricic SM, Srdjenovic BU (2021) Oxidative stress and its role in cancer. J Cancer Res Therapeut 17(1):22–28. https://doi.org/10.4103/jcrt.JCRT_862_16

Jiang Y, Luo W, Wang B, Wang X, Gong P, Xiong Y (2020) Resveratrol promotes osteogenesis via activating SIRT1/FoxO1 pathway in osteoporosis mice. Life Sci 246:117422. https://doi.org/10.1016/j.lfs.2020.117422

Kavanaugh C (2020) RE: petition for authorized health claim for oleic acid in edible oils and a reduction in the risk of coronary heart disease—docket number FDA-2017-Q-08. U.S. food and drug administration, center for food safety and applied nutrition, Washington, D.C. https://www.fda.gov/media/146894/download?attachment. Last accessed 20 Oct 2025

Kim JW, Yoon JS, Park S, Kim H, Kim BY, Lee H, Park SH, Kim SK, Choe JY (2025) Risk of cardiovascular disease associated with long-term use of non-steroidal anti-inflammatory drugs in ankylosing spondylitis. Rheumatology 64(1):261–267. https://doi.org/10.1093/rheumatology/kead684

Landecho MF, Tuero C, Valentí V, Bilbao I, de la Higuera M, Frühbeck G (2019) Relevance of leptin and other Adipokines in obesity-associated cardiovascular risk. Nutrients 11(11):2664. https://doi.org/10.3390/nu11112664

Li W, Ma Y, Yang L (2025) Systematic review and meta-analysis: no evidence that low-dose non-steroidal anti-inflammatory drugs (NSAIDs) reduce the risk of pancreatitis after endoscopic retrograde cholangiopancreatography (ERCP). BMC Gastroenterol 25(1):83. https://doi.org/10.1186/s12876-025-03690-9

Lockyer S, Rowland I (2014) Authorised EU health claims for polyphenols in olive oil. In: Sadler MJ (ed) Foods, nutrients and food ingredients with authorised EU health claims, pp 212–228. Woodhead Publishing, Cambridge. https://doi.org/10.1533/9780857098481.3.212

Loke WM, Hodgson JM, Croft KD (2009) The biochemistry behind the po-tential cardiovascular protection bydietary flavonoids. In: Fraga CG (ed) Plant phenolics and human health: biochemistry, nutrition and pharmacology. Wiley, Hoboken, NJ, pp137–158. https://doi.org/10.1002/9780470531792.ch5

Loke WM, Lai JX, Tan R (2024) Evaluating product demographics, composition and health claims of commercial polyphenol supplements: a review of dosage, bioactive compounds and efficacy. Nutrafoods 2024:607–616. https://doi.org/10.17470/NF-024-0069

Loke WM, Proudfoot JM, McKinley AJ, Needs PW, Kroon PA, Hodgson JM, Croft KD (2008a) Quercetin and its in vivo metabolites inhibit neutrophil-mediated low-density lipoprotein oxidation. J Agri Food Chem 56(10):3609–3615. https://doi.org/10.1021/jf8003042

Loke WM, Proudfoot JM, Stewart S, McKinley AJ, Needs PW, Kroon PA, Hodgson JM, Croft KD (2008b) Metabolic transformation has a profound effect on anti-inflammatory activity of flavonoids such as quercetin: lack of association between antioxidant and lipoxygenase inhibitory activity. Biochem Pharmacol 75(5):1045–1053. https://doi.org/10.1016/j.bcp.2007.11.002

Macready AL, George TW, Chong MF, Alimbetov DS, Jin Y, Vidal A, Spencer JPE, Kennedy OB, Tuohy KM, Minihane AM, Gordon MH, Lovegrove JA, for the FLAVURS Study (2014) Flavonoid-rich fruit and vegetables improve microvascular reactivity and inflammatory status in men at risk of cardiovascular disease—FLAVURS: a randomized controlled trial. Am J Clin Nutr 99(3):479–489. https://doi.org/10.3945/ajcn.113.074237

Malaveille C, Fiorini L, Bianchini M, Davico L, Bertinetti S, Allegro G, Hautefeuille A, Sacerdote C, Vineis P (2004) Randomized controlled trial of dietary intervention: association between level of urinary phenolics and anti-mutagenicity. Mutat Res/genet Toxicol Environ Mutagenes 561(1–2):83–90. https://doi.org/10.1016/j.mrgentox.2004.03.007

Nani A, Murtaza B, Sayed Khan A, Khan NA, Hichami A (2021) Antioxidant and anti-inflammatory potential of polyphenols contained in Mediterranean diet in obesity: molecular mechanisms. Molecules 26(4):985. https://doi.org/10.3390/molecules26040985

Omar A, Omar H, Mohamad A, Haroon S (2025) Non-steroidal anti-inflammatory drugs consumption and awareness about risks during pregnancy in AL-Bayda city, Libya. Alq J Med Appl Sci 119–128.https://doi.org/10.54361/ajmas.258119

Park S, Choi M, Lee M (2021) Effects of anthocyan in supplementation on reduction of obesity criteria: a systematic review and meta-analysis of randomized controlled trials. Nutrients 13(6):2121–2127. https://doi.org/10.3390/nu13062121

Pasparakis M, Haase I, Nestle FO (2014) Mechanisms regulating skin immunity and inflammation. Nat Rev Immunol 14(5):289–301. https://doi.org/10.1038/nri3646

Patanѐ P, Laganà P, Devi P, Vig A, Haddad MA, Natalello S, Cava MA, Ameen SM, Hashim HA (2019) Polyphenols and functional foods from the regulatory viewpoint. J AOAC Int 102(5):1373–1377. https://doi.org/10.5740/jaoacint.19-0130

Petersen SR, Bonnesen K, Grove EL, Pedersen L, Schmidt M (2025) Bleeding risk using non-steroidal anti-inflammatory drugs with anticoagulants after venous thromboembolism: a nationwide Danish study. Eur Heart J 46(1):58–68. https://doi.org/10.1093/eurheartj/ehae736

Pichard V, Guilbaud M, Devaux M, Jaulin N, Journou M, Cospolite M, Garcia A, Ferry N, Michalak-provost S, Gernoux G, Adjali O (2025) Incomplete elimination of viral genomes is associated with chronic inflammation in nonhuman primate livers after AAV-mediated gene transfer. Gene Ther 32(3):287–298, 1–12.https://doi.org/10.1038/s41434-025-00514-z

Ran J, Wang Q, Lu T, Pang X, Liao S, He X (2025) Integrating metabolomics and network analyses to explore mechanisms of Geum japonicum var. chinense against pulmonary fibrosis: involvement of arachidonic acid metabolic pathway. Int J Mol Sci 26(4):1462.https://doi.org/10.3390/ijms26041462

Robinson S (2021) Cardiovascular disease. In: Robinson S (ed) Priorities for health promotion and public health. Routledge/Taylor & Francis Group, Milton Park, UK, pp 355–393

Ruskovska T, Maksimova V, Milenkovic D (2020) Polyphenols in human nutrition: from the in vitro antioxidant capacity to the beneficial effects on cardiometabolic health and related inter-individual variability—an overview and perspective. Br J Nutr 123(3):241–254. https://doi.org/10.1017/S0007114519002733

Santamarina AB, Moreira RG, Mennitti LV, Ferreira YAM, Jucá A, Prado CM, Pisani LP (2025) Low-carbohydrate diet enriched with omega-3 and omega-9 fatty acids modulates inflammation and lipid metabolism in the liver and white adipose tissue of a mouse model of obesity. Nutr Metab Cardiovasc Dis 35(8):103932. https://doi.org/10.1016/j.numecd.2025.103932

Stevenson DE, Hurst RD (2007) Polyphenolic phytochemicals—just an-tioxidants or much more? Cell Molec Life Sci 64:2900–2916. https://doi.org/10.1007/s00018-007-7237-1

Tumbarski Y, Ivanov I, Todorova M, Apostolova S, Tzoneva R, Nikolova K (2025) Phenolic content, antioxidant activity and in vitro anti-inflammatory and antitumor potential of selected Bulgarian propolis samples. Biomedicines 13(2):334. https://doi.org/10.3390/biomedicines13020334

Vetrani C, Maukonen J, Bozzetto L, Della Pepa G, Vitale M, Costabile G, Riccardi G, Rivellese AA, Saarela M, Annuzzi G (2020) Diets naturally rich in polyphenols and/or long-chain n-3 polyunsaturated fatty acids differently affect mi-crobiota composition in high-cardiometabolic-risk individuals. Acta Diabetolog 57(7):853–860. https://doi.org/10.1007/s00592-020-01494-9

Yoon JH, Baek SJ (2005) Molecular targets of dietary polyphenols with anti-inflammatory properties. Yonsei Med J 46(5):585–596

Zamani-Garmsiri F, Emamgholipour S, Rahmani Fard S, Ghasempour G, Jahangard Ahvazi R, Meshkani R (2022) Polyphenols: potential anti-inflammatory agents for treatment of metabolic disorders. Phytother Res 36(1):415–432. https://doi.org/10.1002/ptr.7329

Zhang S, Xu M, Zhang W, Liu C, Chen S (2021) Natural polyphenols in metabolic syndrome: protective mechanisms and clinical applications. Int J Mol Sci 22(11):6110. https://doi.org/10.3390/ijms22116110

# Chapter 10
# Polyphenols as Enhancers of Digestive Processes

**Salvatore Parisi, Sara Petrelli, Carmelo Parisi, Vincenzo Barone, Vito Barone, Giorgia Punzo, Mini Joseph, and Suni Mary Varghese**

**Abstract** The heterogeneous class of phenolics compounds is generally reported to have interesting effects when speaking of human health. This statement is based on a number of recent studies which have demonstrated the relationship between the bioavailability of these compounds in the human diet and beneficial effects on human health. For these reasons at least, the modern industry has considered the use of polyphenols-rich foods and the addition of similar compounds to edible products without a good phenolics profile. The final aim is to use these foods as valid adjuvant strategies against a variety of human diseases. This Chapter, in particular, discusses the matter of digestive mechanisms and the influence of phenolics. In addition, the

S. Parisi (✉)
Department of Nutrition and Food Processing Technology, Lourdes Matha Institute of Hotel Management and Catering Technology, Thiruvananthapuram, Kerala State, India
e-mail: drparisi@inwind.it

S. Petrelli
Capaci (PA), Sicily, Italy
e-mail: sarettap25@gmail.com

C. Parisi · V. Barone · V. Barone · G. Punzo
Palermo, Italy
e-mail: carmelo.parisi2012@libero.it

V. Barone
e-mail: vincenzo.barone2012@gmail.com

V. Barone
e-mail: vitobarone99@gmail.com

G. Punzo
e-mail: punzogiorgia01@gmail.com

M. Joseph
Department of Home Science, Government College for Women, Thiruvananthapuram, Kerala State, India
e-mail: minijoseph66@yahoo.in

S. Mary Varghese
Department of Nutrition and Food Processing Technology, Lourdes Matha Institute of Hotel Management and Catering Technology, Thiruvananthapuram, Kerala State, India
e-mail: sunimary@lmihmct.org

 
R. K. Singla et al. (eds.), *Polyphenols in Health*,
https://doi.org/10.1007/978-3-032-27324-6_10

interaction between these compounds and intestinal microbiota is discussed, also with reference to the inhibition of pathogens. There are several factors which should be taken into account when speaking of polyphenols in the human digestion: role of associated phenolics with dietary fibres; relationship between polysaccharides/polyphenols interaction and the rate of release and absorption of bound and free phenolics, firstly in the small intestine and in the lower gastrointestinal tract; enhancement of microbial fermentation of polyphenols because of the concomitant presence and association of phenolics and dietary fibres; demonstrated influence of polyphenols on the decrease of pathogen agents such as *Clostridium* spp. The consumption of foods and beverages containing notable amount of phenolics is widely recommended: the activity of phenolics favours easy digestion and reduces the production of inflammatory markers.

**Keywords** Dietary fibre · Digestion · Flavonoid · Inflammation · Non-flavonoid · Pathogen · Polyphenol

## 10.1 Polyphenol-Rich Foods and Human Digestive Processes. An Introduction

As discussed in Chap. 6, the heterogeneous class of phenolics compounds, normally defined 'polyphenols', are reported to have interesting effects when speaking of human health. This statement is based on a number of recent studies which have demonstrated the relationship between the bioavailability of these compounds in the human diet and the following features (Arshad et al. 2025; Domingues et al. 2024; Giglio et al. 2018; Hafiz 2025; Kardum et al. 2015; Kupczyk et al. 2025; Issaoui et al. 2020; Liva et al. 2025; Makarewicz et al. 2021; Mechi et al. 2023; Martiniakova et al. 2025; Mendes et al. 2025; Mileo and Miccadesi 2016; Parisi 2019, 2020; Rudrapal et al. 2022; Stoner and Mukhtar 1995; Tillman et al. 2025; Wang and Hu 2023):

(1) Antioxidant behaviour against oxidative stress and inflammatory diseases, including arthritis and metabolic syndrome
(2) Beneficial effects on glucose metabolism and insulin sensitivity
(3) Cardiovascular benefits (restored blood pressure and adequate lipid profiles with specific reference to the low amount of peroxidised lipids)
(4) Good therapeutic advantages when speaking of tumours and associated treatments
(5) Enhanced biodigestibility from natural sources
(6) Positive effects on appetite regulation and fat metabolism
(7) Synergic interaction with gut microbioma.

For these reasons at least, the modern industry has considered the use of polyphenols-rich foods and the addition of similar compounds to edible products without a good phenolics profile. The final aim is to use these foods as valid adjuvant strategies against a variety of human diseases. With specific reference to added foods, the following opportunities have been explored so far (Arzani et al. 2025; Chen et al. 2025; Kumar et al. 2025; Mantravadi et al. 2025; Narayan et al. 2025; Pan et al. 2025; Puspadewi et al. 2025; Solanki et al. 2025; Uti et al. 2025):

(1) Polyphenols can be considered good inhibitory agents against peculiar bacteria in the ambit of intestinal and oral health, and when speaking of natural food preservation
(2) Moreover, added phenolics may be used as nano-carriers in selected pharmaceutical products or novel foods.

Actually, similar strategies should be carefully assessed from the health and regulatory viewpoint (Chaachouay 2025; Karabulut 2025; Kyriacou et al. 2025; Nandabaram et al. 2025; Pavić and Rahelić 2025). Being the relationship between polyphenols consumption and the enhancement of the human digestive processes, a sort of introduction on this topic should be provided. This Chapter, in particular, discusses the matter of digestive mechanisms and the influence of phenolics. In addition, the interaction between these compounds and intestinal microbiota is discussed, also with reference to the inhibition of pathogens. Because of this potentially broad ambit and certain correlations with inflammatory diseases, there are some common points with other Chapters in this book.

## 10.2 Polyphenols and Health Applications by a General Viewpoint. Fibres/Phenolics Agglomeration, Influence on Bioavailability, and Antimicrobial Properties

The broad category of phenolics from vegetable sources is reported to have beneficial effects with reference to different diseases, including oxidative stress and inflammatory problems. The chemical classification of polyphenols (Chap. 6), can be useful when speaking of the comprehension of polyphenols as co-adjuvant agents in the treatment of different diseases (the following list cannot be fully exhaustive) (Abu Rached et al. 2025; Black 2002; Dantzer et al. 2006; Hussain et al. 2016; Pasparakis et al. 2014; Santamarina et al. 2025; Zamani-Garmsiri et al. 2022):

(1) Defence against viruses and other pathogens able to unleash automated defence answers
(2) Restoring action against health damages and superficial wounds
(3) Defence against chemical agents able to suddenly cause irritation (food and environmental allergens, etc.)
(4) Defence from radioactive sources
(5) Treatment of chronic, autoimmune, and low-grade metabolic disorders: insulin resistance, type-2 diabetes, cardiovascular diseases, obesity, severe metabolic syndrome, non-alcoholic fatty liver disease, etc.
(6) Treatment of low-grade and chronic systemic inflammation or 'metainflammation'

(7) Treatment of alcohol and/or tobacco abuse
(8) Treatment of obesity and peculiar anti-dietary behaviours such as prolonged high-calorie consumption (without complications such in the metabolic syndrome, where diabetes and cardiovascular disease have their own weight).

Substantially, inflammation and oxidative stress are normal enough and these effects should be expected. The problem is in the fact that the organism may be not able to carry on effective and rapid answers (inflammation is a positive answer against attacks), also because many different organisms and tissues can be targeted, and peculiar factors (age, sex, etc.), have their importance. Should the answer be insufficient, the resulting inflammation can evolve until severe symptoms are observed (Chalmers et al. 2025; Gasmi et al. 2022; Guo et al. 2025; Harvanová and Duranková 2025; Homey et al. 2006; Pichard et al. 2025; Yoon and Baek 2005):

(a) Modification of blood composition
(b) Fever against different pathogens (viruses, bacteria, etc.)
(c) Well-identifiable diseases such as rheumatoid arthritis or bowel disease.

The use of non-synthetic and natural compounds such as polyphenols with good inhibitory effects appears to be a good solution, also if oxidative stress is observed (Bucciantini et al. 2021; Martín and Ramos 2016; Tumbarski et al. 2025).

With reference to phenolics, the main subdivision in flavonoids and non-flavonoids is widely reported and accepted (Behl et al. 2022; Ganesan and Xu 2017; Nani et al. 2021; Magrone et al. 2019; Matacchione et al. 2020; Zhang et al. 2021). The further subdivision in sub-groups has to be taken into account: the chemical structure has some relationship with associated effects in term of inhibition or enhancement of chemical reactions (by the biochemical viewpoint, the discussion concerns often inhibition or boosting action for several enzymes). Flavonoids (Figs. 6.2, 6.3, 6.4, 6.5, 6.6, 6.7, 6.8, 6.9 and 6.10) comprehend (Baruah 2011; Feng et al. 2017; Mateen et al. 2016; Panche et al. 2016; Petti and Scully 2009):

(1) Anthocyanidins
(2) Chalcones
(3) Flavanols
(4) Flavanones
(5) Flavones
(6) Flavonols
(7) Isoflavones.

In addition, non-flavonoids (Chap. 6) are subdivided as follows, taking into account that the list may be not fully exhaustive (Amiot et al. 2016; Baruah 2011; Gasmi et al. 2022; Mateen et al. 2016; Petti and Scully 2009; Zhang et al. 2021):

(1) Coumarins
(2) Curcominoids
(3) Lignans
(4) Phenolic acids (hydroxybenzoic acids and hydroxycinnamic acids)
(5) Stilbenes

(6) Tannins.

With reference to polyphenols as enhancers of human digestive processes, a premise should be done. First of all, it has to be clarified that each possible action and beneficial effect ascribed to phenolics has to be evaluated on the basis of their effective intake and absorption by the human organism (Nemzer et al. 2025). In other terms (Biéet al. 2023; Lorenzo et al. 2021a, b; Ozdal et al. 2016):

(1) Polyphenols are contained in complex food matrices
(2) After consumption, these matrices, usually of vegetable origin, have to surpass the acid attack in the stomach and subsequently interact with digestive enzymes in the small intestine
(3) After this critical two-steps mechanism, phenolics have to be available for microbial flora in the large intestinal tract
(4) Finally, polyphenols are ready for absorption.

The above-displayed system (Fig. 10.1) shows that digestibility of polyphenols is a multi-steps sequence, and that the amount of really available polyphenols can be different from the original portion of phenolics in the consumed food or beverage. The following factors can reduce this amount (Lorenzo et al. 2021a, b):

(a) Type of food matrix

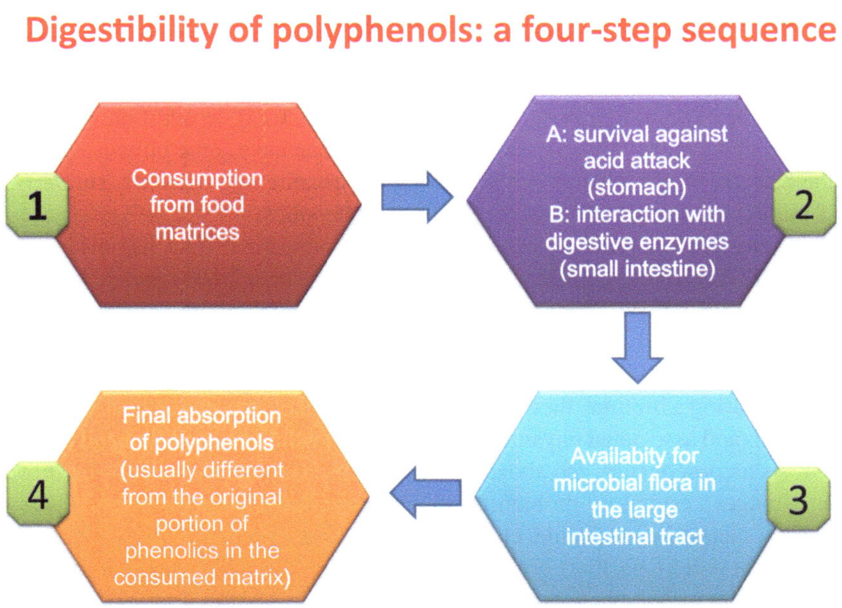


**Fig. 10.1** Digestibility polyphenols is a multi-steps sequence, and the amount of really available polyphenols can be different from the original portion of phenolics in the consumed food or beverage (). In other terms (Biéet al. 2023; Lorenzo et al. 2021a, b; Nemzer et al. 2025; Ozdal et al. 2016). This picture has been created by the authors

(b) Molecular weight of the polyphenol
(c) Molecular structure of the polyphenol, including also conformational features which have to be taken into account when speaking of interactions with enzymatic compounds (position, orientation, etc., are critical pillars in biochemical reactions)
(d) Free or bound nature of the molecule with other macro-compounds
(e) Physicochemical attitude of the polyphenol.

Interestingly, the role of polyphenols is also influenced by the degree of physical interaction with other macromolecules. In fact, the reported amount of bioavailable phenolics concerns free polyphenols which can be normally extracted with organic solvents. These free phenolics are substantially flavonoid aglycones and phenolic acids, with the possibility of some bonds with sugars, and consequent increased bioavailability (De Rossi et al. 2025).

On the other side, there are different 'bound' phenolics which are associated with lignin, cellulose, or proteins, in the vegetable matrix. As a result, similar phenolics cannot be extracted and their agglomerated structure means that degradation can occur only in the large intestine (at the end of the gastrointestinal tract) (De Rossi et al. 2025; Pérez-Jiménez and Torres 2011; Pérez-Jiménez et al. 2013; Rocchetti et al. 2022). The importance of 'non-extractable' polyphenols is critical because these compounds are at disposal of microbes in the lower gastro-intestinal tract; the fermentation occurs, and this fact means that these molecules are not released or absorbed by the human organism. On the contrary, these compounds are metabolised and the resulting mixture of metabolite has some influence on gut microbiota and the healthy state of the human host (Bié et al. 2023; Ding et al. 2020; Fernandes et al. 2023; Jakobek and Matić 2019; Pérez-Jiménez et al. 2013). In addition, the role of vegetable fibres in the human gastrointestinal tract can have some influence because of the attitude of polyphenols to form complex organic aggregations, similarly to their initial state in vegetable sources as bound phenolics.

The variegated population in the human gastrointestinal tract has also critical importance. At present, 12 different divisions (or more than 2,100 species) have been identified so far in this important region of the human body: more than 90% of these identified organisms belong to phyla *Proteobacteria, Firmicutes*, and *Bacteroidetes* (Dieterich et al. 2018; Hugon et al. 2015). Should the composition of gut microbioma be altered during a lifetime, their functions and critical effects on the host's intestinal absorption (and associated physiological functions) would be modified and compromised. As a consequence, the above-mentioned diseases and disorders—cardiovascular damages, respiratory problems, inflammatory bowel, skin diseases, allergenic responses, metabolic syndrome, etc.—would be expected (Nemzer et al. 2025).

For all these reasons, there are several factors which should be taken into account when speaking of polyphenols in the human digestion:

(1) Role of associated phenolics with dietary fibres. In fact, dietary fibres (complex polysaccharides) can 'work' as carriers for antioxidant agents such as polyphenols. In addition, the bioavailability of phenolics for gut microbiota strongly depends on the ability of dietary fibres to transport them until the 'right' place

in the gastrointestinal tract. Because of the great structural diversification of phenolics, the following factors: molecular dimensions and weight; polymerization degree for several molecules; superficial porosity of dietary fibres; steric hindrance and stereochemistry, both for carriers and for phenolics; can have a critical role (Bautista-Ortin et al. 2014; Martins et al. 2022; Chirug et al. 2018; Liu et al. 2020; Rocchetti et al. 2022; Saura-Calixtoet al. 2007–2010; Wong et al. 2016). After all, these compounds—it has to be remembered that currently reported phenolics exceed 8,000 different molecular species—are reported to furnish interesting aggregates with polysaccharide macromolecules in different vegetable matrices such as: bamboo, maize, sugar beet, wheat, spinaches, and tomato peels.

(2) As a result, the polysaccharides/polyphenols interaction (which could concern hydrogen bonds, electrostatic and hydrophobic interactions, etc.) can determine the rate of release and absorption of bound and free phenolics, firstly in the small intestine (no need of carrier fibres at this point) and in the lower gastrointestinal tract (where non-extractable polyphenols need carrier fibres, and these molecules are subsequently released for microbial fermentation). Interestingly, bound polyphenols can be partially demolished in the colon tract by breaking covalent bonds HO-linked aromatic rings and carbohydrates, with additional rupture of glycosidic bonds. These reactions are carried out by enzymes produced by *Clostridium* spp., *Lactobacillus* spp, and *Bifidobacterium* spp. Consequently, obtained metabolites may also be stronger than original phenolics when speaking of general biological activity. Anyway, being fibres/phenolics agglomerations the sum of polysaccharides and bound polyphenols, such as association is a critical point: in other terms, and as reported recently, should dietary fibres be abundant, bound phenolics would be bioavailable in the colon, while the same availability is not demonstrated in the small intestine (at the start of this itinerary in the gastrointestinal tract) (Rocchetti et al. 2022; Rondini et al. 2004; Saura-Calixto 2012; Valli et al. 2012; Vitaglione et al. 2015)

(3) Moreover, the presence of dietary fibres in association with phenolics may enhance microbial fermentation of polyphenols, while free polyphenols could not be fermented with the same results (in vitro experiments). Interestingly, the spreading of *Bacteroidetes* and *Firmicutes* spp is notable against *Clostridium* spp (a good representative for pathogens), and inflammatory molecules are notably reduced, accompanied by reduced inflammatory markers. It should be noted that the enzymatic demolition of non-extractable phenolics can produce phenolics acids such as propionic, cinnamic, and benzoic acids. Involved reactions are cleavage of aromatic rings, dehydroxylation, demethylation, delactonisation, O-deglycosylation, reduction of unsaturated bonds, and ester hydrolysis. These and other metabolites may be subsequently eliminated easily from the body (Appeldoorn et al. 2009; Li et al. 2020; Sánchez-Patán et al. 2015; Tomás-Barberán et al. 2016; Vitaglione et al. 2015).

The last point is interesting also because of the demonstrated influence of polyphenols on the decrease of pathogen agents such as *Clostridium* spp. In particular, the supplementation of catechin and epicatechin (from tea infusions), anthocyanins with selected glicosides, isoflavones, or phenolics acids has been reported to favour the growth of *Bifidobacterium* spp and *Lactobacillus* spp (among other microbes). On the other side, pathogens such as *Clostridium difficile* and *C. perfringens*, *Streptococcus pyogenes* and *S. pneumonia*, *E. coli*, *S. aureus*, *S. typhimurium*, and *Listeria monocytogenes* have been found to be inhibited, although scientific papers demonstrate that some of these microbes such as *E. coli* may be favoured in some occasion (Etxeberria et al. 2015; Hidalgo et al. 2012; Lee et al. 2006; Parkar et al. 2013; Tzouniset al. 2008).

Anyway, it appears that pathogens are generally inhibited in presence of supplemented phenolics in the diet. Chemically speaking, it has been reported that the amount of hydroxyl and methoxyl groups, the—$C_5H_9$—prenyl group, and the demonstrated presence of one or more glycosylations can be critical when speaking of antimicrobial activity. Two examples (De Rossi et al. 2025):

(a) The more glycosilated a specific phenolics compound, the lower the antimicrobial activity
(b) The lower the number of hydroxyl groups, the higher the antimicrobial power because of enhanced lipophilicity and consequent capability of interacting with microbial lipid membranes and finally proteins, with consequent microbial damage.

On the other side, high numbers of hydroxyl groups or glycolisations can give more instability to phenolics (these compounds have antioxidant power), so lowering their importance as effective antimicrobial agents (Zhu et al. 2020).

For these reasons, the consumption of foods and beverages containing notable amount of phenolics (cranberries and blueberries, pomegranate, tea varieties, cocoa, grapes and red wines, etc.), is widely recommended. Recent researches have obtained some result when speaking of basic biochemical reasons which can explain the activity of phenolics in favour of easy digestion and also reduced production of inflammatory markers. It has been found that antimicrobial activity can be considered as the result of different biochemical mechanisms (Bharathiraja et al. 2023; Chun et al. 2005; De Rossi et al. 2025; Li et al. 2019; Wang et al. 2019):

(1) Partial or total modification or demolition of microbial lipid membranes (alteration of permeability)
(2) Demolition of cytoplasmatic matter
(3) Inhibition of cellular pump mechanisms
(4) Chelating attitude to complex needed metals
(5) Reduction of gene expression capability (the microbial cell becomes unable to produce specific proteins. Example: inhibition of the production of cell walls)
(6) Interference with the synthesis of genetic code and of several important proteins
(7) Interference with general cellular pathways, with final cell death.

On the other side, more research is surely needed when speaking of beneficial effects for gut microbiota, and the concomitant antagonistic action for pathogens, including also yeasts and moulds (Abou Baker 2022; Beck and Stengel 2016; Makarewiczet al. 2021; Rocchetti et al. 2022).

## References


Abou Baker DH (2022) An ethnopharmacological review on the therapeutical properties of flavonoids and their mechanisms of actions: a comprehensive review based on up to date knowledge. Toxicol Rep 9:445–469. https://doi.org/10.1016/j.toxrep.2022.03.011

Abu Rached N, Dietrich JW, Ocker L, Stockfleth E, Haven Y, Myszkowski D, Bechara FG (2025) Endotyping insulin-glucose homeostasis in hidradenitis suppurativa: the impact of diabetes mellitus and inflammation. J Clin Med 14(7):2145. https://doi.org/10.3390/jcm14072145

Amiot MJ, Riva C, Vinet A (2016) Effects of dietary polyphenols on metabolic syndrome features in humans: a systematic review. Obes Rev 17(7):573–586. https://doi.org/10.1111/obr.12409

Appeldoorn MM, Vincken JP, Aura AM, Hollman PC, Gruppen H (2009) Procyanidin dimers are metabolized by human microbiota with 2-(3, 4-dihydroxyphenyl) acetic acid and 5-(3, 4-dihydroxyphenyl)-γ-valerolactone as the major metabolites. J Agric Food Chem 57(3):1084–1092. https://doi.org/10.1021/jf803059z

Arshad MT, Maqsood S, Ikram A, Abdullahi MA (2025) Synergistic effects of carrot (Daucus carota L.) Root and Papaya (Carica papaya L.) Extracts in combating inflammation and enhancing immune functions—a new perspective. Food Sci Nutr 13, 6:e70313. https://doi.org/10.1002/fsn3.70313

Arzani V, Soleimani M, Fritsch T, Jacob UM, Calabrese V, Arzani A (2025) Plant polyphenols, terpenes, and terpenoids in oral health. Open Med 20(1):20251183. https://doi.org/10.1515/med-2025-1183

Baruah JB (2011) Chemistry of phenolic compounds: state of the art. Nova Science Publishers, Hauppauge, NY, USA

Bautista-Ortin AB, Cano-Lechuga M, Ruiz-Garcia Y, Gomez-Plaza E (2014) Interactions between grape skin cell wall material and commercial enological tannins. Practical implications. Food Chem 152:558–565. https://doi.org/10.1016/j.foodchem.2013.12.009

Beck S, Stengel J (2016) Mass spectrometric imaging of flavonoid glycosides and biflavonoids in Ginkgo biloba L. Phytochem 130:201–206. https://doi.org/10.1016/j.phytochem.2016.05.005

Behl T, Mehta K, Sehgal A, Singh S, Sharma N, Ahmadi A, Arora S, Bungau S (2022) Exploring the role of polyphenols in rheumatoid arthritis. Crit Rev Food Sci Nutr 62(19):5372–5393. https://doi.org/10.1080/10408398.2021.1924613

Bharathiraja P, Yadav P, Sajid A, Ambudkar SV, Prasad NR (2023) Natural medicinal compounds target signal transduction pathways to overcome ABC drug efflux transporter-mediated multidrug resistance in cancer. Drug Resis Updates 71:101004. https://doi.org/10.1016/j.drup.2023.101004

Bié J, Sepodes B, Fernandes PC, Ribeiro MH (2023) Polyphenols in health and disease: gut microbiota, bioaccessibility, and bioavailability. Compounds 3(1):40–72. https://doi.org/10.3390/compounds3010005

Black PH (2002) Stress and the inflammatory response: a review of neurogenic inflammation. Brain Behav Immun 16(6):622–653. https://doi.org/10.1016/S0889-1591(02)00021-1

Bucciantini M, Leri M, Nardiello P, Casamenti F, Stefani M (2021) Olive polyphenols: antioxidant and anti-inflammatory properties. Antiox 10(7):1044. https://doi.org/10.3390/antiox10071044

Chaachouay N (2025) Synergy, additive effects, and antagonism of drugs with plant bioactive compounds. Drugs Drug Candidates 4(1):4. https://doi.org/10.3390/ddc4010004

Chalmers JD, Metersky M, Aliberti S, Morgan L, Fucile S, Lauterio M, McDonald PP (2025) Neutrophilic inflammation in bronchiectasis. Eur Respir Rev 34(176):240179. https://doi.org/10.1183/16000617.0179-2024

Chen Z, Xiao C, Zhang J, Jian S, Li P, Lin J, He C, Chen Z, Qi Y, Shi J, Chen Q, Chen J, Bo H (2025) The impact of diet on the colonization of beneficial microbes from an ecological perspective. J Agric Food Chem 73(17):10069–10092. https://doi.org/10.1021/acs.jafc.5c02086

Chirug L, Okun Z, Ramon O, Shpigelman A (2018) Iron ions as mediators in pectin-flavonols interactions. Food Hydrocoll 84:441–449. https://doi.org/10.1016/j.foodhyd.2018.06.039

Chun SS, Vattem DA, Lin YT, Shetty K (2005) Phenolic antioxidants from clonal oregano (Origanum vulgare) with antimicrobial activity against Helicobacter pylori. Process Biochem 40(2):809–816. https://doi.org/10.1016/j.procbio.2004.02.018

Dantzer R, Castanon N, Lestage J, Moreau M, Capuron L (2006) Inflammation, sickness behaviour and depression. In: Steptoe A (ed) Depression and physical illness. Cambridge University Press, Cambridge, UK, pp 265–279. https://doi.org/10.1017/CBO9780511544293.013

Das T, Chatterjee N, Capanoglu E, Lorenzo JM, Das AK, Dhar P (2025) The synergistic ramification of insoluble dietary fiber and associated non-extractable polyphenols on gut microbial population escorting alleviation of lifestyle diseases. Food Chem X 18:100697. https://doi.org/10.1016/j.fochx.2023.100697

De Rossi L, Rocchetti G, Lucini L, Rebecchi A (2025) Antimicrobial potential of polyphenols: mechanisms of action and microbial responses—a narrative review. Antiox 14(2):200. https://doi.org/10.3390/antiox14020200

Di Lorenzo C, Colombo F, Biella S, Stockley C, Restani P (2021a) Polyphenols and human health: the role of bioavailability. Nutrients 13(1):273. https://doi.org/10.3390/nu13010273

Di Lorenzo C, Colombo F, Biella S, Stockley C, Restani P (2021b) Polyphenols and human health: the role of bioavailability. Nutr 13(1):273. https://doi.org/10.3390/nu13010273

Dieterich W, Schink M, Zopf Y (2018) Microbiota in the gastrointestinal tract. Med Sci 6(4):116. https://doi.org/10.3390/medsci6040116

Ding Y, Morozova K, Scampicchio M, Ferrentino G (2020) Non-extractable polyphenols from food by-products: current knowledge on recovery, characterisation, and potential applications. Process 8(8):925. https://doi.org/10.3390/pr8080925

Domingues YO, Lemes GA, de Oliveira FL, de Souza TR, Silva B, Bento JAC, Morzelle MC (2024) In vitro simulated gastrointestinal digestion and bioaccessibility of phenolic compounds and antioxidants of soursop (Annona muricata L.) peel and pulp. Lwt 208:116694. https://doi.org/10.1016/j.lwt.2024.116694

Etxeberria U, Arias N, Boqué N, Macarulla MT, Portillo MP, Martínez JA, Milagro FI (2015) Reshaping faecal gut microbiota composition by the intake of trans-resveratrol and quercetin in high-fat sucrose diet-fed rats. J Nutr Biochem 26(6):651–660. https://doi.org/10.1016/j.jnutbio.2015.01.002

Feng W, Hao Z, Li M (2017) Isolation and structure identification of flavonoids. In: Justino GC (ed) Flavonoids, from biosynthesis to human health. Intech Open, Rijeka, Croatia, pp 17–43. https://doi.org/10.5772/67810

Fernandes A, Mateus N, de Freitas V (2023) Polyphenol-dietary fiber conjugates from fruits and vegetables: nature and biological fate in a food and nutrition perspective. Foods 12(5):1052. https://doi.org/10.3390/foods12051052

Ganesan K, Xu B (2017) A critical review on polyphenols and health benefits of black soybeans. Nutrients 9, 5:455. https://doi.org/10.3390/nu9050455

Gasmi A, Mujawdiya PK, Noor S, Lysiuk R, Darmohray R, Piscopo S,Lenchyk L, Antonyak H, Dehtiarova K, Shanaida M, Polishchuk A, Shanaida V, Peana M, Bjørklund G (2022) Polyphenols in metabolic diseases. Mol 27, 19:6280. https://doi.org/10.3390/molecules27196280

Giglio RV, Patti AM, Cicero AF, Lippi G, Rizzo M, Toth PP, Banach M (2018) Polyphenols: potential use in the prevention and treatment of cardiovascular diseases. Curr Pharm des 24(2):239–258. https://doi.org/10.2174/1381612824666180130112652

Guo D, Liu C, Zhu H, Cheng Y, Huo X, Guo Y, Qian H (2025) Food-induced adverse reactions: a review of physiological food quality control, mucosal defense mechanisms, and gastrointestinal physiology. Tox 13(1):61. https://doi.org/10.3390/toxics13010061

Hafiz MS (2025) Anti-diabetic potentials of coffee polyphenols: a narrative review. Curr Nutr Food Sci 21(2):202–211. https://doi.org/10.2174/0115734013292395240219091340

Harvanová G, Duranková S (2025) Inflammatory process: factors inducing inflammation, forms and manifestations of inflammation, immunological significance of the inflammatory reaction. Alergologia Polska-Pol J Allergol 12(1):54–61. https://doi.org/10.5114/pja.2025.147674

Hidalgo M, Oruna-Concha MJ, Kolida S, Walton GE, Kallithraka S, Spencer JP, de Pascual-Teresa S (2012) Metabolism of anthocyanins by human gut microflora and their influence on gut bacterial growth. J Agric Food Chem 60(15):3882–3890. https://doi.org/10.1021/jf3002153

Homey B, Steinhoff M, Ruzicka T, Leung DY (2006) Cytokines and chemokines orchestrate atopic skin inflammation. J Allergy Clin Immunol 118(1):178–189. https://doi.org/10.1016/j.jaci.2006.03.047

Hugon P, Dufour JC, Colson P, Fournier PE, Sallah K, Raoult D (2015) A comprehensive repertoire of prokaryotic species identified in human beings. Lancet Infect Dis 15(10):1211–1219

Hussain T, Tan B, Yin Y, Blachier F, Tossou MC, Rahu N (2016) Oxidative stress and inflammation: what polyphenols can do for us? Oxidative Med Cell Longev 1:7432797. https://doi.org/10.1155/2016/7432797

Issaoui M, Delgado AM, Iommi C, Chammem N (2020) Polyphenols and the mediterranean giet. Springer International Publishing, Cham

Jakobek L, Matić P (2019) Non-covalent dietary fiber-polyphenol interactions and their influence on polyphenol bioaccessibility. Trends Food Sci Technol 83:235–247. https://doi.org/10.1016/j.tifs.2018.11.024

Karabulut G (2025) Tea polyphenols: their role in human health and drug interactions. Dissertation, University of Padova, Italy

Kardum N, Milovanović B, Šavikin K, Zdunić G, Mutavdžin S, Gligorijević T, Spasić S (2015) Beneficial effects of polyphenol-rich chokeberry juice consumption on blood pressure level and lipid status in hypertensive subjects. J Med Food 18(11):1231–1238. https://doi.org/10.1089/jmf.2014.0171

Kumar S, Mukherjee R, Gaur P, Leal É, Lyu X, Ahmad S, Puri P, Chang CM, Raj VS, Pandey RP (2025) Unveiling roles of beneficial gut bacteria and optimal diets for health. Front Microbiol 16:1527755. https://doi.org/10.3389/fmicb.2025.1527755

Kupczyk D, Bilski R, Szeleszczuk Ł, Mądra-Gackowska K, Studzińska R (2025) The role of diet in modulating inflammation and oxidative stress in rheumatoid arthritis, ankylosing spondylitis, and psoriatic arthritis. Nutr 17(9):1603. https://doi.org/10.3390/nu17091603

Kyriacou NM, Gross AS, McLachlan AJ (2025) Green tea catechins as perpetrators of drug pharmacokinetic interactions. Clin Pharmacol Ther 118(1):45–61. https://doi.org/10.1002/cpt.3547

Lee HC, Jenner AM, Low CS, Lee YK (2006) Effect of tea phenolics and their aromatic fecal bacterial metabolites on intestinal microbiota. Res Microbiol 157(9):876–884. https://doi.org/10.1016/j.resmic.2006.07.004

Li J, Liu D, Tian X, Koseki S, Chen S, Ye X, Ding T (2019) Novel antibacterial modalities against methicillin resistant Staphylococcus aureus derived from plants. Crit Rev Food Sci Nutr 59(S1):S153–S161. https://doi.org/10.1080/10408398.2018.1541865

Li H, Christman LM, Li R, Gu L (2020) Synergic interactions between polyphenols and gut microbiota in mitigating inflammatory bowel diseases. Food Funct 11(6):4878–4891. https://doi.org/10.1039/D0FO00713G

Liu X, Le Bourvellec C, Renard CM (2020) Interactions between cell wall polysaccharides and polyphenols: effect of molecular internal structure. Compr Rev Food Sci Food Saf 19(6):3574–3617. https://doi.org/10.1111/1541-4337.12632

Liva K, Panagiotopoulos AA, Foscolou A, Amerikanou C, Vitali A, Zioulis S, Argyri K, Panoutsopoulos GI, Kaliora AC, Gioxari A (2025) High polyphenol extra virgin olive oil and metabolically unhealthy obesity: a scoping review of preclinical data and clinical trials. Clin Pract 15(3):54. https://doi.org/10.3390/clinpract15030054

Magrone T, Magrone M, Russo MA, Jirillo E (2019) Recent advances on the anti-inflammatory and antioxidant properties of red grape polyphenols: in vitro and in vivo studies. Antiox 9(1):35. https://doi.org/10.3390/antiox9010035

Makarewicz M, Drożdż I, Tarko T, Duda-Chodak A (2021) The interactions between polyphenols and microorganisms, especially gut microbiota. Antioxid 10(2):188. https://doi.org/10.3390/antiox10020188

Mantravadi PK, Kovi BS, Reddy SR, Namasivayam GP, Kalesh K, Parthasarathy A (2025) Probing and manipulating the gut microbiome with chemistry and chemical tools. Gut Microbiome 6:e6. https://doi.org/10.1017/gmb.2025.4

Martín MA, Ramos S (2016) Cocoa polyphenols in oxidative stress: potential health implications. J Funct Foods 27:570–588. https://doi.org/10.1016/j.jff.2016.10.008

Martiniakova M, Sarocka A, Penzes N, Biro R, Kovacova V, Mondockova V, Sevcikova A, Ciernikova S, Omelka R (2025) Protective role of dietary polyphenols in the management and treatment of type 2 diabetes mellitus. Nutr 17(2):275. https://doi.org/10.3390/nu17020275

Martins CC, Rodrigues RC, Mercali GD, Rodrigues E (2022) New insights into non-extractable phenolic compounds analysis. Food Res Int 157:111487. https://doi.org/10.1016/j.foodres.2022.111487

Matacchione G, Gurău F, Baldoni S, Prattichizzo F, Silvestrini A, Giuliani A, Pugnaloni A, Espinosa E, Amenta F, Bonafè M, Procopio AD, Rippo MR, Olivieri F, Sabbatinelli J (2020) Pleiotropic effects of polyphenols on glucose and lipid metabolism: focus on clinical trials. Ageing Res Rev 61:101074. https://doi.org/10.1016/j.arr.2020.101074

Mateen S, Moin S, Zafar A, Khan AQ (2016) Redox signalling in rheumatoid arthritis and the preventive role of polyphenols. Clin Chim Acta 463:4–10. https://doi.org/10.1016/j.cca.2016.10.007

Mechi D, Baccouri B, Martín-Vertedor D, Abaza L (2023) Bioavailability of phenolic compounds in californian-style table olives with tunisian aqueous olive leaf extracts. Mol 28(2):707. https://doi.org/10.3390/molecules28020707

Mendes I, Ribeiro MGC, de Souza LF, Rosa CDOB, Hermsdorff HHM, Bressan J (2025) Effect of polyphenol supplementation on adiposity: a systematic review of randomized clinical trials. Curr Nutr Rep 14(1):36. https://doi.org/10.1007/s13668-025-00626-3

Mileo AM, Miccadei S (2016) Polyphenols as modulator of oxidative stress in cancer disease: new therapeutic strategies. Oxid Med Cell Longev 1:6475624. https://doi.org/10.1155/2016/6475624

Nandabaram A, Nelson VK, Mayasa V (2025) Assessment of possible food-drug interactions of pearl millet diet on gliclazide in diabetic rats. Texila Int J Pub Health 13(1):614–628. https://doi.org/10.21522/TIJPH.2013.13.01.Art058

Nani A, Murtaza B, Sayed Khan A, Khan NA, Hichami A (2021) Antioxidant and anti-inflammatory potential of polyphenols contained in Mediterranean diet in obesity: molecular mechanisms. Mol 26(4):985. https://doi.org/10.3390/molecules26040985

Narayan S, Nagpal K, Kumar P (2025) Polyphenol-conjugated polysaccharide nanoplatforms for enhanced therapeutic efficacy. Expert Opin Drug Deliv 22(9):1237–1255. https://doi.org/10.1080/17425247.2025.2514714

Nemzer BV, Al-Taher F, Kalita D, Yashin AY, Yashin YI (2025) Health-improving effects of polyphenols on the human intestinal microbiota: a review. Int J Mol Sci 26(3):1335. https://doi.org/10.3390/ijms26031335

Ozdal T, Sela DA, Xiao J, Boyacioglu D, Chen F, Capanoglu E (2016) The reciprocal interactions between polyphenols and gut microbiota and effects on bioaccessibility. Nutr 8(2):78. https://doi.org/10.3390/nu8020078

Pan X, Duan Y, Liu S, Wang Y, Li Q, Jiang F, Li Y, Huang Z, Su L, Li X, Liu M, Zhou X, Tang H (2025) All-in-one: harnessing multifunctional natural polysaccharide spray hydrogel loaded with polyphenol-metal nanoparticles for fruit preservation. Food Chem 470:142638. https://doi.org/10.1016/j.foodchem.2024.142638

Panche AN, Diwan AD, Chandra SR (2016) Flavonoids: an overview. J Nutr Sci 5:e47. https://doi.org/10.1017/jns.2016.41

Parisi S (2019) Analysis of major phenolic compounds in foods and their health effects. AOAC J 102(5):1354–1355. https://doi.org/10.5740/jaoacint.19-0127

Parisi S (2020) Characterization of major phenolic compounds in selected foods by the technological and health promotion viewpoints. J AOAC 103(4):904–905. https://doi.org/10.1093/jaoacint/qsaa011

Parkar SG, Trower TM, Stevenson DE (2013) Fecal microbial metabolism of polyphenols and its effects on human gut microbiota. Anaerobe 23:12–19. https://doi.org/10.1016/j.anaerobe.2013.07.009

Pasparakis M, Haase I, Nestle FO (2014) Mechanisms regulating skin immunity and inflammation. Nat Rev Immunol 14(5):289–301. https://doi.org/10.1038/nri3646

Pavić E, Rahelić V (2025) Food-drug interactions. In: Banić M, Erceg D (eds) Drug interactions in gastroenterology: a clinical guide. Humana, Cham, Switzerland, pp 245–263. https://doi.org/10.1007/978-3-031-69800-2_15

Pérez-Jiménez J, Torres JL (2011) Analysis of nonextractable phenolic compounds in foods: the current state of the art. J Agric Food Chem 59(24):12713–12724. https://doi.org/10.1021/jf203372w

Pérez-Jiménez J, Díaz-Rubio ME, Saura-Calixto F (2013) Non-extractable polyphenols, a major dietary antioxidant: occurrence, metabolic fate and health effects. Nutr Res Rev 26(2):118–129. https://doi.org/10.1017/S0954422413000097

Petti S, Scully C (2009) Polyphenols, oral health and disease: a review. J Dent 37(6):413–423. https://doi.org/10.1016/j.jdent.2009.02.003

Pichard V, Guilbaud M, Devaux M, Jaulin N, Journou M, Cospolite M, Garcia A, Ferry N, Michalak-provost S, Gernoux G, Adjali O (2025) Incomplete elimination of viral genomes is associated with chronic inflammation in nonhuman primate livers after AAV-mediated gene transfer. Gene Ther 32, 3:287–298, 1–12.https://doi.org/10.1038/s41434-025-00514-z

Puspadewi R, Milanda T, Muhaimin M, Chaerunisaa AY (2025) Nanoparticle-encapsulated plant polyphenols and flavonoids as an enhanced delivery system for anti-acne therapy. Pharm 18(2):209. https://doi.org/10.3390/ph18020209

Rocchetti G, Gregorio RP, Lorenzo JM, Barba FJ, Oliveira PG, Prieto MA, Simal-Gandara J, Mosele JI, Motilva MJ, Tomas M, Patrone V, Capanoglu E, Lucini L (2022) Functional implications of bound phenolic compounds and phenolics–food interaction: a review. Compr Rev Food Sci Food Saf 21(2):811–842. https://doi.org/10.1111/1541-4337.12921

Rondini L, Peyrat-Maillard MN, Marsset-Baglieri A, Fromentin G, Durand P, Tomé D, Prost M, Berset C (2004) Bound ferulic acid from bran is more bioavailable than the free compound in rat. J Agric Food Chem 52(13):4338–4343. https://doi.org/10.1021/jf0348323

Rudrapal M, Khairnar SJ, Khan J, Dukhyil AB, Ansari MA, Alomary MN et al (2022) Dietary polyphenols and their role in oxidative stress-induced human diseases: insights into protective effects, antioxidant potentials and mechanism (s) of action. Front Pharmacol 13:806470.https://doi.org/10.3389/fphar.2022.806470

Sánchez-Patán F, Barroso E, Van de Wiele T, Jiménez-Girón A, Martín-Alvarez PJ, Moreno-Arribas MV, Martínez-Cuesta MC, Peláez C, Requena T, Bartolomé B (2015) Comparative in vitro fermentations of cranberry and grape seed polyphenols with colonic microbiota. Food Chem 183:273–282. https://doi.org/10.1016/j.foodchem.2015.03.061

Santamarina AB, Moreira RG, Mennitti LV, Ferreira YAM, Jucá A, Prado CM, Pisani LP (2025) Low-carbohydrate diet enriched with omega-3 and omega-9 fatty acids modulates inflammation and lipid metabolism in the liver and white adipose tissue of a mouse model of obesity. Nutr Metab Cardiovasc Dis 35(8):103932. https://doi.org/10.1016/j.numecd.2025.103932

Saura-Calixto F (2012) Concept and health-related properties of nonextractable polyphenols: the missing dietary polyphenols. J Agric Food Chem 60(45):11195–11200. https://doi.org/10.1021/jf303758j

Saura-Calixto F, Serrano J, Goñi I (2007–2010) Intake and bioaccessibility of total polyphenols in a whole diet. Food Chem 101(2):492–501. https://doi.org/10.1016/j.foodchem.2006.02.006

Saura-Calixto F, Pérez-Jiménez J, Touriño S, Serrano J, Fuguet E, Torres JL, Goni I (2010) Proanthocyanidin metabolites associated with dietary fibre from in vitro colonic fermentation and proanthocyanidin metabolites in human plasma. Mol Nutr Food Res 54(7):939–946. https://doi.org/10.1002/mnfr.200900276

Solanki R, Rawat L, Tabasum S, Pal S, Patel S, Sabarwal A (2025) A comprehensive review of anti-cancer mechanisms of polyphenol honokiol and nano carrier-based approaches to enhance its therapeutic potential. Phytochem Rev 1–27. https://doi.org/10.1007/s11101-025-10090-0

Stoner GD, Mukhtar H (1995) Polyphenols as cancer chemopreventive agents. J Cell Biochem 59(S22):169–180. https://doi.org/10.1002/jcb.240590822

Tillman L, Rieres JM, Ahjem E, Bishop-Guest F, McGrath M, Hatrick H, Pranjol MZI (2025) Thinking outside the therapeutic box: the potential of polyphenols in preventing chemotherapy-induced endothelial dysfunction. Cells 14(8):566. https://doi.org/10.3390/cells14080566

Tomás-Barberán FA, Selma MV, Espín JC (2016) Interactions of gut microbiota with dietary polyphenols and consequences to human health. Curr Opin Clin Nutr Metab Care 19(6):471–476. https://doi.org/10.1097/MCO.0000000000000314

Tumbarski Y, Ivanov I, Todorova M, Apostolova S, Tzoneva R, Nikolova K (2025) Phenolic content, antioxidant activity and in vitro anti-inflammatory and antitumor potential of selected bulgarian propolis samples. Biomed 13(2):334. https://doi.org/10.3390/biomedicines13020334

Tzounis X, Vulevic J, Kuhnle GG, George T, Leonczak J, Gibson GR, Kwik-Uribe C, Spencer JP (2008) Flavanol monomer-induced changes to the human faecal microflora. Brit J Nutr 99(4):782–792. https://doi.org/10.1017/S0007114507853384

Uti DE, Alum EU, Atangwho IJ, Ugwu OPC, Egbung GE, Aja PM (2025) Lipid-based nano-carriers for the delivery of anti-obesity natural compounds: advances in targeted delivery and precision therapeutics. J Nanobiotechnol 23(1):1–37. https://doi.org/10.1186/s12951-025-03412-z

Valli V, Gómez-Caravaca AM, Di Nunzio M, Danesi F, Caboni MF, Bordoni A (2012) Sugar cane and sugar beet molasses, antioxidant-rich alternatives to refined sugar. J Agric Food Chem 60(51):12508–12515. https://doi.org/10.1021/jf304416d

Vitaglione P, Mennella I, Ferracane R, Rivellese AA, Giacco R, Ercolini D, Gibbons SM, La Storia A, Gilbert JA, Jonnalagadda S, Thielecke F, Gallo M, Scalfi L, Fogliano V (2015) Whole-grain wheat consumption reduces inflammation in a randomized controlled trial on overweight and obese subjects with unhealthy dietary and lifestyle behaviors: role of polyphenols bound to cereal dietary fiber. Am J Clin Nutr 101(2):251–261. https://doi.org/10.3945/ajcn.114.088120

Wang K, Hu S (2023) The synergistic effects of polyphenols and intestinal microbiota on osteoporosis. Front Immunol 14:1285621. https://doi.org/10.3389/fimmu.2023.1285621

Wang Q, Cao J, Yu H, Zhang J, Yuan Y, Shen X, Li C (2019) The effects of EGCG on the mechanical, bioactivities, cross-linking and release properties of gelatin film. Food Chem 271:204–210. https://doi.org/10.1016/j.foodchem.2018.07.168

Wong X, Carrasco-Pozo C, Escobar E, Navarrete P, Blachier F, Andriamihaja M, Annaig Lan A, Tomé D, Cires MJ, Pastene E, Gotteland M (2016) Deleterious effect of p-cresol on human colonic epithelial cells prevented by proanthocyanidin-containing polyphenol extracts from fruits and proanthocyanidin bacterial metabolites. J Agric Food Chem 64(18):3574–3583. https://doi.org/10.1021/acs.jafc.6b00656

Yoon JH, Baek SJ (2005) Molecular targets of dietary polyphenols with anti-inflammatory properties. Yonsei Med J 46(5):585–596

Zamani-Garmsiri F, Emamgholipour S, Rahmani Fard S, Ghasempour G, Jahangard Ahvazi R, Meshkani R (2022) Polyphenols: potential anti-inflammatory agents for treatment of metabolic disorders. Phytother Res 36(1):415–432. https://doi.org/10.1002/ptr.7329

Zhang S, Xu M, Zhang W, Liu C, Chen S (2021) Natural polyphenols in metabolic syndrome: protective mechanisms and clinical applications. Int J Mol Sci 22(11):6110. https://doi.org/10.3390/ijms22116110

Zhu L, Li W, Deng Z, Li H, Zhang B (2020) The composition and antioxidant activity of bound phenolics in three legumes, and their metabolism and bioaccessibility of gastrointestinal tract. Foods 9, 12:1816. https://doi.org/10.3390/foods9121816

# Chapter 11
# Polyphenols in Human Health: Safety Evaluation, Toxicity, and Pharmacological Interactions

**Arup Dey, Manish Kumar Gautam, and Mayank Gangwar**

**Abstract** Polyphenols are a wide family of plant-derived chemicals found in fruits, vegetables, tea, and wine that are renowned for their antioxidant, anti-inflammatory, anti-cancer, and cardioprotective effects. The potential use of these chemicals in nutraceuticals and functional meals has garnered a lot of interest lately. Concerns regarding the safety, toxicity, and possible interactions with pharmaceutical medications are brought up by the growing usage of extracts and supplements high in polyphenols. The toxicological profiles of common polyphenols, including resveratrol, epigallocatechin gallate, quercetin, and curcumin, are evaluated in this study based on recent in vitro, in vivo, and clinical data. Numerous polyphenols exhibit therapeutic benefits at low to moderate dosages; nevertheless, cytotoxicity, oxidative stress, and organ-specific toxicity have been associated with large dosages or prolonged exposure. Furthermore, polyphenols have the ability to either activate or inhibit cytochrome P450 enzymes and drug transporters, which might change the pharmacokinetics of drugs and result in interactions between herbs and drugs. While modest whole food eating is typically safe, longer-term supplementation, standardised dosage, and population-specific responses require more thorough research. This article emphasises the necessity of thorough safety, toxicity evaluations, and interactions to guarantee the secure use of polyphenols in the promotion of health and the prevention of illness.

**Keywords** Dietary polyphenol · Drug–polyphenol interaction · Safety study · Polyphenol · Toxicity evaluation

---

A. Dey
Faculty of Pharmaceutical Science, Assam down town University, Guwahati, Assam, India
e-mail: arupdey2088@gmail.com

M. K. Gautam · M. Gangwar (✉)
Clinical Studies and Trials Unit, Division of Development Research, Indian Council of Medical Research, New Delhi, India
e-mail: mayank.gangwar@icmr.gov.in

M. K. Gautam
e-mail: manish.g@adtu.in



R. K. Singla et al. (eds.), *Polyphenols in Health*,
https://doi.org/10.1007/978-3-032-27324-6_11

## Abbreviations

| | |
|---|---|
| MTT | 3-(4,5-Dimethylthiazol-2-yl)-2,5-diphenyltetrazolium bromide (assay) |
| AE | Adverse events |
| $V_z/F$ | Apparent volume of distribution after oral administration |
| AUOC | Area under and over the curve |
| $AUC_0$–t | Area under the curve from time zero to last measurable concentration |
| AUC | Area under the curve |
| BD | *Bis in die* (twice daily) |
| CYP2C8 | Cytochrome P450 2C8 isoenzyme |
| CYP2D6 | Cytochrome P450 2D6 isoenzyme |
| CYP3A4 | Cytochrome P450 3A4 isoenzyme |
| CYP | Cytochrome P450 |
| DNA | Deoxyribonucleic acid |
| ECG | Electrocardiogram |
| EGCG | Epigallocatechin gallate |
| EFSA | European Food Safety Authority |
| GAE | Gallic acid equivalent |
| GSE | Grape seed extract |
| GSPE | Grape seed polyphenolic extract |
| $IC_{50}$ | Half maximal inhibitory concentration |
| HED | Human equivalent dose |
| HepG2 | Human hepatocellular carcinoma cell line |
| HL-60 | Human promyelocytic leukemia cell line |
| $H_2O_2$ | Hydrogen peroxide |
| ·OH | Hydroxyl radical |
| Fe | Iron |
| $C_{max}$ | Maximum plasma concentration |
| MRNA | Messenger ribonucleic acid |
| NOAEL | No observed adverse effect level |
| OPP | Oil palm phenolics |
| OD | *Omni die* (once daily) |
| ORS | Oral rehydration solution |
| OATP | Organic anion transporting polypeptide |
| OCT | Organic cation transporter |
| OECD | Organisation for Economic Co-operation and Development |
| FEEDAP | Panel on Additives and Products or Substances used in Animal Feed |
| P-gp | P-glycoprotein |
| PCR | Polymerase chain reaction |
| H9c2 | Rat embryonic cardiomyocyte cell line |
| RSC96 | Rat Schwann cell line |
| ROS | Reactive oxygen species |
| SAE | Serious adverse events |
| SHR | Spontaneously hypertensive rat |
| UV | Ultraviolet |

## 11.1 Introduction

Dietary polyphenols are secondary metabolites of plants that have a polyphenol structure. They may be found in a variety of meals and drinks, including fruits, vegetables, cereals, tea, and red wine (Aravind et al. 2021). Dietary polyphenols offer a range of health advantages to people (Farha et al. 2022; Wan et al. 2021). Adults' daily consumption of polyphenols was around 1,193 $\pm$ 510 mg, with 200–300 mg of polyphenols from diet included in every 100 g of fresh produce (Li et al. 2023; Perez-Jimenez et al. 2011). The potential use of dietary polyphenols as functional food components in the creation of nutraceutical and functional foods has garnered a lot of interest lately. However, most dietary polyphenols have no effect in the small colon; instead, they may accumulate in the large intestine, where the gut bacteria may further break them down into easily absorbed, health-promoting low-molecular-weight phenolic metabolites (Cardona et al. 2013). These substances gained notoriety because they are widely found in plant-based diets, and there is strong evidence linking their intake to an increased risk of diabetes, cardiovascular disease, and cancer. Diets rich in polyphenols may lower the risk of a number of age-related chronic illnesses, according to both epidemiological and clinical data (Debelo et al. 2020). The 'biochemical scavenger theory', which holds that polyphenolic substances neutralise free radicals by creating stable chemical complexes, is the most widely accepted explanation for these advantages (Sroka et al. 2003). By producing hydrogen peroxide ($H_2O_2$), polyphenols also protect against oxidative damage. This action can help when speaking of the regulation of immune response processes, including cellular growth (Saeidnia and Abdollahi 2013; Sroka et al. 2003). However, the vast bulk of the data is derived from in vitro models, and it is uncertain if these pathways apply to people (Auclair et al. 2010; Habas et al. 2017). Additionally, recent research has clarified how absorption pharmacokinetics affect polyphenols' effectiveness as antioxidants alongside various potentially health-promoting processes; these physicochemical characteristics of the molecules could account for inconsistent results in the literature as well as the differing effects seen in both animal and human experiments (Jaiswal and Abu-Ghannam 2013; Selby-Pham et al. 2018). Consumption of polyphenols has been linked with several complications. It has been demonstrated that drinks with polyphenolic plant extracts have adverse consequences, especially for people with degenerative illnesses, elevated blood pressure, thyroid issues, epileptic seizures, or cardiovascular disease (Cory et al. 2018). By interacting with thiamine and iron during the digestive process, before they are absorbed, dietary polyphenols have been demonstrated to decrease their absorption. Furthermore, interactions between drugs and reaction-related enzymes or drug carrier proteins can alter their activity, which, depending on the circumstance, can either enhance or reduce bioavailability (Zhang and Tsao 2016). As an illustration, low iron status may result from polyphenols' iron-chelating and iron-absorption-inhibiting properties (Hurrell and Egli 2010). This result might be detrimental to communities that already have inadequate iron storage and consume grains like sorghum, beans, and millet that are high in phytates, which further prevent iron absorption. When infants are fed soy-based formula, isoflavones may have an effect on their long-term growth and pubertal development (Kim et al. 2011; Nguyen et al. 2011).

## 11.2 Safety of Polyphenols

The safety of polyphenols in illness prevention and treatment has not been well investigated, but it is nevertheless crucial to promote their usage for human health. Consuming green tea and its extracts on a regular basis seems to be usually harmless (Shen et al. 2017). More research may be necessary to ensure the safe use of concentrated extracts, including dosages of particular elements taken in solid dosage form, especially in the form of the typical infusion (Hu et al. 2018). Moderate dosages of resveratrol administered directly appear to be both safe and cardioprotective. Therefore, in an animal trial, the administration of a resveratrol supplement (Longevinex) did not appear to have any negative consequences, suggesting positive benefits and safe usage (Sangeetha et al. 2013). However, grape seed extract showed anti-inflammatory and antioxidant qualities and appears to be safe for healthy rats, even at large repeated dosages (Charradi et al. 2018). As demonstrated in a recent study employing curcumin self-nanomicellising solid dispersion aimed at Alzheimer's therapy, the effectiveness of polyphenols can be increased by their incorporation into novel pharmaceutical products that target them specifically while seemingly avoiding side effects (Parikh et al. 2018). However, like with medical pomegranate products for cancer, inadequate regulatory restrictions on commercial polyphenol supplements and nonpharmaceutical formulations raise concerns about their safe usage (Vlachojannis et al. 2015).

The bark of the cinnamon plant, *Cinnamomum zeylanicum,* is used as a spice in cooking and has been shown to have therapeutic and health advantages. The primary bioactive class of components in cinnamon bark is represented by polyphenols (PP). However, for safe human consumption, toxicity data and tolerable and safe dosage levels of oral exposure to PP-based standardised cinnamon bark extract (PP-CZ) are essential. The scientists assessed PP-CZ mutagenicity, genotoxic potential, acute oral, and sub-chronic oral toxicities in vivo. In accordance with Organisation for Economic Co-operation and Development (OECD) Tests 423 and 408, respectively, the effects of oral therapy with a single- and 90-day recurrent dosage were assessed in rates. Effects on haematology, biochemistry, histology, body weight, food and drink consumption, and organ weight were documented. The chromosomal aberrations (OECD Test 473) and Ames (OECD No. 471) tests were used to assess mutagenicity and genotoxicity. This suggests a 'median lethal dose' >2,000 mg/kg, as a single oral dose of PP-CZ did not result in death or treatment-related adverse consequences. A sub-chronic dose of 500 mg/kg/day for 90 days was also shown to be safe in rats, indicating a 'human equivalent dose' (HED) of 4.8 g/day and a 'no observed adverse effect level' of 500 mg/kg. Additionally, experiments conducted in vitro revealed that PP-CZ exhibited neither mutagenicity nor genotoxicity. In rats, PP-CZ demonstrated a strong safety profile devoid of genotoxicity or mutagenicity (Thakurdesai et al. 2024).

Relatively few researchers have looked at the potential toxicity of polyphenols, with the majority of studies focusing on their preventive properties against illness or harmful pharmaceuticals. Rats or mice were given an oral dose of 0.5 or 2 g/kg body weight of a grape seed proanthocyanidin extract, and rats were given a dose of 60 g/kg diet of punicalagin, an ellagitannin found in pomegranate juice, without experiencing any acute toxicity. However, when rats were given large dosages of quercetin (2 or 4%), chronic nephropathy was seen. The study found no influence on survival times; however, mice's life expectancy was dramatically lowered when quercetin (0.1%) was added to their diet (Mennen et al. 2005).

According to another study, in order to monitor cardiovascular activity remotely, the researchers implanted telemetry transmitters in eight naïve beagle dogs as part of a safety pharmacology study at Ridglan Farms. The dogs were then given one oral dose of the polyphenolic compound resveratrol at a dosage of either 0, 200, 400, or 1,000 mg/kg. After a one-week washout period, weekly dosing was maintained until each animal got a single pill administration of resveratrol at every dose level. Weekly body weight measurements, daily clinical observations, daily mortality/moribundity observations, and general toxicity assessments were also included in the study. At least two hours before and 24 h after each resveratrol dosage, telemetry was used to continually measure body temperature and cardiovascular performance. Cardiovascular parameters were monitored, including electrocardiography, pulse, diastolic, systolic, and mean arterial pressure. Resveratrol may enhance cardiovascular health, according to study findings. The study used a balanced Latin-Square crossover design to administer resveratrol orally as pills and found no treatment-related effects in any of the examined parameters. No animals passed away throughout the trial, and no negative effects were observed in any of the animals during physical tests and clinical assessments. The comparison between mean body weights and total body weights revealed no treatment-related changes. The delivery of resveratrol had no effect on the body temperature, pulse, arterial pressure, or electrocardiogram (ECG) measures. The authors have not disclosed any safety information in their study (Johnson et al. 2011).

Apple polyphenol extract is utilised as a food additive and nutritional supplement in Japan because of its potent antioxidant activity and wide range of physiological effects. A detailed study on the safety evaluation of Applephenon1, a polyphenol extract made from unripe apples, reported that only a modest mutagenicity effect was observed at the level of 2,500 mg/plate (Ames test). On the other hand, mutagenicity was not assessed when speaking of chromosomal aberration and the micronucleus tests. Finally, no detectable effects at the level of 2,000 mg/kg were reported, showing that this apple phenolic compound can be evaluated as safe and healthy-effective, in the ambit of a normal nutritional intake (Shoji et al. 2004).

Most people agree that drugs derived from natural sources are harmless, even in high concentrations. The toxicity of natural plant extracts is therefore not well documented in the available research. In addition, the safety of consuming large amounts of polyphenols as food additives or dietary supplements over an extended period of time is unclear and requires further research. Therefore, identifying any possible cytotoxicity requires knowing the right amounts of polyphenol and other natural substances to utilise as ingredients in food (Ofosu et al. 2020).

The most part of studies concerning polyphenols aimed to determine the protective effects of polyphenols against diseases or toxic drugs, and relatively few investigators have examined their possible toxicity. No acute toxicity was reported after oral administration of a grape seed proanthocyanidin extract at a dose of 0.5 or 2 g/kg body weight to rats or mice or after administration of punicalagin (an ellagitannin present in pomegranate juice) at a dose of 60 g/kg diet to rats. However, chronic nephropathy was observed in rats when high doses of quercetin (2 or 4%) were added to their diet. No effect on survival times was observed in that study, whereas the addition of quercetin (0.1%) to the diet of mice significantly reduced their life expectancy (Yamakoshi et al. 2002).

According to a reported study, rats given large doses of 2 and 4 g/kg of proanthocyanin-enriched extracts from grape seeds orally did not experience acute or sub-chronic toxicity. On the fourteenth day of clinical monitoring, the fatal dosage in the acute trial was greater than 4 g/kg. However, after a 90-day sub-chronic toxicity investigation, the 'no observed adverse effect level'(NOAEL) was determined to be 2 g/kg of grape seed extract (GSE), which is around 240 times the expected daily consumption of proanthocyanin by humans. In a 90-day sub-chronic toxicity investigation, grape seed polyphenolic extract (GSPE) demonstrated great tolerance and was deemed safe in animal models fed both extract and chow. Following therapy with dosages ranging from 200 to 2,150 mg/kg/day, no discernible side effects were noted (Yamakoshi et al. 2002). Dried grape extract is safe and effective for use as a flavouring agent in animal feed for all animal species and categories (with the exception of dogs) at a maximum inclusion rate of 100 mg per kg of complete feed, according to the findings of the EFSA, the Panel on Additives and Products or Substances used in Animal Feed (FEEDAP). During this focus, the Panel stressed the fact that there are no consumer safety issues (EFSA FEEDAP 2014).

To assess oil palm phenolics' (OPP) efficacy and safety in healthy participants, a phase I- single-blind clinical experiment was carried out (Fairus et al. 2018). 25 healthy human volunteers participated in this single-blind study, receiving either control treatments or 450 mg gallic acid equivalent (GAE)/day of OPP for 60 days. On days 1, 30, and 60, fasting blood and urine samples were taken three times during the whole study, also including a medical check-up. During the control and OPP treatment periods, no major adverse effects (AE) and significant adverse effects (SAE) were observed, with the final evaluation of OPP as a safe supplement and possibly able to lower cholesterol amounts (Fairus et al. 2018). As shown in Fig. 11.1, natural polyphenols possess diverse structural reasons that contribute to their distinct biological functions.

To assess the safety of 'LifeDrops' solutions, an aqueous formulation that contains punicalagin-α and punicalagin-β from pomegranate fruit and EGCG and epicatechin gallate derived from green tea leaves, a randomised, double-blind clinical experiment was carried out. 44 babies and kids between the ages of 6 months and 5 years participated in the trial. They received 'LifeDrops' solution twice a day for a maximum

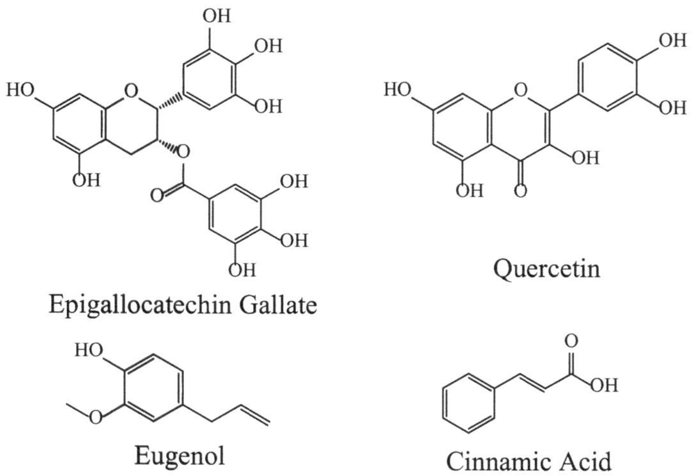


**Fig. 11.1** Chemical structures of selected natural polyphenols. BKchem version 0.13.0, 2009 (http://bkchem.zirael.org/index.html), has also been used for drawing this structure. This picture has been created by the authors

of 4 days without experiencing any negative physiological effects or adverse occurrences. In order to keep them hydrated during the research, patients received extra oral rehydration solution (ORS) as needed. Venous blood was used to assess the total amount of blood, minerals, the condition of the liver, and creatinine. 95% assurance intervals were used to compare the mean values. The measured blood factors either stayed constant throughout the course of the four days or rose or fell at the identical rates in the two groups. As a result, it was demonstrated that children and newborns can safely use 'LifeDrops' solution, according to the study findings (Sarker et al. 2015). The safety of natural products and formulations containing polyphenols has been well studied in a variety of in vitro, in vivo, and clinical contexts. These safety assessments are thoroughly summarised in Table 11.1, which also includes model systems, dose ranges, and observed toxicological results. These findings provide credence to the possibility of a safe therapeutic use when used within advised bounds.

## 11.3 Toxicity of Polyphenols

In 2024, a group of researchers used the ethanolic extract of *Euphorbia hirta L.*, which contains a lot of polyphenols, to do a 3-(4,5-dimethylthiazol-2-yl)-2,5-diphenyltetrazolium bromide (MTT) assay experiment on human hepatocellular carcinoma cell line (HepG2) cells, which are human liver cancer cells. They stated that HepG2 human liver cancer cell lines were used to evaluate the cytotoxic effect of *Euphorbia hirta L.*-ethanolic extract. Using the MTT test at 24 h, the inhibitory

**Table 11.1** Safety evaluation of polyphenol-containing natural products and formulations

| Polyphenol/product | Source/compound | Study type | Dosage | Model/subjects | Key safety findings | References |
|---|---|---|---|---|---|---|
| Green tea/extract | Green tea polyphenols | Human data | Regular consumption | Humans | Safe with moderate use; further data needed on extract dosages | Hu et al. (2018), Shen et al. (2017) |
| Resveratrol | Longevinex supplement | Animal (in vivo) | Not specified | Rats | No adverse effects; safe and cardioprotective | Sangeetha et al. (2013) |
| Grape seed extract | Proanthocyanidins | Animal (in vivo) | High repeated doses | Rats | Safe, anti-inflammatory, and antioxidant | Charradi et al. (2018) |
| Curcumin (nanodispersion) | Curcumin formulation | Animal/ tech-based (in vivo) | Not specified | Alzheimer's model | Targeted delivery enhances efficacy with no side effects | Parikh et al. (2018) |
| Pomegranate polyphenols | Medical pomegranate | General remark | – | – | Safety concerns due to regulatory issues with supplements | Vlachojannis et al. (2015) |
| Cinnamon bark extract (PP-CZ) | Cinnamomumzeylanicum | In vivo + in vitro | 2000 mg/kg (acute), 500 mg/kg/day (sub-chronic); HED ~ 4.8 g/day | Rats | No genotoxicity/ mutagenicity; NOAEL 500 mg/kg/day | Thakurdesai et al. (2024) |
| Quercetin | Flavonoid | Animal (in vivo) | 2–4% in diet | Rats | Chronic nephropathy; reduced lifespan in mice at 0.1% | Mennen et al. (2005) |
| Resveratrol (cardiac safety) | Pure compound | Beagle dog crossover study | 200–1000 mg/kg | Dogs | No adverse effects on cardiovascular, ECG, body weight, temp | Johnson et al. (2011) |

(continued)

**Table 11.1** (continued)

| Polyphenol/product | Source/compound | Study type | Dosage | Model/subjects | Key safety findings | References |
|---|---|---|---|---|---|---|
| Applephenon® | Unripe apple extract | Toxicity + genotoxicity | 2500 mg/plate; 2000 mg/kg | Rats | No toxic effects observed; slight mutagenicity at very high dose | Shoji et al. ( 2004) |
| General polyphenols | Various plant sources | General review | – | – | Need for more data on long-term safety in high doses | Ofosu et al. (2020) |
| Grape Seed polyphenol extract (GSPE) | Grapes | Sub-chronic toxicity | 2–4 g/kg; NOAEL = 2 g/kg | Rats | Safe with high tolerance; no observed toxicity | EFSA FEEDAP (2014), Yamakoshi et al. (2002) |
| Oil Palm Phenolics (OPP) | Palm (gallic acid equivalents) | Phase I clinical trial | 450 mg GAE/day for 60 days | 25 healthy humans | Safe, no major adverse events; improved lipid profile | Fairus et al. (2018) |
| LifeDrops solution | Pomegranate + Green Tea (EGCG) | RCT – Clinical Trial | Twice daily for 4 days | 44 children (6 mo – 5 yrs) | No adverse physiological effects; safe in pediatric population | Sarker et al. (2015) |

concentration needed for 50% cell death (HepG2-alf maximal inhibitory concentration, $IC_{50}$) was found to be 200 μg/mL of *Euphorbia hirta L.* ethanolic extract by graphing cell viability under treated and untreated conditions (Palanisamy et al. 2024).

In 2023, researchers examined the effects of eight *Camellia sinensis*-derived green tea and eight *Camellia* black tea samples on rat Schwann cell lines (RSC96) and H9c2 embryonic heart cells. The findings showed a negative correlation between $IC_{50}$ values and polyphenol content, indicating that higher levels of polyphenol were associated with greater toxicity. At concentrations beginning at 2 mg/ml, green tea samples dramatically reduced RSC96-cell viability, demonstrating a dose-dependent impact. Likewise, at 3.0 mg/ml, two black tea samples decreased cell viability. At 0.5 mg/ml, H9c2 cell viability started to decrease, and three black tea samples showed noticeable toxicity starting at 1.0 mg/ml. Overall, the study showed that the polyphenol level of tea samples affected their toxicity (Wu et al. 2023).

By examining the total area under and over the curve (AUOC) in relation to the initial state, the researchers assessed the in vitro toxicity of 22 polyphenol-rich drugs. The three main cellular markers nuclear size, mitochondrial membrane potential, and cell membrane integrity were used to rank these substances. Among the phenolics they studied, naringin showed the least amount of harm, according to their findings. Additionally, kale leaf extract, wasted hop extract, and a commercial extract (Aronox) showed little cytotoxicity. As a positive control, kaempferol, on the other hand, demonstrated obvious harmful effects on nuclear and mitochondrial architecture. Furthermore, extracts from hollyhock flower, walnut husk, and buckthorn bark were shown to have minimal effect on the nucleus but to be extremely harmful to the mitochondria and cell membrane (Boncler et al. 2017).

Glei and coworkers et al. investigated the effects of functional foods high in polyphenols, particularly anthocyanins from black carrots and catechins from green tea, on human colon cancer cells. The findings demonstrated that both sources' water extracts severely decreased cell viability and multiplication while damaging DNA. Given that the whole extracts were more hazardous than their active ingredients, compound combinations may have additive or synergistic harmful effects. The study also discovered that neither black carrot nor green tea extracts offered defence against hydrogen peroxide-induced oxidative stress, highlighting the need to evaluate whole food extracts on an individual basis (Glei et al. 2003).

In a research published in 2005, Murias et al. examined the cytotoxic effects of resveratrol, a naturally occurring polyphenol, and its hydroxylated derivatives on human promyelocytic leukemia cell line (HL-60) cells. The findings indicated that hydroxylated resveratrol derivatives were much more harmful than resveratrol itself, especially those with ortho-dihydroxy structures. This enhanced cytotoxic impact was attributed to these polyphenolic chemicals' prooxidant behaviour, which facilitated redox cycling and the creation of semiquinone intermediates, leading to a higher generation of reactive oxygen species (ROS). These ROS destroy essential cellular components, cause oxidative stress, and trigger apoptosis. The work emphasises how, when properly optimised, polyphenols may play two roles in cancer treatment (Murias et al. 2005).

The safety and pharmacokinetic profile of green tea polyphenols were assessed in a clinical investigation following four weeks of daily oral administration of either EGCG or polyphenol E. Figures 6.1 and 6.11 (Chap. 6), and 11.1 show some of these natural compounds. Finding out how long-term ingestion of these polyphenols impacts the skin's susceptibility to erythema caused by ultraviolet (UV) radiation was the main goal. A two-week lead-in phase was followed by a random assignment of healthy volunteers with Fitzpatrick skin types II or III to receive either EGCG 500 mg *omni die*(OD), 400 mg *bis in die*(BD), 800 mg Polyphenon E OD, 400 mg Polyphenon E BD, or a placebo. Eight people participated in each group, and the study ran for four weeks. During the therapy period, migraines, fainting, and muscular pains were among the reported adverse effects, along with digestive issues such as bloating, sickness, heartburn, and stomach discomfort. Throughout the experiment, no appreciable alterations in haematological markers or blood chemistry were noted, despite these minor side effects. After four weeks of green tea polyphenol therapy at a dose of 800 mg once daily, the area under the plasma EGCG concentration–time curve increased by more than 60%. Following repeated green tea polyphenol administration at a dose of 400 mg twice a day, no appreciable alterations in the pharmacokinetics of EGCG were seen. UV-induced erythema was not prevented by four weeks of green tea polyphenol therapy at the chosen dosage and dosing schedule (Chow et al. 2003). Table 11.2 summarises the toxicity profiles of several natural polyphenols measured in human and cellular models. To evaluate their therapeutic potential, our results emphasise safety limits and dose-dependent cytotoxic effects.

## 11.4 Interactions of Polyphenols

The highly reactive nature of polyphenols' functional groups allows them to interact with a variety of chemicals, including medications, potentially changing their therapeutic effects. These interactions have an impact on drug transporters such as P-glycoprotein, organic anion transporting polypeptides (OATP), organic cation transporters (OCT), and drug-metabolising enzymes (mainly cytochrome P450 (CYP) isoenzymes such as CYP3A4 isoenzyme (CYP3A4) and CYP2D6 isoenzyme (CYP2D6), influencing drug absorption, distribution, metabolism, and elimination. These enzymes can be activated and inhibited by polyphenols, which can lead to increased drug toxicity or concentration or decreased therapeutic effectiveness. For medications with a limited therapeutic window, such as digoxin, cyclosporine A, and warfarin, this is especially risky. The precise mechanism by which they stimulate or prevent the enzyme is often unclear, although this is somewhat recognised. For instance, patients are advised against consuming grapefruit juice while taking specific drugs. To prevent any negative effects from food-drug or herb–drug combos, it is crucial that patients and medical practitioners understand these interactions (Duda-Chodak and Tarko 2023).

Another study looked at how polyphenols and iron (Fe) interact, specifically with regard to Fenton and Fenton-like reactions, which are frequently employed

**Table 11.2** Toxicity assessment of natural polyphenols in cellular and human models

| Polyphenol/product | Source/compound | Study type | Dosage | Model/subjects | Key findings | Duration | Mechanism of toxicity | dose dependency | Assay type method | References |
|---|---|---|---|---|---|---|---|---|---|---|
| *Euphorbia hirta* extract | Polyphenol-rich ethanolic extract | In vitro | 200 µg/mL | HepG2 (human liver cancer) | IC50 at 200 µg/mL; moderate cytotoxicity | 24 h | Cytotoxicity | Dose-dependent | MTT assay | Palanisamy et al. (2024) |
| *Camellia sinensis* (green and black tea) | Green and black tea polyphenols | In vitro | 2–3 mg/mL | RSC96 Schwann cells, H9c2 cardiomyocytes | Toxicity correlated with polyphenol content; decreased cell viability | Dose-dependent | Cytotoxicity | Dose-dependent | MTT assay | Wu et al. (2023) |
| Multiple plant extracts | 22 polyphenol-rich compounds | In vitro | Various | Cell-based toxicity screening | Naringin least toxic; kaempferol and some highly cytotoxic extracts | Not specified | Mitochondrial and membrane damage | Varied | AUOC ranking system | Boncler et al. (2017) |
| Black carrot and green tea | Anthocyanins and catechins | In vitro | Not specified | Human colon cancer cells | Decreased viability; DNA damage; synergistic toxicity | Not specified | DNA damage and oxidative stress | Additive/synergistic | Functional food extract test | Glei e al. (2003) |

(continued)

**Table 11.2** (continued)

| Polyphenol/product | Source/compound | Study type | Dosage | Model/subjects | Key findings | Duration | Mechanism of toxicity | dose dependency | Assay type method | References |
|---|---|---|---|---|---|---|---|---|---|---|
| Resveratrol and derivatives | Hydroxylated resveratrol | In vitro | Not specified | HL-60 leukemia cells | Derivatives more cytotoxic due to ROS production | Not specified | Oxidative stress and apoptosis | Structure-dependent | ROS-mediated apoptosis | Murias et al. (2005) |
| Green tea polyphenols (EGCG, Polyphenon E) | EGCG/polyphenon E | Clinical trial | 400–800 mg/day | Healthy adults | Mild GI and neurological AE; no major lab changes | 4 weeks | Mild systemic side effects | Mild, not dose-limiting | PK and safety study | Chow et al. (2003) |

in the breakdown of organic contaminants. The authors will concentrate on how polyphenols transform Feions back into Fe ions in the iron-chelating regeneration cycle, which is crucial for Fenton chemistry. The effectiveness and adaptability of Fenton-like reactions are enhanced by these interactions, which significantly increase hydroxyl radical (·OH) production even at high pH values. Additionally, it locks at how the Fe-polyphenol complexes behave in oxidative settings and provides experimental evidence for their dual function as iron chelators and reductants in complicated oxidative processes (Pan et al. 2022).

The covalent connections between polyphenols and proteins during food processing were studied by Zhang and coworkers, with an emphasis on the processes of production, techniques of measurement, bioactive effects, and applications. They clarified that under processing circumstances like heating, enzymatic activity, or pH changes, polyphenols particularly those with catechol or galloyl structures oxidise into reactive quinones. These quinones create permanent covalent connections with the residues of nucleophilic amino acids in proteins, such as lysine, cysteine, histidine, and tyrosine. Protein solubility, emulsifying qualities, digestibility, and allergenicity are all impacted by these interactions, which have a substantial impact on protein structure and function. These complexes are investigated using analytical methods such as mass spectrometry, electrophoresis, and spectroscopy. The improved bioactivities of covalent polyphenol–protein complexes include hypoglycemic, anti-inflammatory, gut microbiota-modulating, and antioxidant properties. They have the potential to be used in functional biomaterials, delivery methods, and food additives. The authors also point out some of the existing issues, such as regulating response settings, enhancing quantification precision, and assessing in vivo safety and effectiveness (Zhang et al. 2024).

Many fruits, vegetables, teas, and wines include polyphenols, particularly flavonoids, which make up a sizable amount of the bioactive substances found in food. Drug-metabolising enzymes, especially cytochrome P450 isoforms like CYP3A4, CYP1A1, CYP1A2, and others, are known to interact with these flavonoids. Given its significant function in the metabolism of several medications, CYP3A4 has been the subject of the most research among them. Through inhibition or activation of CYP3A4, flavonoids can modify medication absorption, distribution, and excretion. Depending on their metabolism and conversion into glucuronidated or sulfated forms, flavonols such as quercetin and kaempferol, for example, can either raise or lower medication concentrations in vivo while inhibiting CYP3A4 activity in vitro. Apigenin and chrysin are two flavones able to show strong CYP3A4 inhibition. Apigenin also increases medication bioavailability by lowering drug efflux through P-glycoprotein inhibition (Choi et al. 2011).

Other subclasses of polyphenols also exhibit noteworthy interactions with CYP3A4. There are well-established food-drug interactions because flavanones, such as naringenin (found in grapefruit juice), are recognised inhibitors. CYP3A4 is inhibited to varied degrees by isoflavones, anthocyanins, stilbenoids (like resveratrol), lignans (like silymarin), and phenolic acids (like caffeic acid and gallic acid) via a variety of routes, including competitive and non-competitive inhibition or even mechanism-based inactivation. Particularly for medications with limited therapeutic indices, these interactions may result in changed therapeutic results or elevated drug toxicity. In order to prevent harmful drug–nutrient interactions, it is important to carefully consider taking polyphenol-rich foods or supplements at the same time as traditional medications (Basheer and Kerem 2015).

This study examined the effects of myricetin, a dietary flavonoid, on the metabolism of amodiaquine, a CYP2C8 substrate, in Wistar rats. Rats were given oral amodiaquine (60 mg/kg) after receiving an intravenous dosage of myricetin (5 mg/kg). For the purpose of determining pharmacokinetic parameters, serial blood samples were obtained. The findings demonstrated that myricetin had a strong in vivo suppression of CYP2C8-mediated metabolism by considerably raising the plasma exposure as area under the curve (AUC) of amodiaquine and decreasing its clearance. This demonstrates that myricetin can change medication metabolism at physiologically relevant dosages, highlighting the possibility of substantial drug–polyphenol interactions (Bhatt et al. 2022).

The study on the interaction between naringenin and rivaroxaban was carried out in 2023 using in vitro and in vivo models. Mixed inhibition for the rivaroxaban metabolism was shown by naringenin in rat liver microsomes, with an $IC_{50}$ of 38.89 μM, Ki of 54.91 μM, and αKi of 73.33 μM. In vivo, naringenin pre-treatment of Sprague–Dawley rats at 10 mg/kg/day for 2 weeks significantly enhanced the exposure to rivaroxaban with reference to: area under the curve from time zero to last measurable concentration ($AUC_0$–t), and maximum plasma concentration ($C_{max}$). On the other side, parameters such as apparent volume of distribution after oral administration ($V_z$/F) suggested a measurable decline. This indicated a strong inhibitory effect of naringenin on the clearance of rivaroxaban, which may enhance the anticoagulant activity and risk of bleeding from the drug. The study indicates that possible drug–nutrient interactions are to be monitored in clinical practice, particularly with respect to foods or supplements containing high levels of polyphenols (Shi et al 2021).

In another study, Tan and coworkers looked into, in rats with spontaneous hypertension (SHR), oral EGCG (10 mg/kg/day for 13 days) dramatically decreased the bioavailability of the non-metabolised β-blocker nadolol. While EGCG pretreatment had no effect on elimination half-life or renal clearance, it did result in a ~51% drop in AUC and a ~53% decrease in plasma $C_{max}$ when compared to control. The EGCG-nadolol group reduced antihypertensive effectiveness, was shown by their higher systolic blood pressure. According to real-time polymerase chain reaction (PCR), EGCG dramatically decreased messenger ribonucleic acid (mRNA) levels of important intestinal and hepatic drug uptake transporters, including Oct1, P-glycoprotein (P-gp), and Oatp1a5. The decrease ranged from around 2.6-fold for the liver to eight-fold for the ileum. These results raise concerns about herb–drug interactions in hypertensive patients because they imply that EGCG decreases nadolol absorption through transporter regulation, resulting in lower systemic drug exposure and impaired pharmacodynamic impact (Tan et al. 2021). Table 11.3 provides an overview of the pharmacological interactions of polyphenols and their related biological functions. Representative molecular structures from many polyphenol subclasses are shown in

Figs. 6.2, 6.3, 6.4, 6.5, 6.6, 6.7, 6.8, 6.9, 6.10 and 6.11 (Chap. 6), which sheds light on their chemical variety and their functions.

## 11.5 Conclusions

Polyphenols have outstanding therapeutic potential and play a major role in the positive health impacts of plant-based diets. Laboratory and preclinical research have provided substantial evidence for their anti-inflammatory, antioxidant, and disease-modifying qualities. However, new research indicates that natural substances might have negative effects at large dosages or after extended usage, casting doubt on the notion that they are always safe. Modification of important metabolic enzymes, organ-specific toxicity, DNA damage, and cytotoxicity all demonstrate the complex bioactivity of polyphenols. Research on animals and clinical trials shows that although polyphenols such as EGCG, resveratrol, and grape seed extract are typically safe at modest quantities, certain polyphenols might cause unexpected physiological effects, particularly in sensitive people or when taken with pharmaceuticals.

Their potential to alter medication metabolism and transport, which might increase or decrease therapeutic effectiveness or safety, is a serious worry. Patients with chronic illnesses or those following polypharmacy regimens should pay special attention to this. Additionally, the bioavailability and responsiveness to polyphenols can be influenced by differences in human metabolism, gut microbiota composition, and genetic background. Strong clinical data and tailored intake recommendations are therefore essential. The current focus of polyphenols in therapeutic and preventative medicine is on striking a balance between maximising their advantages and reducing any possible hazards by using them carefully and intelligently.

In conclusion, polyphenols are bioactive substances produced from plants that have a variety of health-promoting properties, such as anti-inflammatory, antioxidant, and disease-preventive properties. Nevertheless, mounting data indicate that their safety varies depending on the situation and dose. Long-term usage, particularly through supplements, or high dosages, may result in harmful drug–nutrient interactions or cytotoxicity. Certain polyphenols can potentially cause toxicity or impair the effectiveness of treatment by interfering with medication metabolism and nutrient absorption. Their widespread safety is supported by historic use, but more research is needed for contemporary uses, especially when it comes to isolated or concentrated forms. Therefore, to guarantee their safe inclusion into medicinal and nutritional practices, future research ought to focus on thorough toxicological evaluations, formulation standardisation, and rigorous pharmacokinetic profiling.

**Table 11.3** Mechanistic insights into polyphenol interactions with drugs and biological systems

| Polyphenol type | Mechanism of interaction | Affected enzymes/ transporters | Affected drugs | Effect on drug | Outcome | References |
|---|---|---|---|---|---|---|
| General polyphenols | Inhibition or activation of drug metabolism | CYP450 (CYP3A4, CYP2D6), P-gp, OATP, OCT | Digoxin, cyclosporine A, warfarin | Altered metabolism and transport | Increased toxicity or reduced efficacy | Duda-Chodak and Tarko (2023) |
| Polyphenols (Fenton context) | Iron chelation and redox cycling | $Fe^{2+}/Fe^{3+}$ (Fenton chemistry) | Not drug-specific | Enhanced ·OH radical generation | Boosts pollutant degradation efficiency in AOP | Pan et al. (2022) |
| Catechol/Galloyl polyphenols | Covalent bonding during food processing | N/A (targets protein residues: Lys, Cys, His, Tyr) | Not available | Alters protein solubility, emulsification, and digestibility | Bioactive benefits (e.g., antioxidant, hypoglycemic effects) | Zhang et al. (2024) |
| Flavonoids (quercetin, kaempferol) | CYP3A4 inhibition | CYP3A4, P-gp | Various CYP3A4-metabolised drugs | Modifies plasma drug levels | Decreased or increased exposure | Choi et al. (2011) |
| Apigenin, chrysin | Strong CYP3A4 and P-gp inhibition | CYP3A4, P-gp | Various | Inhibits metabolism and efflux | Increased bioavailability | Choi et al. (2011) |
| Naringenin (grapefruit) | CYP3A4 inhibition (mixed-type) | CYP3A4 | Rivaroxaban | ↓ CL, ↑ AUC, ↑ $C_{max}$ | Enhanced bleeding risk due to elevated anticoagulant exposure | Basheer and Kerem (2015), Shi et al. (2021) |
| Isoflavones, anthocyanins, stilbenoids, lignans, phenolic acids | Competitive/ non-comp inhibition of CYP3A4 | CYP3A4 | Various | Drug level fluctuations | Potential toxicity or therapeutic failure | Basheer and Kerem (2015) |

(continued)

**Table 11.3** (continued)

| Polyphenol type | Mechanism of interaction | Affected enzymes/ transporters | Affected drugs | Effect on drug | Outcome | References |
|---|---|---|---|---|---|---|
| Myricetin | CYP2C8 inhibition | CYP2C8 | Amodiaquine | ↑ AUC, ↓ CL | Strong interaction at physiologic levels | Bhatt et al. (2022) |
| EGCG (green tea catechin) | Downregulation of uptake transporters | Oct1, P-gp, Oatp1a5 (mRNA suppression) | Nadolol | ↓ AUC by ~51%, ↓ $C_{max}$ by ~53% | Decreased drug absorption and effect | Tan et al. (2021) |

## References


Aravind SM, Wichienchot S, Tsao R, Ramakrishnan S, Chakkaravarthi S (2021) Role of dietary polyphenols on gut microbiota, their metabolites and health benefits. Food Res Int 142:110189. https://doi.org/10.1016/j.foodres.2021.110189

Auclair S, Chironi G, Milenkovic D, Hollman PCH, Renard CMGC, Mégnien JL Gariepy J, Paul JL, Simon A, Scalbert A (2010) The regular consumption of a polyphenol-rich apple does not influence endothelial function: a randomised double-blind trial in hypercholesterolemic adults. Eur J Clin Nutr 64(10):1158–1165. https://doi.org/10.1038/ejcn.2010.135

Basheer L, Kerem Z (2015) Interactions between CYP3A4 and dietary polyphenols. Oxid Med Cell Longev 2015:854015. https://doi.org/10.1155/2015/854015

Bhatt S, Manhas D, Kumar V, Gour A, Sharma K, Dogra A, Ojha PK, Nandi U (2022) Effect of myricetin on cyp2c8 inhibition to assess the likelihood of drug interaction using in silico, in vitro, and in vivo approaches. ACS Omega 7(15):13260–13269. https://doi.org/10.1021/acsomega.2c00726

Boncler M, Golanski J, Lukasiak M, Redzynia M, Dastych J, Watala C (2017) A new approach for the assessment of the toxicity of polyphenol-rich compounds with the use of high content screening analysis. PLoS ONE 12(6):e0180022. https://doi.org/10.1371/journal.pone.0180022

Cardona F, Andrés-Lacueva C, Tulipani S, Tinahones FJ, Queipo-Ortuño MI (2013) Benefits of polyphenols on gut microbiota and implications in human health. J Nutr Biochem 24(8):1415–1422. https://doi.org/10.1016/j.jnutbio.2013.05.001

Charradi K, Mahmoudi M, Bedhiafi T, Jebari K, El May MV, LimamF AE (2018) Safety evaluation, anti-oxidative and anti-inflammatory effects of subchronically dietary supplemented high dosing grape seed powder (GSP) to healthy rat. Biomed Pharmacother 107:534–546. https://doi.org/10.1016/j.biopha.2018.08.031

Choi JS, Piao YJ, Kang KW (2011) Effects of quercetin on the bioavailability of doxorubicin in rats: role of CYP3A4 and P-gp inhibition by quercetin. Arch Pharmacal Res 34(4):607–613. https://doi.org/10.1007/s12272-011-0411-x

Chow HS, Cai Y, Hakim IA, Crowell JA, Shahi F, Brooks CA, Dorr RT, Hara Y, Alberts DS (2003) Pharmacokinetics and safety of green tea polyphenols after multiple-dose administration of epigallocatechin gallate and polyphenon E in healthy individuals. Clin Cancer Res 9(9):3312–3319

Cory H, Passarelli S, Szeto J, Tamez M, Mattei J (2018) The role of polyphenols in human health and food systems: a mini-review. Front Nutr 5:87. https://doi.org/10.3389/fnut.2018.00087

Debelo H, Li M, Ferruzzi MG (2020) Processing influences on food polyphenol profiles and biological activity. Curr Opin Food Sci 32:90–102. https://doi.org/10.1016/j.cofs.2020.03.001

Duda-Chodak A, Tarko T (2023) Possible side effects of polyphenols and their interactions with medicines. Mol 28(6):2536. https://doi.org/10.3390/molecules28062536

EFSA FEEDAP (2014) Scientific opinion on the safety and efficacy of tannic acid when used as feed flavouring for all animal species. EFSA J 12(10):3828.https://doi.org/10.2903/j.efsa.2014.3828

Fairus S, Leow SS, Mohamed IN, Tan YA, Sundram K, Sambanthamurthi R (2018) A phase I single-blind clinical trial to evaluate the safety of oil palm phenolics (OPP) supplementation in healthy volunteers. Sci Rep 8(1):8217. https://doi.org/10.1038/s41598-018-26384-7

Farha AK, Gan RY, Li HB, Wu DT, Atanasov AG, Gul K, Zhang JR, Yang QQ, Corke H (2022) The anticancer potential of the dietary polyphenol rutin: current status, challenges, and perspectives. Crit Rev Food Sci Nutr 62(3):832–859. https://doi.org/10.1080/10408398.2020.1829541

Glei M, Matuschek M, Steiner C, Böhm V, Persin C, Pool-Zobel BL (2003) Initial in vitro toxicity testing of functional foods rich in catechins and anthocyanins in human cells. Toxicol Vitro 17(5–6):723–729. https://doi.org/10.1016/S0887-2333(03)00099-7

Habas K, Brinkworth MH, Anderson D (2017) Diethylstilbestrol induces oxidative DNA damage, resulting in apoptosis of spermatogonial stem cells in vitro. Toxicol 382:117–121. https://doi.org/10.1016/j.tox.2017.03.013

Hu J, Webster D, Cao J, Shao A (2018) The safety of green tea and green tea extract consumption in adults—results of a systematic review. Reg Toxicol Pharmacol 95:412–433. https://doi.org/10.1016/j.yrtph.2018.03.019

Hurrell R, Egli I (2010) Iron bioavailability and dietary reference values. Am J Clin Nutr 91(5):1461S-1467S. https://doi.org/10.3945/ajcn.2010.28674F

Jaiswal AK, Abu-Ghannam N (2013) Degradation kinetic modelling of color, texture, polyphenols and antioxidant capacity of York cabbage after microwave processing. Food Res Int 53(1):125–133. https://doi.org/10.1016/j.foodres.2013.04.007

Johnson WD, Morrissey RL, Usborne AL, Kapetanovic I, Crowell JA, Muzzio M, McCormick DL (2011) Subchronic oral toxicity and cardiovascular safety pharmacology studies of resveratrol, a naturally occurring polyphenol with cancer preventive activity. Food Chem Toxicol 49(12):3319–3327. https://doi.org/10.1016/j.fct.2011.08.023

Kim J, Kim S, Huh K, Kim Y, Joun H, Park M (2011) High serum isoflavone concentrations are associated with the risk of precocious puberty in Korean girls. Clin Endocrinol 75(6):831–835. https://doi.org/10.1111/j.1365-2265.2011.04127.x

Li W, Chen H, Xu B, Wang Y, Zhang C, Cao Y, Xing X (2023) Research progress on classification, sources and functions of dietary polyphenols for prevention and treatment of chronic diseases. J Future Foods 3(4):289–305. https://doi.org/10.1016/j.jfutfo.2023.03.001

Mennen LI, Walker R, Bennetau-Pelissero C, Scalbert A (2005) Risks and safety of polyphenol consumption. Am J Clin Nutr 81(1):326S-329S. https://doi.org/10.1093/ajcn/81.1.326S

Murias M, Jäger W, Handler N, Erker T, Horvath Z, Szekeres T, Nohl H, Gille L (2005) Antioxidant, prooxidant and cytotoxic activity of hydroxylated resveratrol analogues: structure-activity relationship. Biochem Pharmacol 69(6):903–912. https://doi.org/10.1016/j.bcp.2004.12.001

Nguyen RH, Umbach DM, Parad RB, Stroehla B, Rogan WJ, Estroff JA (2011) US assessment of estrogen-responsive organ growth among healthy term infants: piloting methods for assessing estrogenic activity. Pediatr Radiol 41:633–642. https://doi.org/10.1007/s00247-010-1895-0

Ofosu FK, Daliri EBM, Elahi F, Chelliah R, Lee BH, Oh DH (2020) New insights on the use of polyphenols as natural preservatives and their emerging safety concerns. Front Sustain Food Sys 4:525810. https://doi.org/10.3389/fsufs.2020.525810

Palanisamy S, Srinivas K, Rao MVB (2024) Evaluation of the in vitro antioxidant and cytotoxic effects of Euphorbia hirta ethanolic extract. Asian J BiolLife Sci 13(1):139. https://doi.org/10.5530/ajbls.2024.13.18

Pan Y, Qin R, Hou M, Xue J, Zhou M, Xu L, Zhang Y (2022) The interactions of polyphenols with Fe and their application in Fenton/Fenton-like reactions. Sep Purif Technol 300:121831. https://doi.org/10.1016/j.seppur.2022.121831

Parikh A, Kathawala K, Li J, Chen C, Shan Z, Cao X, Garg S (2018) Curcumin-loaded self-nanomicellizing solid dispersion system: Part II: In vivo safety and efficacy assessment against behavior deficit in Alzheimer disease. Drug Deliv Transl Res 8:1406–1420. https://doi.org/10.1007/s13346-018-0570-0

Perez-Jimenez J, Fezeu L, Touvier M, Arnault N, Manach C, Hercberg S, Scalbert A (2011) Dietary intake of 337 polyphenols in French adults. Am J Clin Nutr 93(6):1220–1228. https://doi.org/10.3945/ajcn.110.007096

Saeidnia S, Abdollahi M (2013) Antioxidants: friends or foe in prevention or treatment of cancer: the debate of the century. Toxicol Appl Pharmacol 271(1):49–63. https://doi.org/10.1016/j.taap.2013.05.004

Sangeetha MK, Vallabi DE, Sali VK, Thanka J, Vasanthi HR (2013) Sub-acute toxicity profile of a modified resveratrol supplement. Food Chem Toxicol 59:492–500. https://doi.org/10.1016/j.fct.2013.06.037

Sarker SA, Sultana S, Pietroni M, Dover A (2015) Safety of a bioactive polyphenol dietary supplement in pediatric subjects with acute diarrhoea. Int J Pediatr 2015:387159. https://doi.org/10.1155/2015/387159

Selby-Pham SNB, Howell KS, Dunshea FR, Ludbey J, Lutz A, Bennett L (2018) Statistical modelling coupled with LC-MS analysis to predict human upper intestinal absorption of phytochemical mixtures. Food Chem 245:353–363. https://doi.org/10.1016/j.foodchem.2017.10.102

Shen CL, Brackee G, Song X, Tomison MD, Finckbone V, Mitchell KT, Tang L, Chyu MC, Dunn DM, Wang JS (2017) Safety evaluation of green tea polyphenols consumption in middle-aged ovariectomized rat model. J Food Sci 82(10):2192–2205. https://doi.org/10.1111/1750-3841.13745

Shi HF, Zhao FL, Chen H, Zhou Q, Geng PW, Zhou YF, Wu HL, Chong J, Wang F, Dai DP, Yang JF, Wang SH (2021) Naringenin has an inhibitory effect on rivaroxaban in rats both in vitro and in vivo. Pharmacol ResPerspect 9(3):e00782. https://doi.org/10.1002/prp2.782

Shoji T, Akazome Y, Kanda T, Ikeda M (2004) The toxicology and safety of apple polyphenol extract. Food Chem Toxicol 42(6):959–967. https://doi.org/10.1016/j.fct.2004.02.008

Sroka Z, Cisowski W (2003) Hydrogen peroxide scavenging, antioxidant and anti-radical activity of some phenolic acids. Food Chem Toxicol 41(6):753–758. https://doi.org/10.1016/S0278-6915(02)00329-0

Tan HJ, Ling WC, Chua AL, Lee SK (2021) Oral epigallocatechin gallate reduces intestinal nadolol absorption via modulation of Oatp1a5 and Oct1 transcriptional levels in spontaneously hypertensive rats. Phytomed 90:153623. https://doi.org/10.1016/j.phymed.2021.153623

Thakurdesai PA, Deshpande PO, Bhaskaran S (2024) Preclinical toxicity assessment of polyphenols-based standardized extract of Cinnamomum zeylanicum bark. CurrRes Nutr Food SciJ 12(2):579–590. https://doi.org/10.12944/CRNFSJ.12.2.8

Vlachojannis C, Zimmermann BF, Chrubasik-Hausmann S (2015) Efficacy and safety of pomegranate medicinal products for cancer. EvidBased Complement Altern Med 2015:258598. https://doi.org/10.1155/2015/258598

Wan MLY, Co VA, El-Nezami H (2021) Dietary polyphenol impact on gut health and microbiota. Crit Rev Food Sci Nutr 61(4):690–711. https://doi.org/10.1080/10408398.2020.1744512

Wu MH, Liu JY, Tsai FL, Syu JJ, Yun CS, Chen LY, Ye JC (2023) The adverse and beneficial effects of polyphenols in green and black teas in vitro and in vivo. Int J Med Sci 20(10):1247-1255. https://doi.org/10.7150/ijms.85521

Yamakoshi J, Saito M, Kataoka S, Kikuchi M (2002) Safety evaluation of proanthocyanidin-rich extract from grape seeds. Food Chem Toxicol 40(5):599–607. https://doi.org/10.1016/S0278-6915(02)00006-6

Zhang H, Tsao R (2016) Dietary polyphenols, oxidative stress and antioxidant and anti-inflammatory effects. Curr Opin Food Sci 8:33–42. https://doi.org/10.1016/j.cofs.2016.02.002

Zhang K, Huang J, Wang D, Wan X, Wang Y (2024) Covalent polyphenols-proteins interactions in food processing: formation mechanisms, quantification methods, bioactive effects, and applications. Front Nutr 11:1371401. https://doi.org/10.3389/fnut.2024.1371401

# Chapter 12
# Polyphenols for Weight Management and Obesity

**D. K. Meera, Vincenzo Barone, Carmelo Parisi, Sara Petrelli, Sony George, Mini Joseph, and Suni Mary Varghese**

**Abstract** Polyphenols, a large and diverse group of plant-derived compounds, play a promising role in weight management and the prevention of obesity-related metabolic disorders. These bioactive molecules, abundant in foods such as tea, coffee, cocoa, berries, citrus fruits, and whole grains, influence body weight regulation through multiple interrelated mechanisms involving appetite control, lipid metabolism, and gut microbiota modulation. Polyphenols affect appetite by interacting with the gut–brain axis and modulating key satiety and hunger hormones, including leptin, ghrelin, and peptide YY, thereby enhancing fullness and reducing energy intake. At the same time, they contribute to the regulation of fat metabolism by inhibiting lipogenesis, stimulating fatty acid oxidation, and improving lipid profiles—manifested

D. K. Meera (✉)
Department of Home Science, Government College for Women, Thiruvananthapuram, Kerala, India
e-mail: meerasunil@gmail.com

V. Barone · C. Parisi
Palermo, Italy
e-mail: vincenzo.barone2012@gmail.com

C. Parisi
e-mail: carmelo.parisi2012@libero.it

S. Petrelli
Capaci, PA, Italy
e-mail: sarettap25@gmail.com

S. George
Department of Food Technology, TKM Institute of Technology, Kollam, India
e-mail: sony.g@tkmit.ac.in

M. Joseph
Department of Food and Nutrition, Government College for Women, Thiruvananthapuram, Kerala, India
e-mail: minijoseph66@yahoo.in

S. M. Varghese
Department of Nutrition and Food Processing Technology, Lourdes Matha Institute of Hotel Management and Catering Technology, Thiruvananthapuram, Kerala, India
e-mail: sunimary@lmihmct.org



R. K. Singla et al. (eds.), *Polyphenols in Health*,
https://doi.org/10.1007/978-3-032-27324-6_12

by reduced triglyceride levels and increased HDL cholesterol. Compounds such as catechins, resveratrol, chlorogenic acid, and citrus flavonoids have shown significant potential in preventing adipose tissue accumulation and improving metabolic flexibility. Incorporating polyphenol-rich foods into daily dietary patterns, such as green tea, coffee, cocoa, fruits, vegetables, whole grains, legumes, nuts, herbs, and spices, provides a natural and sustainable strategy for weight management. These foods not only supply potent antioxidants but also exert synergistic effects on appetite regulation and lipid metabolism. Although long-term human studies are needed to establish standardized dosages and efficacy, current evidence supports the inclusion of polyphenol-rich foods as part of a balanced lifestyle to promote healthy weight control, enhance metabolic health, and reduce the risk of chronic diseases associated with obesity.

**Keywords** Appetite regulation · Fat metabolism · Functional foods · Gut microbiota · Polyphenol · Weight management

## Abbreviations

| | |
|---|---|
| AMPK | Adenosine monophosphate-activated protein kinase |
| BMI | Body mass index |
| CCAAT | Cytosine-cytosine-adenosine-adenosine-thymidine |
| C/EBPα | CCAAT/enhancer-binding protein alpha |
| CGA | Chlorogenic acid |
| GC | Decaffeinated green coffee phenolic extract |
| EGC | Epigallocatechin |
| EGCG | Epigallocatechin-3-gallate |
| GLP-1 | Glucagon-like peptide-1 |
| HDL | High-density lipoprotein |
| LDL | Low-density lipoprotein |
| BG | Oat β-glucans |
| PYY | Peptide YY |
| PPARγ | Peroxisome proliferator-activated receptor gamma |
| RCT | Randomized controlled trial |
| SIRT1 | Sirtuin 1 |
| VAS | Visual analogue scale |

## 12.1 Effects on Appetite Regulation and Fat Metabolism

### *12.1.1 Polyphenols and Appetite*

Polyphenols, naturally occurring compounds found in various plant-based foods, have garnered attention for their potential role in appetite regulation. Research indicates that these compounds can influence appetite through mechanisms involving the gut–brain axis and modulation of hormones related to satiety. By interacting with gut microbiota and producing beneficial metabolites, polyphenols may enhance feelings of fullness and reduce overall caloric intake. This emerging understanding positions polyphenols as promising agents in the management of appetite and, consequently, body weight.

Two extensively studied polyphenols include decaffeinated green coffee phenolic extract (GC) and oat β-glucans (BG). Their effects have been analyzed subjectively using visual analog scales (VAS). In addition, the variables such as actual food intake, post-prandial appetite, and satiety hormones have been collected. The dataset involved 29 overweight volunteers for 8 weeks. On the basis of obtained VAS results and subsequent analysis, GC was not effective enough in comparison with GC/BG when speaking of hunger reduction. On the other side, food intakes were judged as data without statistically significant differences. Substantially, some differences between nutraceuticals were observed with concern to acute and regular effects on appetite and satiety; at the same time, there were several similarities between subjective data and recorded feelings by volunteers, meaning that GC and GC/BG may reduce hunger but GC/BG appears more efficient in this restricted ambit (Redondo-Puente et al. 2021).

Plant-derived phenolic-rich foods exert positive effects on satiety. In this specific ambit, phenolic acids appear to be able to influence safety and hunger-related mechanisms. This evidence has been reported to be reliable when speaking of in vitro, in vivo, and in silico studies concerning the effects of the following phenolic acids: chlorogenic acid (CGA), ferulic acid (FA), gallic acid (GA), and protocatechuic acid. Basically, it has been reported that these acids are able to interact with several hormones, including also their specific receptors, and hunger/satiety-related enzymatic processes. The reason appears to be strictly linked with the chemical skeleton of these phenolic acids, and specifically with the position of hydroxyl groups (Zuñiga-Martínez et al. 2024). In effect, the notable presence of hydroxyl groups in the chemical structure (Chaps. 6 and 12) can be reliably considered as one of the main factors which contribute to high reactivity and biological significance of phenolics in general, and phenolic acids in this restricted ambit (Rentzsch et al. 2009; Zhang et al. 2021). This fact can also explain the notable interest of pharmaceutical industries when speaking of good and beneficial effects against different human diseases and disorders, on condition that these compounds are really available and bioaccessible for human consumption (Mechi et al. 2023; Neilson et al. 2017; Park 2014; Tihana et al. 2019). The significance of these phenolic acids and polyphenols on a general level is easily reported on these bases (Arshad et al. 2025; Domingues et al. 2024;

Giglio et al. 2018; Hafiz 2025; Kardum et al. 2015; Kupczyk et al. 2025; Issaoui et al. 2020; Liva et al. 2025; Makarewicz et al. 2021; Martiniakova et al. 2025; Mechi et al. 2023; Mendes et al. 2025; Mileo and Miccadei 2016; Parisi 2019, 2020; Rudrapal et al. 2022; Stoner and Mukhtar 1995; Tillman et al. 2025; Wang and Hu 2023).

A systematic review of bioactive phytochemicals as food supplements to facilitate dieting and to improve patients' adherence to reducing food and caloric intake was studied (Stuby et al. 2019). The basic aim was to evaluate critically the effects on human behavior (with safety considerations concerning human consumption) of different bioactive phytochemicals when speaking of suppression of appetite and the concomitant increase of satiety. Because of the variety of possible phytochemicals and the consequent broad choice with relation to vegetable sources, this study considered extracts including mixtures phytochemicals from *Camellia sinensis* (also known as green tea), *Capsicum annuum*, and *Coffea* species. This choice is comprehensible because tea plants (green, oolong, and black types) are a major source of polyphenols (Ganugapati et al. 2011; Karabulut 2025; Kyriacou et al. 2025). Green tea infusions can deliver a large amount of monomeric catechins such as epigallocatechin gallate (EGCG), epigallocatechin (EGC), epicatechin gallate, and epicatechin. Black tea contains theaflavins and thearubigins. Consequently, the choice of infusions from these tea sources is a good option, taking into account that polyphenols yield is significantly influenced by some variables, including the brew strength, leaf grade, and steep time (Bhagwat et al. 2014). In addition, roasted coffee is a leading dietary source of chlorogenic acids (esters of caffeic and quinic acids), including 5-caffeoylquinic acid and dicaffeoylquinic acids. Coffee can give a relevant amount of chlorogenic acids including 5-caffeoylquinic acid and related dicaffeoylquinic acids. Actually, roasting may lead to reduction of these acids, also generating melanoidins: these compounds help in the binding of phenolics and hence contribute to the antioxidant capacity (Gu et al. 2004).

With reference to the systematic review, it is notable to consider that the claimed positive effects on hunger/appetite (in terms of measurable, not perceivable, reduction) and on satiety induction or enhancement were not assessed with reliability. In detail, it was reported that treatment effects were not revealed with sufficient clarity, and this result was assessed for all tested extracts in different experiments. On the other side, some adverse effects were observed, but the real significance was not important enough (no serious events). In summary, these extracts—and contained compounds—were judged to be unable to show real and reliable results in terms of suppression or reduction of appetite and/or satiety enhancement or induction (Stuby et al. 2019).

Another research on polyphenol-rich blackcurrant extracts has reported that the supposed good effects on the limitation of post-prandial glucose amounts (with concomitant suppression or reduction of appetite) were not clearly assessed when speaking of healthy human participants. In detail, observed effects were minimal, according to the authors (Pinto et al. 2023).

A randomized, crossover, placebo-controlled study looked into the effect of polyphenol-rich fruits in the appetite of 20 participants and found no change in appetite or appetite modulating hormones (Smith et al. 2019).

### 12.1.2 Polyphenols and Fat Metabolism

Dietary polyphenols are metabolized by microorganisms to produce small active molecules, which further influence gut microbes to regulate host lipid metabolism.

Growing evidence indicated that polyphenols are involved in lipid metabolism and may prevent obesity development (Adriouch et al. 2018). Animal studies with the administration of polyphenol-rich plant extract showed improved plasma lipid levels and endotoxaemia (Aires et al. 2019).

In another randomized, double-blind, placebo-controlled clinical trial, it was found that dietary polyphenols might reduce cardiovascular disease risk and atherosclerosis via lowering triglyceride levels. In this clinical study, the beneficial effects of polyphenols contained in Goishi tea on blood lipid profiles were tested on more than seventy participants showing low-density lipoprotein (LDL) cholesterol $\geq$ 120 mg/mL. The research was carried out dividing randomly the number of participants in two different groups for a total duration of 12 weeks. The comparison between the two groups (the difference was the polyphenol intake: the first group consumed Goishi tea containing 122 mg of polyphenols, while the second group consumed another drink containing 1/10 of polyphenol intake when compared with the first group) showed reliably that the higher the consumption of Goishi tea, the higher the augment of high-density lipoprotein (HDL) cholesterol with consequent decrease of triglycerides. In addition, such a cumulative effect was notable if treated subjects showed a body mass index < 25 kg/m$^2$, suggesting that polyphenols contained in Goishi tea may positively reduce arteriosclerosis and cardiovascular risks (Ishida et al. 2018).

Citrus flavonoids are polyphenolic compounds with significant biological properties. These fruits and juices supply flavanones such as naringenin, hesperidin, nobiletin, and tangeretin. Related concentrations can vary depending on grape variety, climate, and winemaking (Zamora-Ros et al. 2014). Consequently, citrus flavonoids have emerged as potential therapeutics for the treatment of metabolic dysregulation. Rodent diets containing citrus flavonoids have been reported to have positive effects on dyslipidemia, insulin resistance, and hepatic steatosis. It seems that similar effects are dependent on two mechanisms: (1) the inhibition of the synthesis of hepatic fatty acids, and (2) the enhanced oxidation of fatty acids. In attotion, it has been reported that anti-inflammatory effects of citrus flavonoids take place in the adipose tissues, the aorta, the liver, and the kidney, suggesting that the mentioned actions are extremely localized and not by chance. Actually, such a multifaceted effect is dependent on mechanisms which have not been fully clarified at present, even if several targets appear to be clear. The problem is to export the therapeutic significance of these results on another level, when speaking of human subjects and required efficacy, dose, bioavailability, and safety requirements (Mulvihill et al. 2016).

### *12.1.3 Polyphenols and Weight Management*

Polyphenols have been suggested to reduce body weight and modify body composition through different mechanisms. These effects have been extensively studied in animals and in vitro and to a lesser extent in humans. A systematic literature research (Farhat 2024) on 19 studies on the effect of polyphenols and weight management found that only nine studies showed a clear diminution of body weight in examined subjects. The authors opined that further research is needed before assessing polyphenols as efficient weight-reduction agents when speaking of overweight and obese human subjects. This reflection considers phenolics as coadjuvant agents in weigh-reduction strategies, also stating that a long experimental duration exceeding 12 months would be needed (Farhat 2024).

Another study by Martínez-Rodríguez (2024) performed with the aim of assessing the effects of a botanical-based blend comprised lemon verbena and hibiscus extracts on weight management, body fat, appetite, lipid, and glucose profiles was considered. In detail, involved volunteers—expression of an overweight population—showed a clear and significant reduction of weight after three months: the average result was approximately 3.4% on total weight, with the concomitant diminution of body fat mass (6% loss when compared with initial evaluation, in average). Additionally, cholesterol appeared to be lowered by up to 7%, and volunteers in the supplemented-diet group declared increased satisfaction after meal consumption, in comparison with the placebo group. Consequently, the daily intake of similar polyphenol-containing mixtures may be recommended for overweight patients (Martínez-Rodríguez 2024).

On the other side, the scientific debate on the real role and associated effects of phenolics in human subjects has not reached unanimous evaluation. A meta-analysis of 100 articles based on human, clinical, and epidemiological studies has recently evaluated the role of several bioactive polyphenols (including caffeine, gallic acid, ascorbic acid, oleuropein, and other well-known phenolics) from vegetable sources as also described in Chap. 2, with reference to the ambit of weight management strategies. On the one side, the authors considered the role of polyphenols with some favor on a general level, but also stated that claimed effects on hunger, satiety, enhanced oxidation of fatty acids, etc. are not reliably evident. Otherwise, research should be extremely limited in scope, target, and object of the investigation, such as a recent work on the efficacy and safety effects of Sinetrol Xpur (polyphenolic citrus dry extract) in weight management. Briefly, the study demonstrated that treated human volunteers showed a significant reduction of waist and hip circumference and abdominal fat when compared with the placebo group. In addition, oxidative stress was evaluated by means of chemical profiles of malondialdehyde (decrease), superoxide dismutase (increase), and glutathione (increase), and without adverse effects or damages to critical organs. In summary, the authors reported that Sinetrol Xpur can be used in weight management strategies, also with interesting antidiabetes and anti-inflammatory answers (Dallas et al. 2014).

## 12.2 Key Polyphenol Compounds

Polyphenols comprehend diverse classes of plant secondary metabolites that differ in structure, physicochemical properties, and biological actions (Chaps. 6 and 10). Table 12.1 contain the principal polyphenol families and representative compounds that have been most extensively investigated for effects relevant to weight management and obesity.

### *12.2.1 Bioavailability and Metabolism Considerations*

Most polyphenols are extensively metabolized (phase II conjugation, microbial catabolism) and circulate primarily as glucuronides, sulfates, and microbial metabolites rather than parent aglycones. Bioactivity in vivo therefore often reflects metabolites and host–microbiome co-metabolites rather than the native compounds (Cardona et al. 2013).

Poor and variable oral bioavailability, rapid clearance, and inter-individual differences in microbiota composition are major constraints on translating in vitro potency into clinical efficacy. Strategies to enhance bioavailability (formulations, co-administration with bioavailability enhancers, prodrugs) are active areas of research (Crozier et al. 2009).

### *12.2.2 Typical Doses Used in Human Studies*

With relation to polyphenols and typical doses used in human studies, the following list can be offered here:

(a) EGCG/green tea extracts: common doses are in the range of 300–1000 mg/day of catechins, often combined with caffeine (Hursel and Westerterp-Plantenga 2013).
(b) Resveratrol: clinical trials have employed 150 mg/day to 2 g/day; metabolic effects in humans often require sustained dosing and results are mixed (Baur and Sinclair 2006).
(c) Curcumin: studies commonly use 500–2000 mg/day of curcumin or standardized turmeric extracts; bioavailability enhancers (piperine, nanoparticles) are commonly considered (Panahi et al. 2016).
(d) Chlorogenic acid/green coffee: doses range widely. In general, green coffee extract trials often used 200–800 mg/day of total extract standardized to CGA (Greenberg et al. 2006).

**Table 12.1** Polyphenols in foods and their health benefits

| Class | Representative compounds | Main sources | Mechanistic relevance |
|---|---|---|---|
| Flavan-3-ols (catechins) | Epigallocatechin-3-gallate (EGCG), epicatechin, epigallocatechin | Green tea, black tea (oxidized forms), cocoa | Thermogenesis and increased fat oxidation; inhibition of intestinal lipid absorption; AMPK activation (Dulloo et al. 1999; Hursel and Westerterp-Plantenga 2013) |
| Flavonols | Quercetin, kaempferol, myricetin | Onions, apples, berries, tea | Modulation of adipogenesis; anti-inflammatory and antioxidant signaling; effects on energy expenditure and insulin sensitivity (Boots et al. 2008) |
| Anthocyanins | Cyanidin, delphinidin, malvidin glycosides | Berries, red grapes, purple vegetables | Attenuation of adipocyte differentiation; modulation of gut microbiota; improvements in lipid handling and glucose homeostasis (Wallace and Giusti 2015) |
| Stilbenes | Resveratrol | Red wine, grapes, peanuts (minor) | SIRT1/AMPK activation; mitochondrial biogenesis; improved insulin sensitivity and fatty acid oxidation (Baur and Sinclair 2006; Timmers et al. 2011) |
| Phenolic acids | Chlorogenic acid (CGA), caffeic acid, ferulic acid | Coffee (green beans), fruits, whole grains | Inhibition of glucose absorption; modulation of hepatic lipid metabolism and AMPK signaling; pancreatic lipase inhibition (Greenberg et al. 2006) |
| Curcuminoids | Curcumin (diferuloylmethane) | Turmeric (*Curcuma longa*) | Anti-inflammatory actions; down-regulation of adipogenesis; improvements in insulin signaling and adipokine profiles (Panahi et al. 2016) |
| Isoflavones | Genistein, daidzein | Soy and soy products | Estrogen receptor-mediated and non-genomic effects on adipocyte differentiation, lipid metabolism, and energy balance (Messina 2016) |

(continued)

**Table 12.1** (continued)

| Class | Representative compounds | Main sources | Mechanistic relevance |
|---|---|---|---|
| Proanthocyanidins/ condensed tannins | Oligomeric procyanidins (grape seed extract), cranberry protocatechuic acid | Grapes, apples, cocoa, berries | Inhibition of digestive enzymes (pancreatic lipase, α-amylase); modulation of gut microbiota; attenuation of adipose inflammation (Spencer et al. 2013) |
| Other notable compounds | Oleuropein, hesperidin, naringenin | Olive leaves/ fruit; citrus fruits | Modulates lipid metabolism and inflammation; improves lipid and glucose metabolism; potential effects on adipocyte biology |

### *12.2.3 Safety and Interactions*

Polyphenols are generally safe at dietary levels; higher supplemental doses can cause adverse effects (e.g., gastrointestinal upset, liver enzyme changes with high-dose green tea extracts), and may interact with drugs (cytochrome P450 modulation, antiplatelet effects). Safety profiles depend on specific compound, dose, and formulation and require evaluation in clinical contexts (Hursel and Westerterp-Plantenga 2013; Panahi et al. 2016).

## 12.3 Potential Benefits in Obesity

This section synthesizes mechanistic rationale and clinical evidence for polyphenols as adjuncts to weight management and obesity prevention/treatment. Evidence spans molecular, preclinical, and clinical domains; overall, polyphenols show multiple biologically plausible pathways that could favorably influence energy balance and adiposity, but clinical effects in humans are modest and heterogeneous.

### *12.3.1 Mechanistic Pathways Relevant to Obesity. Energy Expenditure and Thermogenesis*

Some polyphenols (notably green tea catechins, EGCG, and resveratrol) increase energy expenditure and fat oxidation via sympathetic stimulation, inhibition of catechol-O-methyltransferase (prolonging norepinephrine action), and activation of adenosine monophosphate-activated protein kinase (AMPK) and Sirtuin 1 (SIRT1) pathways that enhance mitochondrial function and uncoupling protein expression in adipose and muscle (Dulloo et al. 1999; Baur and Sinclair 2006).

### 12.3.2 Mechanistic Pathways Relevant to Obesity. Appetite Regulation and Satiety Hormones

Polyphenols may modulate gut hormone secretion such as glucagon-like peptide-1 (GLP-1) and peptide YY, and central appetite pathways, though human evidence is limited and inconsistent. Modest reductions in subjective appetite or energy intake have been reported with some extracts in acute settings.

### 12.3.3 Mechanistic Pathways Relevant to Obesity: Adipogenesis and Lipogenesis

In vitro and animal models, inhibition of preadipocyte differentiation, down-regulation of peroxisome proliferator-activated receptor gamma (PPARγ), and cytosine-cytosine-adenosine-adenosine-thymidine (CCAAT)/enhancer-binding protein alpha (C/EBPα) reduced lipogenesis and increased lipolysis. These results have been observed and reported with quercetin, resveratrol, curcumin, and anthocyanins (Boots et al. 2008; Panahi et al. 2016).

### 12.3.4 Mechanistic Pathways Relevant to Obesity: Insulin Sensitivity and Glucose Homeostasis

Improvements in insulin signaling and glucose uptake (via AMPK activation, reduced oxidative stress, and inflammation) can limit adipose expansion and ectopic lipid deposition. Several polyphenols have shown favorable effects on insulin sensitivity in animal studies and some human trials (Timmers et al. 2011; Greenberg et al. 2006).

### 12.3.5 Mechanistic Pathways Relevant to Obesity: Gut Microbiota Modulation

Polyphenols and their metabolites alter microbial composition and function, promoting taxa associated with leanness and increased short-chain fatty acid production; microbiota changes mediate metabolic effects in animal models and likely contribute to inter-individual variability in responses (Cardona et al. 2013).

In detail:

(a) Intestinal lipid and carbohydrate digestion/absorption: Certain polyphenols inhibit pancreatic lipase, α-amylase, and α-glucosidase, reducing nutrient absorption and post-prandial excursions (Spencer et al. 2013).

(b) Anti-inflammatory and antioxidant effects: By lowering adipose inflammation and systemic inflammatory markers, polyphenols may ameliorate the chronic inflammatory state of obesity that contributes to insulin resistance and dysregulated energy balance (Boots et al. 2008; Panahi et al. 2016).

## 12.4 Summary of Evidence from Human Studies

### *12.4.1 Aggregate Findings*

Meta-analyses and systematic reviews generally report small but statistically significant reductions in body weight, body mass index (BMI), or waist circumference with certain polyphenol-rich extracts (e.g., green tea catechins, green coffee/chlorogenic acid, curcumin, anthocyanin-rich extracts) when used as adjuncts to diet/exercise or as supplements over weeks to months. Effect sizes are modest (often ~ 0.5–2.0 kg body weight reduction vs. placebo), and heterogeneity is high (Hursel and Westerterp-Plantenga 2013; Panahi et al. 2016; Greenberg et al. 2006).

#### 12.4.1.1 Green Tea Catechins (EGCG)

Several randomized controlled trials (RCT) and pooled analyses indicate modest increases in energy expenditure and small reductions in body weight/fat, particularly when combined with caffeine and in subjects with higher baseline BMI or when weight-loss behavioral measures are concurrently applied (Dulloo et al. 1999).

#### 12.4.1.2 Resveratrol

Promising metabolic effects in animal models (improved mitochondrial function, insulin sensitivity) have not consistently translated to robust weight loss in humans. Some trials show improvements in metabolic biomarkers but limited direct impact on body weight (Baur and Sinclair 2006; Timmers et al. 2011).

#### 12.4.1.3 Curcumin

Trials and meta-analyses report modest reductions in body weight and BMI and improvements in inflammatory markers and insulin resistance indices, though heterogeneity in formulations and varying bioavailability complicate interpretation (Panahi et al. 2016).

#### 12.4.1.4 Chlorogenic Acid/Green Coffee Extract

Small RCT and meta-analyses have reported modest weight loss with standardized green coffee extracts; some methodological concerns about trial quality and blinding have been raised (Greenberg et al. 2006).

#### 12.4.1.5 Anthocyanins, Quercetin, and Isoflavones

Evidence is mixed; some trials suggest improvements in body composition, waist circumference, or metabolic markers, but overall effect sizes on weight are small and inconsistent across studies (Wallace and Giusti 2015; Boots et al. 2008; Messina 2016).

#### 12.4.1.6 Limitations of the Current Evidence

The following limitations and exceptions should be taken into account:

(a) Heterogeneity in preparations and doses: Studies use whole foods, crude extracts, standardized fractions, or purified compounds at widely varying doses, making cross-study comparisons difficult.
(b) Bioavailability and formulation differences: Differences in absorption-enhancing strategies (e.g., piperine with curcumin, nano formulations) influence outcomes and limit generalizability.
(c) Study quality and duration: Many trials are small, short (weeks to a few months), or lack rigorous controls and dietary/exercise standardization.
(d) Inter-individual variability: Genetic background, baseline metabolic state, habitual diet, and gut microbiota composition modify responses.
(e) Outcome relevance: Clinical significance of modest weight changes is unclear; improvements in metabolic health markers may occur independently from large weight loss and might be clinically relevant.

## 12.5 Safety Considerations and Clinical Use

At dietary levels from whole foods, polyphenols are safe and provide multiple health benefits beyond weight management.

High-dose supplements can produce adverse effects (e.g., hepatotoxicity reported with some concentrated green tea extracts), and interactions with medications (anticoagulants, antihypertensives, CYP-metabolized drugs) should be considered. Clinicians should evaluate benefit–risk on an individual basis (Hursel and Westerterp-Plantenga 2013; Panahi et al. 2016).

Current evidence supports considering polyphenol-rich foods (tea, berries, coffee, citrus, soy, cocoa, turmeric-containing foods, olives) as part of a healthy dietary pattern for weight management and metabolic health, rather than relying on supplements as monotherapy.

## 12.6 Recommendations for Future Research

Large, well-powered RCT of standardized, well-characterized polyphenol formulations with adequate duration and clinically relevant endpoints (body composition, metabolic outcomes, long-term maintenance) are required.

Studies linking pharmacokinetics (parent compounds and metabolites), microbiome composition, and metabolic responses will clarify responder phenotypes and mechanisms.

Comparative effectiveness trials assessing polyphenol interventions as adjuncts to behavioral weight-loss programs are warranted.

## 12.7 Concluding Summary

Polyphenols represent a biologically plausible, multi-targeted class of compounds with mechanisms that can favorably influence energy balance, adipocyte biology, and metabolic health. In humans, most polyphenol interventions produce modest reductions in body weight and improvements in metabolic markers, with substantial heterogeneity. Emphasizing polyphenol-rich whole foods within caloric control and lifestyle programs is reasonable; therapeutic expectation for supplements should be cautious pending more rigorous long-term clinical data.

## References

Adriouch S, Kesse-Guyot E, Feuillet T, Touvier M, Olié V, Andreeva V, Hercberg S, Galan P, Fezeu LK (2018) Total and specific dietary polyphenol intakes and 6-year anthropometric changes in a middle-aged general population cohort. Int J Obes 42(3):310–317. https://doi.org/10.1038/ijo.2017.227

Aires V, Labbé J, Deckert V, Pais de Barros JP, Boidot R, Haumont M, Maquart G, Guern NL, Masson D, Prost-Camus E, Prost M, Lagrost L (2019) Healthy adiposity and extended lifespan in obese mice fed a diet supplemented with a polyphenol-rich plant extract. Sci Rep 9:9134. https://doi.org/10.1038/s41598-019-45600-6

Arshad MT, Maqsood S, Ikram A, Abdullahi MA (2025) Synergistic effects of carrot (*Daucus carota* L.) root and papaya (*Carica papaya* L.) extracts in combating inflammation and enhancing immune functions—a new perspective. Food Sci Nutr 13(6):e70313. https://doi.org/10.1002/fsn3.70313

Baur JA, Sinclair DA (2006) Therapeutic potential of resveratrol: the in vivo evidence. Nat Rev Drug Discov 5(6):493–506. https://doi.org/10.1038/nrd2060

Bhagwat S, Haytowitz DB, Holden JM (2014) USDA database for the flavonoid content of selected foods, release 3.1. U.S. Department of Agriculture, Agricultural Research Service, Washington, D.C. Available https://www.ars.usda.gov/ARSUserFiles/80400535/Data/Flav/Flav_R03-1.pdf. Accessed 13 Oct 2025

Boots AW, Haenen GRMM, Bast A (2008) Health effects of quercetin: from antioxidant to nutraceutical. Eur J Pharmacol 585(2–3):325–337. https://doi.org/10.1016/j.ejphar.2008.03.008

Cardona F, Andrés-Lacueva C, Tulipani S, Tinahones FJ, Queipo-Ortuño MI (2013) Benefits of polyphenols on gut microbiota and implications in human health. J Nutr Biochem 24(8):1415–1422. https://doi.org/10.1016/j.jnutbio.2013.05.001

Crozier A, Jaganath IB, Clifford MN (2009) Dietary phenolics: chemistry, bioavailability and effects on health. Nat Prod Rep 26(8):1001–1043. https://doi.org/10.1039/B802662A

Dallas C, Gerbi A, Elbez Y, Caillard P, Zamaria N, Cloarec M (2014) Clinical study to assess the efficacy and safety of a citrus polyphenolic extract of red Orange, grapefruit, and Orange (Sinetrol-XPur) on weight management and metabolic parameters in healthy overweight individuals. Phytother Res 28(2):212–218. https://doi.org/10.1002/ptr.4981

Domingues YO, Lemes GA, de Oliveira FL, de Souza TR, Silva B, Bento JAC, Morzelle MC (2024) In vitro simulated gastrointestinal digestion and bioaccessibility of phenolic compounds and antioxidants of soursop (*Annona muricata* L.) peel and pulp. Lwt 208:116694. https://doi.org/10.1016/j.lwt.2024.116694

Dulloo AG, Duret C, Rohrer D, Girardier L, Mensi N, Fathi M, Philippe C, Vandermander J (1999) Efficacy of a green tea extract rich in catechin polyphenols and caffeine in increasing 24-h energy expenditure and fat oxidation in humans. Am J Clin Nutr 70(6):1040–1045. https://doi.org/10.1093/ajcn/70.6.1040

Farhat G (2024) Polyphenols in obesity and weight management: are they worth further research? An umbrella review. Nutr Bull 49(2):126–131. https://doi.org/10.1111/nbu.12667

Ganugapati J, Mukkavalli S, Sahithi A (2011) Docking studies of green tea flavonoids as insulin mimetics. Int J Comp App 30(4):48–52

Giglio RV, Patti AM, Cicero AF, Lippi G, Rizzo M, Toth PP, Banach M (2018) Polyphenols: potential use in the prevention and treatment of cardiovascular diseases. Curr Pharm Des 24(2):239–258. https://doi.org/10.2174/1381612824666180130112652

Greenberg JA, Boozer CN, Geliebter A (2006) Coffee, diabetes, and weight control. Am J Clin Nutr 84(4):682–693. https://doi.org/10.1093/ajcn/84.4.682

Gu L, Kelm MA, Hammerstone JF, Beecher G, Holden J, Haytowitz D, Prior RL (2004) Concentrations of proanthocyanidins in common foods and estimations of normal consumption. J Nutr 134(3):613–617. https://doi.org/10.1093/jn/134.3.613

Hafiz MS (2025) Anti-diabetic potentials of coffee polyphenols: a narrative review. Curr Nutr Food Sci 21(2):202–211. https://doi.org/10.2174/0115734013292395240219091340

Hursel R, Westerterp-Plantenga MS (2013) Catechin- and caffeine-rich teas for control of body weight in humans. Am J Clin Nutr 98(6):1682S–1693S. https://doi.org/10.3945/ajcn.113.058396

Ishida N, Iizuka M, Kataoka K, Okazaki M, Shiraishi K, Yagi Y, Jobu K, Yokota J, Oishi M, Moriyama H, Shimamura T, Matsumura Y, Ukeda H, Miyamura M (2018) Improvement of blood lipid profiles by Goishi tea polyphenols in a randomised, double-blind, placebo-controlled clinical study. Int J Food Sci Nutr 69(5):598–607. https://doi.org/10.1080/09637486.2017.1386629

Issaoui M, Delgado AM, Iommi C, Chammem N (2020) Polyphenols and the Mediterranean diet. Springer International Publishing, Cham

Karabulut G (2025) Tea polyphenols: their role in human health and drug interactions. Dissertation, University of Padova, Italy

Kardum N, Milovanović B, Šavikin K, Zdunić G, Mutavdžin S, Gligorijević T, Spasić S (2015) Beneficial effects of polyphenol-rich chokeberry juice consumption on blood pressure level and

lipid status in hypertensive subjects. J Med Food 18(11):1231–1238. https://doi.org/10.1089/jmf.2014.0171
Kupczyk D, Bilski R, Szeleszczuk Ł, Mądra-Gackowska K, Studzińska R (2025) The role of diet in modulating inflammation and oxidative stress in rheumatoid arthritis, ankylosing spondylitis, and psoriatic arthritis. Forum Nutr 17(9):1603. https://doi.org/10.3390/nu17091603
Kyriacou NM, Gross AS, McLachlan AJ (2025) Green tea catechins as perpetrators of drug pharmacokinetic interactions. Clin Pharmacol Ther 118(1):45–61. https://doi.org/10.1002/cpt.3547
Liva K, Panagiotopoulos AA, Foscolou A, Amerikanou C, Vitali A, Zioulis S, Argyri K, Panoutsopoulos GI, Kaliora AC, Gioxari A (2025) High polyphenol extra virgin olive oil and metabolically unhealthy obesity: a scoping review of preclinical data and clinical trials. Clin Pract 15(3):54. https://doi.org/10.3390/clinpract15030054
Makarewicz M, Drożdż I, Tarko T, Duda-Chodak A (2021) The interactions between polyphenols and microorganisms, especially gut microbiota. Antioxid 10(2):188. https://doi.org/10.3390/antiox10020188
Martínez-Rodríguez A, Martínez-Olcina M, Vicente-Martínez M, Asencio-Mas N, Navarro P, Caturla N, Jones J (2024) Effectiveness of a polyphenol-enriched blend on weight management and metabolic syndrome-related parameters in healthy overweight adults. Appl Sci 14(9):3882. https://doi.org/10.3390/app14093882
Martiniakova M, Sarocka A, Penzes N, Biro R, Kovacova V, Mondockova V, Sevcikova A, Ciernikova S, Omelka R (2025) Protective role of dietary polyphenols in the management and treatment of type 2 diabetes mellitus. Forum Nutr 17(2):275. https://doi.org/10.3390/nu17020275
Mechi D, Baccouri B, Martín-Vertedor D, Abaza L (2023) Bioavailability of phenolic compounds in Californian-style table olives with Tunisian aqueous olive leaf extracts. Mol 28(2):707. https://doi.org/10.3390/molecules28020707
Mendes I, Ribeiro MGC, de Souza LF, Rosa CDOB, Hermsdorff HHM, Bressan J (2025) Effect of polyphenol supplementation on adiposity: a systematic review of randomized clinical trials. Curr Nutr Rep 14(1):36. https://doi.org/10.1007/s13668-025-00626-3
Messina M (2016) Soy and health update: evaluating the evidence. Forum Nutr 8(12):754. https://doi.org/10.3390/nu8120754
Mileo AM, Miccadei S (2016) Polyphenols as modulator of oxidative stress in cancer disease: new therapeutic strategies. Oxidative Med Cell Longev 1:6475624. https://doi.org/10.1155/2016/6475624
Mulvihill E, Burke A, Huff M (2016) Citrus flavonoids as regulators of lipoprotein metabolism and atherosclerosis. Ann Rev Nutr 36:275–299. https://doi.org/10.1146/annurev-nutr-071715-050718
Neilson AP, Goodrich KM, Ferruzzi MG (2017) Bioavailability and metabolism of bioactive compounds from foods. In: Coulston AM, Boushey CJ, Ferruzzi MG, Delahanty LM (eds) Nutrition in the prevention and treatment of disease. Academic Press, Cambridge, pp 301–319. https://doi.org/10.1016/B978-0-12-802928-2.00015-1
Panahi Y, Hosseini MS, Khalili N, Naimi E, Soflaei SS, Muhammed Majeed M, Sahebkar A (2016) Effects of curcumin on body weight, waist circumference, and lipid profile in overweight or obese subjects: a randomized controlled trial and meta-analysis. Forum Nutr 32(10):1116–1122. https://doi.org/10.1016/j.nut.2016.03.018
Park J (2014) Flaxseed secoisolariciresinol diglucoside and visceral obesity: a closer look at its chemical properties, absorption, metabolism, bioavailability, and effects on visceral fat, lipid profile, systemic inflammation, and hypertension. In: Watson RR (ed) Nutrition in the prevention and treatment of abdominal obesity book, 1st edn. Academic Press, San Diego, pp 317–328
Parisi S (2019) Analysis of major phenolic compounds in foods and their health effects. AOAC J 102(5):1354–1355. https://doi.org/10.5740/jaoacint.19-0127

Parisi S (2020) Characterization of major phenolic compounds in selected foods by the technological and health promotion viewpoints. J AOAC 103(4):904–905. https://doi.org/10.1093/jaoacint/qsaa011
Pinto AR, Hobden MD, Brown K, Farrimond J, Targett D, Corpe CP, Ellis PR, Todorova Y, Socha K, Bahsoon S, Haworth C, Marcel M, Nie X, Hall WL (2023) Acute effects of drinks containing blackcurrant and citrus (poly)phenols and dietary fibre on postprandial glycaemia, gut hormones, cognitive function and appetite in healthy adults: two randomised controlled trials. Food Funct 14(22):10163–10176. https://doi.org/10.1039/D3FO03085G
Redondo-Puente M, Mateos R, Seguido MA, García-Cordero J, González S, Tarradas RM, Bravo-Clemente L, Sarriá B (2021) Appetite and satiety effects of the acute and regular consumption of green coffee phenols and green coffee phenol/oat β-glucan nutraceuticals in subjects with overweight and obesity. Foods 10(11):2511. https://doi.org/10.3390/foods10112511
Rentzsch M, Wilkens A, Winterhalter P (2009) Non-flavonoid phenolic compounds. In: Moreno Arribas MV, Polo MC (eds) Wine chemistry and biochemistry. Springer, New York, NY, USA, pp 509–527
Rudrapal M, Khairnar SJ, Khan J, Dukhyil AB, Ansari MA, Alomary MN, Fahad M, Alshabrmi FM, Palai S, Deb PK, Devi R (2022) Dietary polyphenols and their role in oxidative stress-induced human diseases: insights into protective effects, antioxidant potentials and mechanism(s) of action. Front Pharmacol 13:806470. https://doi.org/10.3389/fphar.2022.806470
Smith TJ, Karl JP, Wilson MA, Whitney CC, Barrett A, Farhadi NF, Chen CYO, Montain SJ (2019) Glycaemic regulation, appetite and ex vivo oxidative stress in young adults following consumption of high-carbohydrate cereal bars fortified with polyphenol-rich berries. Brit J Nutr 121(9):1026–1038. https://doi.org/10.1017/S0007114519000394
Spencer JPE, Vauzour D, Rendeiro C (2013) Flavonoids and cognition: the molecular mechanisms underlying their behavioural effects. Arch Biochem Biophys 492(1–2):1–9. https://doi.org/10.1016/j.abb.2009.10.003
Stoner GD, Mukhtar H (1995) Polyphenols as cancer chemopreventive agents. J Cell Biochem 59(S22):169–180. https://doi.org/10.1002/jcb.240590822
Stuby J, Gravestock I, Wolfram E, Pichierri G, Steurer J, Burgstaller JM (2019) Appetite-suppressing and satiety-increasing bioactive phytochemicals: a systematic review. Forum Nutr 11(9):9. https://doi.org/10.3390/nu11092238
Tihana Ž, Abdelkebir R, Alcantara C, Carmen M, García-Pérez JV, Meléndez-Martínez AJ, Re A, Lorenzo JM, Barba FJ (2019) From extraction of valuable compounds to health promoting benefits of olive leaves through bioaccessibility, bioavailability and impact on gut microbiota. Trends Food Sci Technol 83:63–77. https://doi.org/10.1016/j.tifs.2018.11.005
Tillman L, Rieres JM, Ahjem E, Bishop-Guest F, McGrath M, Hatrick H, Pranjol MZI (2025) Thinking outside the therapeutic box: the potential of polyphenols in preventing chemotherapy-induced endothelial dysfunction. Cells 14(8):566. https://doi.org/10.3390/cells14080566
Timmers S, Konings E, Bilet L, Houtkooper RH, van de Weijer T, Goossens GH, Schrauwen P et al (2011) Calorie restriction-like effects of resveratrol on energy metabolism and metabolic profile in obese humans. Cell Metab 14(5):612–622
Wallace TC, Giusti MM (2015) Anthocyanins: nature's health-promoting pigments. Foods 8(11):550. https://doi.org/10.3390/foods8110550
Wang K, Hu S (2023) The synergistic effects of polyphenols and intestinal microbiota on osteoporosis. Front Immunol 14:1285621. https://doi.org/10.3389/fimmu.2023.1285621
Zamora-Ros, Knaze V, Rothwell JA, Hémon B, Moskal A, Overvad K, Tjønneland A, Kyrø C, Fagherazzi G,, Boutron-Ruault MC, Touillaud M, Katzke V, Kühn T, Boeing H, Förster J, Trichopoulou A, Valanou E, Peppa E, Palli D, Agnoli C, Ricceri F, Tumino R, Santucci de Magistris M, Peeters PHM, Bueno-de-Mesquita HB, Engeset D, Skeie G, Hjartåker A, Menéndez V, Agudo A, Molina-Montes E, Huerta JM, Barricarte A, Amiano P, Sonestedt E, Nilsson LM, Landberg R, Key TJ, Khaw KT, Wareham NJ, Lu Y, Slimani N, Romieu I, Riboli E, Scalbert A (2014) Dietary polyphenol intake in Europe: the European prospective investigation into cancer

and nutrition (EPIC) study. Eur J Nutr 55(4):1359–1375. https://doi.org/10.1007/s00394-015-0950-x

Zhang S, Xu M, Zhang W, Liu C, Chen S (2021) Natural polyphenols in metabolic syndrome: protective mechanisms and clinical applications. Int J Mol Sci 22(11):6110. https://doi.org/10.3390/ijms22116110

Zuñiga-Martínez BS, Domínguez-Avila JA, Montiel-Herrera M, Villegas-Ochoa MA, Robles-Sánchez RM, Ayala-Zavala JF, Viuda-Martos M, González-Aguilar GA (2024) Consumption of plant-derived phenolic acids modulates hunger and satiety responses due to chemical interactions with Enteroendocrine mediators. Foods 13(22):3640. https://doi.org/10.3390/foods13223640

# Chapter 13
# Biological Activities and Metabolism of Xanthones

**Ai Sze Wee, Hin Yee Thew, Bey Hing Goh, and Kooi Yeong Khaw**

**Abstract** The heterogeneous class of phenolic compounds commonly found in vegetable products is generally reported to have interesting effects when speaking of human health. This statement is based on a number of recent studies which have demonstrated the relationship between the bioavailability of these compounds in the human diet and beneficial effects on human health. For these reasons at least, the modern industry has considered the use of polyphenols-rich foods and the addition of similar compounds to edible products without a good phenolic profile. The final aim is to use these foods as valid adjuvant strategies against a variety of human diseases. This chapter, in particular, discusses the matter of digestive mechanisms and the influence of phenolics. In addition, the interaction between these compounds and intestinal microbiota is discussed, also with reference to the inhibition of pathogens. There are several factors which should be taken into account when speaking of polyphenols in the human digestion: role of associated phenolics with dietary fibres; relationship between polysaccharides/polyphenols interaction and the rate of release and absorption of bound and free phenolics, firstly in the small intestine and in the lower gastrointestinal tract; enhancement of microbial fermentation of polyphenols because of the concomitant presence and association of phenolics and dietary fibres; and demonstrated influence of polyphenols on the decrease of pathogen agents such as *Clostridium* spp. The consumption of foods and beverages containing notable amounts of phenolics is widely recommended: the activity of phenolics favours easy digestion and reduces the production of inflammatory markers.

---

A. S. Wee (✉)
Faculty of Medicine, Nursing and Health Sciences, SEGi University, Kota Damansara, Selangor, Malaysia
e-mail: sandrawee@segi.edu.my

H. Y. Thew · B. H. Goh
Sunway Biofunctional Molecules Discovery Centre (SBMDC), School of Medical and Life Sciences, Subang Jaya, Malaysia

B. H. Goh
Faculty of Health, Australian Research Centre in Complementary and Integrative Medicine, University of Technology Sydney, Ultimo, NSW, Australia

K. Y. Khaw
School of Pharmacy, Monash University Malaysia, Bandar Sunway, Subang Jaya, Malaysia

 
R. K. Singla et al. (eds.), *Polyphenols in Health*,
https://doi.org/10.1007/978-3-032-27324-6_13

**Keywords** Antimicrobial activity · Anticancer activity · Anti-inflammatory activity · Antioxidant activity · Mangosteen · Neuroprotective activity · Xanthone

## Abbreviations

| | |
|---|---|
| ADMET | Absorption Distribution Metabolism Excretion and Toxicity |
| AMPK | Adenosine monophosphate-activated protein kinase |
| AD | Alzheimer's disease |
| CPX | Caged polyprenylated xanthone |
| CYP | Cytochrome P450 |
| MAPK | Mitogen-activated protein kinase |
| NF-κB | Nuclear factor kappa B |
| Nrf2 | Nuclear factor-erythroid 2-related factor 2 |
| PD | Parkinson's disease |
| HM-1 | 1-hydroxyl-2,3,5-trimethoxyxanthone |
| MPTP | 1-methyl-4-phenyl-1,2,3,6-tetrahydropyridine |

## 13.1 Introduction

Xanthones are a structurally diverse class of polyphenolic compounds widely distributed in medicinal plants (Table 13.1). They have attracted considerable interest due to their broad spectrum of biological activities. Despite promising in vitro bioactivities, the clinical application of xanthones is often limited by poor oral bioavailability and extensive metabolic transformation. Structural features such as glycosylation, methoxylation, and prenylation significantly influence both their biological activity and metabolic fate. This chapter provides a comprehensive overview of the biological activities of xanthones and their metabolic pathways, highlighting their pharmacological potential and the challenges associated with their therapeutic development.

## 13.2 Biological Activities of Xanthones

Xanthones exhibit a wide range of biological activities, making them promising candidates for therapeutic development. Their structural diversity enables them to interact with multiple molecular targets, conferring antimicrobial, anticancer, antioxidant, anti-inflammatory, and neuroprotective effects (Table 13.1).

**Table 13.1** Structural classes of xanthones and their predominant metabolic pathways

| Xanthone subclass | Structural characteristics | Representative compounds | Dominant metabolic system | Major metabolic reactions | Expected bioavailability |
|---|---|---|---|---|---|
| C-glycosylxanthones | C–C glycosidic linkage, polyhydroxylated | Mangiferin | | Deglycosylation, phase II conjugation | Low to moderate |
| Prenylated xanthones | One or more prenyl chains | α-Mangostin, gartanin | Hepatic CYP450 enzymes | Hydroxylation, oxidation, conjugation | Low |
| Methoxylated xanthones | Methoxy substitutions | HM-1 | Hepatic CYP450 enzymes | Demethylation, Hydroxylation | Moderate |
| Caged polyprenylated xanthones | Polycyclic caged scaffold with prenyl groups | Gambogenic acid | Predicted CYP450 metabolism | Predicted oxidation and inhibition | Unknown |

## 13.2.1 Antifungal Activity

Fungal infections, caused by both human and plant pathogens, remain a significant health and agricultural challenge. Natural xanthones have emerged as promising antifungal agents due to their diverse chemical structures and bioactivities. Several xanthones isolated from the *Hypericum* species have demonstrated notable antifungal potential. For example, 5-hydroxy-1-methoxyxanthone, 6-deoxyjacareubin, and 1,5-dihydroxyxanthone, obtained from the dichloromethane extract of *Hypericum brasiliense* stems and roots, inhibited the growth of the phytopathogenic fungus *Cladosporium cucumerinum* in a thin-layer chromatography bioautographic assay (Rocha et al. 1994). In a related study, six xanthones isolated from *Hypericum roeperanum*, including 2-deprenyl-reediaxanthone B, 5-O-methyl-2-deprenylrheediaxanthone B, calcinoxanthone D, roeperanone, 5-O-demethylpaxanthonin, and 5-O-methylisojacareubin, were evaluated against *Candida albicans* and *Cladosporium cucumerinum* (Rath et al. 1996). Most compounds selectively inhibited *C. albicans* at 1 μg/mL, a potency comparable to amphotericin B but lower than miconazole.

Xanthones derived from mangosteen (*Garcinia mangostana*) also exhibited antifungal activity. In detail, α-mangostin inhibited *Candida albicans* and displayed stronger activity than clotrimazole and nystatin (Kaomongkolgit et al. 2009). It also inhibited *Colletotrichum gloeosporioides*, with a greater effect on spore germination than on mycelial growth (Ye et al. 2020). γ-mangostin exhibited activity against *Cryptococcus neoformans* in both in vitro assays and a *Caenorhabditis elegans* infection model (Li et al. 2026). Overall, these findings indicate that xanthones from diverse plant sources possess broad-spectrum antifungal potential, supporting their continued investigation as natural antifungal agents.

## 13.2.2 Antibacterial Activity

Naturally occurring xanthones have been reported to exhibit broad-spectrum antibacterial activity against Gram-positive and Gram-negative bacteria. Prenylated xanthones such as jacareubin and 1,3,5,6-tetrahydroxy-2-prenyl xanthone from *Calophyllum brasiliense* displayed antibacterial activity against methicillin-resistant and methicillin-sensitive *Staphylococcus aureus* strains, with moderate effects on *Escherichia coli* (Yasunaka et al. 2005). Fungal-derived xanthones, including trihydroxyxanthone derivatives from lignicolous fungus BCC 28210 (Yoiprommarat et al. 2020), as well as xanthones from *Cratoxylum sumatranum* (Tantapakul et al. 2016) and *Aspergillus* species (Zhu et al. 2018), have also shown activity against various Gram-positive bacteria, including *Bacillus*, *Micrococcus*, and *Staphylococcus* species.

Xanthones from *Garcinia* species continue to demonstrate potent antibacterial effects. For instance, nineteen xanthones isolated from the young fruits and flowers of cowa fruit (*G. cowa*) inhibited *Bacillus cereus, B. subtilis, S. aureus, Salmonella typhimurium, E. coli,* and *Pseudomonas aeruginosa* (Sriyatep et al. 2015). Xanthones derived from mangosteen (*G. mangostana*), including α-, β-, and γ-mangostin, have demonstrated broad antibacterial activity. These compounds showed inhibitory activity against Gram-positive bacteria, including *S. aureus*—methicillin-resistant strains (Iinuma et al. 1996), as well as *Clostridium perfringens* (Liu et al. 2023) and various mycobacterial species (Suksamrarn et al. 2003). Xanthones from *Garcinia smeathmannii,* including ananixanthone and cheffouxanthone, exhibited activity against *Enterococcus faecalis,* while rubraxanthone from Asam Kandis (*G. dioica*) was active against both methicillin-resistant and methicillin-sensitive strains of *S. aureus.* These studies collectively highlight xanthones as a diverse and potent class of antibacterial agents with activity across multiple bacterial genera.

### 13.2.3 Anticancer Activity

Naturally occurring xanthones have been reported to exhibit selective cytotoxicity against a variety of human cancer cell lines. Xanthones isolated from *Aspergillus versicolor*, including sterigmatocystin and dihydrosterigmatocystin, demonstrated in vitro cytotoxicity against colorectal, ovarian, lung, brain, and melanoma cell lines, with sterigmatocystin showing greater potency and stronger activity against liver and lung cancer cells (Lee et al. 2010). Similarly, two xanthone derivatives from the endophytic fungus ZSU-H16, obtained from the mangrove *Avicennia marina*, displayed notable cytotoxicity against oral carcinoma and multidrug-resistant oral carcinoma cells (Huang et al. 2010).

Prenylated xanthones derived from mangosteen (*G. mangostana*) have also shown significant anticancer potential. A novel prenylated xanthone isolated from mangosteen pericarp demonstrated moderate to strong cytotoxicity across brain, gastric, prostate, lung, and nasopharyngeal cancers, acting through apoptosis induction (Yang et al. 2017). Ananixanthone and caloxanthone B, isolated from the stem bark of *Calophyllum* species, exhibited moderate to strong activity against leukaemia cells, likely via protein kinase inhibition (Zamakshshari et al. 2019). Additional xanthone derivatives, such as those isolated from the propolis of the stingless bee *Lisotrigona furva*, showed weak to moderate cytotoxicity against lung, liver, and oral carcinoma cells (Oanh et al. 2021). Overall, these findings highlight xanthones as a structurally diverse class of compounds with promising anticancer activity, supporting further exploration of their mechanisms and therapeutic potential.

### 13.2.4 Antioxidant Activity

Natural xanthones have been shown to exhibit significant antioxidant activity. Ten xanthones isolated from *Polygala japonica*, including gartanin, 8-desoxygartanin,

2-hydroxyxanthone, 2,7-dihydroxyxanthone, 1,5,6-trihydroxyxanthone, 1,6,7-trihydroxyxanthone, norathyriol, 1,2,5-trihydroxyxanthone, isomangiferin, and mangiferin, demonstrated measurable electron-transfer-based antioxidant capacity using the ferric-reducing antioxidant power assay (Li et al. 2018).

In addition, several xanthones have been reported to modulate nuclear factor-erythroid 2-related factor 2 (Nrf2) signalling, a key pathway in cellular defence against oxidative stress. For example, mangiferin activated the Nrf2/antioxidant response elements pathway in neuronal, hepatic, gastric, and hematopoietic models, mitigating oxidative stress, inflammation, and DNA damage (Liu et al. 2017; Mahmoud-Awny et al. 2015; Pan et al. 2016; Zhang et al. 2015). Similarly, tovophyllin A protected against acetaminophen-induced oxidative stress in murine liver models through Nrf2 activation (Ibrahim et al. 2018). Other xanthones, including a-mangostin and y-mangostin, were shown to exert antioxidant effects via Nrf2 activation in retinal and liver cells (Fang et al. 2016; Wang et al. 2018). These findings highlight the potent antioxidant potential of natural xanthones, both through direct electron-transfer mechanisms and through modulation of cellular defence pathways.

### 13.2.5 Anti-Inflammatory Activity

Xanthones have demonstrated significant anti-inflammatory activity across both cellular and animal models. Several xanthones isolated from *Hypericum patulum* callus tissue, including demethylpaxantonin, patulone, garcinone B, tripteroside, and 1,3,5,6-tetrahydroxyxanthone, were shown to inhibit prostaglandin release and nuclear factor kappa B (NF-κB)-dependent transcription (Yamakuni et al. 2006). Additionally, α-mangostin inhibited nuclear translocation of p65 and reduced activation of key NF-κB pathway proteins in human fibroblast-like synoviocytes derived from rheumatoid arthritis patients (Zuo et al. 2018). Beyond joint inflammation, a xanthone derivative from *Cudrania tricuspidata* suppressed microglial activation and reduced lipopolysaccharide-induced pro-inflammatory mediators via inhibition of mitogen-activated protein kinase (MAPK) and NF-κB signalling, accompanied by induction of heme oxygenase-1 (Ko et al. 2023). In vivo, mangiferin attenuated neuroinflammation in a rat subarachnoid haemorrhage model by reducing oxidative stress, inflammasome activation, and NF-κB signalling (Wang et al. 2017). Overall, xanthones exhibit anti-inflammatory activity across both cellular and animal models.

### 13.2.6 Neuroprotective Activity

Natural xanthones have emerged as promising bioactive compounds across a wide range of neurodegenerative disease models, highlighting their broad neuroprotective properties. In Alzheimer's disease (AD) models, xanthones from *Polygala* species regulated tau aggregation, apoptosis, and inflammatory signalling in neuronal cells (Çalış et al. 2023), while garcinone D inhibited amyloid-β42 oligomerisation and alleviated amyloid-induced paralysis in transgenic *Caenorhabditis elegans*, demonstrating in vivo efficacy (Thew et al. 2024). Additional mangosteen-derived xanthones showed cholinesterase inhibition, antioxidant activity, and protection against amyloid and glutamate-induced neurotoxicity, with cognitive improvement observed in rodent models (Abdallah et al. 2021; Khaw et al. 2014; Sethiya and Mishra 2014; Thew et al. 2025; Wang et al. 2016).

In Parkinson's disease (PD) models, α-mangostin conferred neuroprotection in rotenone-treated SH-SY5Y cells by reducing oxidative stress, inhibiting α-synuclein aggregation, preserving mitochondrial function, and suppressing apoptosis (Hao et al. 2017), while tovophyllin A protected dopaminergic neurons and improved behavioural outcomes in toxin-induced cellular and 1-methyl-4-phenyl-1,2,3,6-tetrahydropyridine (MPTP) mouse models (Huang et al. 2020). Beyond AD and PD, mangiferin attenuated neurotoxicity in a 3-nitropropionic acid-induced Huntington's disease rat model, supporting its protective effects against mitochondrial dysfunction and neuronal degeneration (Lum et al. 2023). In amyotrophic lateral sclerosis-models, α-mangostin restored energy homeostasis and enhanced autophagic clearance of mutant SOD1 aggregates in microglia via adenosine monophosphate-activated protein kinase (AMPK) activation. Collectively, these studies demonstrate that xanthones exhibit neuroprotective effects across diverse experimental models of neurodegenerative diseases.

## 13.3 Metabolism of Xanthones

Xanthones undergo extensive metabolic transformation, which significantly affects their bioavailability and therapeutic potential. Metabolism is influenced by structural characteristics such as glycosylation, methoxylation, and prenylation, which determine whether metabolism is primarily gut microbiota-dependent or mediated by host enzymes.

### *13.3.1 Mangiferin, a C-Glycosylxanthone*

Mangiferin is a natural C-glycosylxanthone with diverse metabolic activities but intrinsically low bioavailability. It is primarily absorbed in the small intestine via passive diffusion, with varying efficiency across intestinal segments. Despite this absorption, mangiferin's low solubility and poor transmembrane permeability significantly limit its uptake into systemic circulation (Mei et al. 2021).

The initial metabolic step involves deglycosylation, where mangiferin is hydrolysed to its aglycone, norathyriol (1,3,6,7-tetrahydroxyxanthone). This reaction is predominantly mediated by gut microbial enzymes, as the C-glucosidic bond in mangiferin is resistant to mammalian digestive enzymes. Recent metagenomic and metabolomic analyses demonstrate a moderate prevalence of gut bacteria capable of converting mangiferinto norathyriol, identifying two previously uncultured strains, CAKRHR01 *sp934339005* and *Enterococcus sp900549885*, as likely contributors to this metabolic process. Notably, the abundance of CAKRHR01 *sp934339005* was positively correlated with the conversion of mangiferin to norathyriol, highlighting inter-individual variability in microbial-dependent metabolic capacity (Bunt et al. 2025).

Deglycosylation is notably pronounced under carbohydrate-depleted conditions, suggesting that the activation of bacterial C-glycoside-cleaving pathways represents an adaptive response to low-nutrient niches in the colon. Importantly, this microbial conversion constitutes a prodrug-like activation step, as norathyriol exhibits significantly higher apparent bioavailability (approximately 30.4%) than mangiferin and serves as the primary substrate for subsequent metabolism (Guo et al. 2018).

Following absorption, norathyriol undergoes phase II-metabolism in both intestinal and hepatic tissues, as well as through gut microbial conjugative pathways. These metabolic reactions include glucuronidation, sulphation, methylation, and secondary glycosylation, generating a diverse spectrum of conjugated metabolites that enhance solubility, facilitate systemic distribution, and promote efficient elimination (Zhou et al. 2023).

### *13.3.2 Xanthones from Garcinia Mangostana*

Xanthones are a structurally diverse class of polyphenolic compounds predominantly found in the pericarp of *G. mangostana* (mangosteen), a tropical fruit native to Southeast Asia and traditionally used for treating inflammation, infections, and skin ailments. Among the various xanthones identified, α-mangostin and gartanin are the most abundant and pharmacologically significant molecules.

Both in silico and in vitro studies indicate that α-mangostin and gartanin undergo extensive biotransformation mediated by multiple cytochrome P450 (CYP) isoforms. Phase I-metabolism primarily involves hydroxylation and oxidation, while phase II-metabolism includes glucuronidation and sulphation. Liquid chromatography/tandem mass spectrometry analyses have confirmed tri-oxidation products for α-mangostin and sulphated conjugates for gartanin, consistent with computational 'Absorption, Distribution, Metabolism, Excretion, and Toxicity' (ADMET) predictions (Rukthong et al. 2020).

In murine models of dextran sulphate sodium-induced ulcerative colitis, administration of *G. mangostana* pericarp extract and α-mangostin restored suppressed hepatic CYP1A1, CYP1A2, CYP2E1, CYP2C29, CYP2D9, CYP3A11, and CYP3A13 expression and enzyme activities to near-normal levels, highlighting the potential of xanthones to modulate CYP-dependent metabolic responses under inflammatory conditions (Nopwinyoowong et al. 2023).

While pre-clinical studies indicate that xanthones can influence CYP activity, clinical pharmacokinetic data suggest that systemic exposure remains limited.

In a study of healthy adults consuming mangosteen juice with a high-fat meal, both free and conjugated xanthones were detected in serum. α-mangostin reached maximum concentrations (approximately 113 nmol/L) while urinary recovery of xanthones accounted for only approximately 2% of the ingested dose over 24 h, and this fact is indicative of limited absorption and extensive first-pass metabolism (Chitchumroonchokchai et al. 2012).

Furthermore, xanthone-containing fruits such as mangosteen and pineapple can inhibit multiple CYP isoforms, suggesting potential food–drug interactions (Foti et al. 2009; Hidaka et al. 2008). Overall, α-mangostin and gartanin undergo host-mediated metabolic transformations that significantly shape pharmacokinetic profiles, systemic exposure, and therapeutic potential.

### 13.3.3 Other Xanthones

1-hydroxyl-2,3,5-trimethoxyxanthone (HM-1) is a major compound extracted from *Hakea elliptica* D. Don, a Tibetan medicinal plant. HM-1 undergoes phase I-metabolism via hepatic cytochrome P450 enzymes, with demethylation of methoxy groups and hydroxylation of aromatic rings as the major biotransformation pathways, as demonstrated using pooled human liver microsomes and recombinant CYP isoforms, particularly CYP3A4 and CYP2C8 (Feng et al. 2014).

Caged polyprenylated xanthones (CPX) are a unique subclass of xanthones characterised by a highly oxygenated tricyclic caged scaffold and prenyl substituents that increase lipophilicity. They are primarily found in the resin of *G. hanburyi*, a tropical plant used in traditional medicine for anti-inflammatory, anticancer, and antimicrobial activities. No experimental metabolism studies have been reported for CPX at present. Therefore, absorption, biotransformation, and excretion remain unknown. This fact represents a clear knowledge gap in understanding CPX pharmacokinetics and metabolism. In silico predictions suggest that most CPX are CYP3A4 substrates, and some (e.g. gambogenic acid) may also act as CYP3A4 inhibitors, indicating potential drug–drug interactions, although these predictions remain unvalidated (He et al. 2023).

## 13.4 Conclusion

Xanthones represent a highly versatile class of natural compounds with extensive biological activities, including antifungal, antibacterial, anticancer, antioxidant, anti-inflammatory, and neuroprotective effects. Their therapeutic potential is, however, influenced by complex metabolic transformations, poor bioavailability, and potential food–drug interactions. Structural modifications such as glycosylation, methoxylation, and prenylation play a critical role in determining both activity and metabolism. Further studies integrating in vitro, in vivo, and clinical investigations are warranted to optimise xanthone pharmacokinetics and maximise their translational potential as therapeutic agents.

## References

Abdallah HM, El Sayed NS, Sirwi A, Ibrahim SRM, Mohamed GA, Abdel Rasheed NO (2021) Mangostanaxanthone IV ameliorates streptozotocin-induced neuro-inflammation, amyloid deposition, and tau hyperphosphorylation via modulating PI3K/Akt/GSK-3β pathway. Biology 10(12):1298. https://doi.org/10.3390/biology10121298

Bunt D, Schwalbe M, Hayeeawaema F, El Aidy S (2025) Gut microbiota-mediated conversion of mangiferin to norathyriol alters short chain fatty acid and urate metabolism. Gut Microb 17(1):2508422. https://doi.org/10.1080/19490976.2025.2508422

Çalış İ, Becer E, Ünlü A, Uğurlu Aydın Z, Hanoğlu A, Vatansever HS, Dönmez AA (2023) Comparative phytochemical studies on the roots of *polygala azizsancarii* and *P. Peshmenii* and neuroprotective activities of the two xanthones. Phytochemistry 210:113650. https://doi.org/10.1016/j.phytochem.2023.113650

Chitchumroonchokchai C, Riedl KM, Suksumrarn S, Clinton SK, Kinghorn AD, Failla ML (2012) Xanthones in mangosteen juice are absorbed and partially conjugated by healthy adults. J Nutr 142(4):675–680. https://doi.org/10.3945/jn.111.156992

Fang Y, Su T, Qiu X, Mao P, Xu Y, Hu Z, Zhang Y, Zheng X, Xie P, Liu Q (2016) Protective effect of alpha-mangostin against oxidative stress induced-retinal cell death. Sci Rep 6:21018. https://doi.org/10.1038/srep21018

Feng R, Zhou X, Tan XS, Or PM, Hu T, Fu J, Ma JY, Huang M, He CY, Shi JG, Che CT, Yeung JH, Wang Y (2014) *In vitro* identification of cytochrome P450 isoforms responsible for the metabolism of 1-hydroxyl-2,3,5-trimethoxy-xanthone purified from *Halenia elliptica* D. Don. Chem Biol Interact 210:12–19. https://doi.org/10.1016/j.cbi.2013.12.008

Foti RS, Pearson JT, Rock DA, Wahlstrom JL, Wienkers LC (2009) In vitro inhibition of multiple cytochrome P450 isoforms by xanthone derivatives from mangosteen extract. Drug Metab Dispos 37(9):1848–1855. https://doi.org/10.1124/dmd.109.028043

Guo X, Cheng M, Hu P, Shi Z, Chen S, Liu H, Shi H, Xu Z, Tian X, Huang C (2018) Absorption, metabolism, and pharmacokinetics profiles of Norathyriol, an Aglycone of Mangiferin, in rats by HPLC-MS/MS. J Agric Food Chem 66(46):12227–12235. https://doi.org/10.1021/acs.jafc.8b03763

Hao XM, Li LD, Duan CL, Li YJ (2017) Neuroprotective effect of α-mangostin on mitochondrial dysfunction and α-synuclein aggregation in rotenone-induced model of Parkinson's disease in differentiated SH-SY5Y cells. J Asian Nat Prod Res 19(8):833–845. https://doi.org/10.1080/10286020.2017.1339349

He R, Jia B, Peng D, Chen W (2023) Caged Polyprenylated Xanthones in *Garcinia hanburyi* and the biological activities of them. Drug Des Devel Ther 17:3625–3660. https://doi.org/10.2147/DDDT.S426685

Hidaka M, Nagata M, Kawano Y, Sekiya H, Kai H, Yamasaki K, Okumura M, Arimori K (2008) Inhibitory effects of fruit juices on cytochrome P450 2C9 activity *in vitro*. Biosci Biotechnol Biochem 72(2):406–411. https://doi.org/10.1271/bbb.70511

Huang Y, Sun L, Zhu S, Xu L, Liu S, Yuan C, Guo Y, Wang X (2020) Neuroprotection against Parkinson's disease through the activation of Akt/GSK3β signaling pathway by Tovophyllin A. Front Neurosci 14:723. https://doi.org/10.3389/fnins.2020.00723

Huang Z, Yang R, Guo Z, She Z, Lin Y (2010) A new xanthone derivative from mangrove endophytic fungus no. ZSU-H16. Chem Nat Comp 46(3):348–351. https://doi.org/10.1007/s10600-010-9614-5

Ibrahim SRM, El-Agamy DS, Abdallah HM, Ahmed N, Elkablawy MA, Mohamed GA (2018) Protective activity of tovophyllin A, a xanthone isolated from Garcinia mangostana pericarps, against acetaminophen-induced liver damage: role of Nrf2 activation. Food Funct 9(6):3291–3300. https://doi.org/10.1039/c8fo00378e

Iinuma M, Tosa H, Tanaka T, Asai F, Kobayashi Y, Shimano R, Miyauchi K (1996) Antibacterial activity of xanthones from guttiferaeous plants against methicillin-resistant *Staphylococcus aureus*. J Pharm Pharmacol 48(8):861–865. https://doi.org/10.1111/j.2042-7158.1996.tb03988.x

Kaomongkolgit R, Jamdee K, Chaisomboon N (2009) Antifungal activity of alpha-mangostin against Candida albicans. J Oral Sci 51(3):401–406. https://doi.org/10.2334/josnusd.51.401

Khaw KY, Choi SB, Tan SC, Wahab HA, Chan KL, Murugaiyah V (2014) Prenylated xanthones from mangosteen as promising cholinesterase inhibitors and their molecular docking studies. Phytomed 21(11):1303–1309. https://doi.org/10.1016/j.phymed.2014.06.017

Ko W, Baek JS, Liu Z, Dong L, Kim N, Lee H, Yoon CS, Kim NY, Kim SC, Lee DS (2023) Anti-inflammatory activity of 1,6,7-Trihydroxy-2-(1,1-dimethyl-2-propenyl)-3-methoxyxanthone isolated from *Cudrania tricuspidata* via NF-κB, MAPK, and HO-1 signaling pathways in lipopolysaccharide-stimulated RAW 264.7 and BV2 cells. Mol 28(21):7299. https://doi.org/10.3390/molecules2821729

Lee YM, Li H, Hong J, Cho HY, Bae KS, Kim MA, Kim DK, Jung JH (2010) Bioactive metabolites from the sponge-derived fungus *Aspergillus versicolor*. Arch Pharm Res 33(2):231–235. https://doi.org/10.1007/s12272-010-0207-4

Li X, Chen S, Wu Y, Liu F, Jumai A, Zhu B, Xiong B, Qiu SX (2026) Γ-Mangostin from *Garcinia mangostana* (L.) exerts the antifungal activity against *Cryptococcus neoformans* by inhibiting ribosome biogenesis. J Ethnopharmacol 358:120986. https://doi.org/10.1016/j.jep.2025.120986

Li X, Jiang Q, Chen B, Luo X, Chen D (2018) Structure-activity relationship and prediction of the electron-transfer potential of the Xanthones series. ChemistryOpen 7(9):730–736. https://doi.org/10.1002/open.201800108

Liu Y, Zhu J, Qu S, Shen J, Zhu K (2023) Plant-derived Xanthones against Clostridial enteric infections. Antibiot 12(2):232. https://doi.org/10.3390/antibiotics12020232

Liu YW, Cheng YQ, Liu XL, Hao YC, Li Y, Zhu X, Zhang F, Yin XX (2017) Mangiferin Upregulates Glyoxalase 1 through activation of Nrf2/ARE signaling in central neurons cultured with high glucose. Mol Neurobiol 54(6):4060–4070. https://doi.org/10.1007/s12035-016-9978-z

Lum PT, Sekar M, Seow LJ, Shaikh MF, Arulsamy A, Retinasamy T, Gan SH, Gnanaraj C, Esa NM, Ramachawolran G, Subramaniyan V, Chinni SV, Wu YS (2023) Neuroprotective potency of mangiferin against 3-nitropropionic acid induced Huntington's disease-like symptoms in rats: possible antioxidant and anti-inflammatory mechanisms. Front Pharmacol 14:1189957. https://doi.org/10.3389/fphar.2023.1189957

Mahmoud-Awny M, Attia AS, Abd-Ellah MF, El-Abhar HS (2015) Mangiferin mitigates gastric ulcer in ischemia/ Reperfused rats: involvement of PPAR-γ, NF-κB and Nrf2/HO-1 signaling pathways. PLoS One 10(7):e0132497. https://doi.org/10.1371/journal.pone.0132497

Mei S, Perumal M, Battino M, Kitts DD, Xiao J, Ma H, Chen X (2021) Mangiferin: a review of dietary sources, absorption, metabolism, bioavailability, and safety. Crit Rev Food Sci Nutr 63(18):3046–3064. https://doi.org/10.1080/10408398.2021.1983767

Nopwinyoowong N, Chatuphonprasert W, Tatiya-aphiradee N, Jarukamjorn K (2023) *Garcinia mangostana* pericarp and α-mangostin normalized the expression of colonic and renal cytochrome P450 genes and abated the expression of renal inflammatory genes in dextran sulfate sodium-induced ulcerative colitis in mice. J Appl Pharm Sci 13(7):251–257. https://doi.org/10.7324/JAPS.2023.132717

Oanh VTK, Thoa HT, Hang NTM, Phuong DTL, Lien NTP, Popova M, Trusheva B, Bankova V, Le TN (2021) New dihydrochromene and xanthone derivatives from Lisotrigona furva propolis. Fitoter 149:104821. https://doi.org/10.1016/j.fitote.2020.104821

Pan CW, Pan ZZ, Hu JJ, Chen WL, Zhou GY, Lin W, Jin LX, Xu CL (2016) Mangiferin alleviates lipopolysaccharide and D-galactosamine-induced acute liver injury by activating the Nrf2 pathway and inhibiting NLRP3 inflammasome activation. Eur J Pharmacol 770:85–91. https://doi.org/10.1016/j.ejphar.2015.12.006

Rath G, Potterat O, Mavi S, Hostettmann K (1996) Xanthones from Hypericum roeperanum. Phytochem 43(2):513–520. https://doi.org/10.1016/0031-9422(96)00284-1

Rocha L, Marston A, Kaplan MA, Stoeckli-Evans H, Thull U, Testa B, Hostettmann K (1994) An antifungal gamma-pyrone and xanthones with monoamine oxidase inhibitory activity from Hypericum brasiliense. Phytochem 36(6):1381–1385. https://doi.org/10.1016/s0031-9422(00)89727-7

Rukthong P, Sereesongsang N, Kulsirirat T, Boonnak N, Sathirakul K (2020) *In vitro* investigation of metabolic fate of α-mangostin and gartanin via skin permeation by LC-MS/MS and in silico evaluation of the metabolites by ADMET predictor™. BMC Complement Med Ther 20(1):359. https://doi.org/10.1186/s12906-020-03144-7

Sethiya NK, Mishra SH (2014) Investigation of Mangiferin, as a promising natural polyphenol Xanthone on multiple targets of Alzheimer's disease. J Biolog Act Prod Nat 4(2):111–119. https://doi.org/10.1080/22311866.2014.921121

Sriyatep T, Siridechakorn I, Maneerat W, Pansanit A, Ritthiwigrom T, Andersen RJ, Laphookhieo S (2015) Bioactive prenylated xanthones from the young fruits and flowers of *Garcinia cowa*. J Nat Prod 78(2):265–271. https://doi.org/10.1021/np5008476

Suksamrarn S, Suwannapoch N, Phakhodee W, Thanuhiranlert J, Ratananukul P, Chimnoi N, Suksamrarn A (2003) Antimycobacterial activity of prenylated xanthones from the fruits of Garcinia mangostana. Chem Pharm Bull 51(7):857–859. https://doi.org/10.1248/cpb.51.857

Tantapakul C, Maneerat W, Sripisut T, Ritthiwigrom T, Andersen RJ, Cheng P, Cheenpracha S, Raksat A, Laphookhieo S (2016) New Benzophenones and Xanthones from *Cratoxylum sumatranum* ssp. neriifolium and their antibacterial and antioxidant activities. J Agric Food Chem 64(46):8755–8762. https://doi.org/10.1021/acs.jafc.6b03643

Thew HY, Boon Keat K, Tan YC, Ong YS, Parat MO, Murugaiyah V, Goh BH, Khaw KY (2024) Probing the anti-Aβ42 aggregation and protective effects of prenylated xanthone against Aβ42-induced toxicity in transgenic Caenorhabditis elegans model. Chem Biol Interact 394:110978. https://doi.org/10.1016/j.cbi.2024.110978

Thew HY, Lim WM, Ong YS, Ong LK, Parat MO, Goh BH, Khaw KY (2025) Correction: Garcinone D mitigates amyloid β42-induced neurotoxicity: unravelling mechanisms of neuroprotection. Sci Rep 15(1):36902. https://doi.org/10.1038/s41598-025-24970-0

Wang A, Li D, Wang S, Zhou F, Li P, Wang Y, Lin L (2018) γ-Mangostin, a xanthone from mangosteen, attenuates oxidative injury in liver via NRF2 and SIRT1 induction. J Funct Foods 40:544–553. https://doi.org/10.1016/j.jff.2017.11.047

Wang SN, Li Q, Jing MH, Alba E, Yang XH, Sabaté R, Han YF, Pi RB, Lan WJ, Yang XB, Chen JK (2016) Natural Xanthones from *Garcinia mangostana* with multifunctional activities for the therapy of Alzheimer's disease. Neurochem Res 41(7):1806–1817. https://doi.org/10.1007/s11064-016-1896-y

Wang Z, Guo S, Wang J, Shen Y, Zhang J, Wu Q (2017) Nrf2/HO-1 mediates the neuroprotective effect of mangiferin on early brain injury after subarachnoid hemorrhage by attenuating mitochondria-related apoptosis and neuroinflammation. Sci Rep 7(1):11883. https://doi.org/10.1038/s41598-017-12160-6

Yamakuni T, Aoki K, Nakatani K, Kondo N, Oku H, Ishiguro K, Ohizumi Y (2006) Garcinone B reduces prostaglandin E2 release and NF-kappaB-mediated transcription in C6 rat glioma cells. Neurosci Lett 394(3):206–210. https://doi.org/10.1016/j.neulet.2005.10.023

Yang R, Li P, Li N, Zhang Q, Bai X, Wang L, Xiao Y, Sun L, Yang Q, Yan J (2017) Xanthones from the pericarp of *Garcinia mangostana*. Molecules 22(5):683. https://doi.org/10.3390/molecules22050683

Yasunaka K, Abe F, Nagayama A, Okabe H, Lozada-Pérez L, López-Villafranco E, Muñiz EE, Aguilar A, Reyes-Chilpa R (2005) Antibacterial activity of crude extracts from Mexican medicinal plants and purified coumarins and xanthones. J Ethnopharmacol 97(2):293–299. https://doi.org/10.1016/j.jep.2004.11.014

Ye H, Wang Q, Zhu F, Feng G, Yan C, Zhang J (2020) Antifungal activity of alpha-Mangostin against *Colletotrichum gloeosporioides* in vitro and in vivo. Molecules 25(22):5335. https://doi.org/10.3390/molecules25225335

Yoiprommarat S, Kongthong S, Choowong W, Boonyuen N, Isaka M, Bunyapaiboonsri T (2020) Xanthones from a lignicolous freshwater fungus (BCC 28210). Nat Prod Res 34(9):1233–1237. https://doi.org/10.1080/14786419.2018.1556655

Zamakshshari NH, Ee GCL, Ismail IS, Ibrahim Z, Mah SH (2019) Cytotoxic xanthones isolated from *Calophyllum depressinervosum* and *Calophyllum buxifolium* with antioxidant and cytotoxic activities. Food Chem Toxicol 133:110800. https://doi.org/10.1016/j.fct.2019.110800

Zhang B, Zhao J, Li S, Zeng L, Chen Y, Fang J (2015) Mangiferin activates the Nrf2-ARE pathway and reduces etoposide-induced DNA damage in human umbilical cord mononuclear blood cells. Pharm Biol 53(4):503–511. https://doi.org/10.3109/13880209.2014.927890

Zhou H, Song S, Lan X, Li Y, Yuan X, Yang J, Li M, Cao T, Zhang J (2023) Comprehensive profiling of Mangiferin metabolites in vivo and in vitro based on the drug metabolite clusters analytical strategy. ACS Omega 8(11):9934–9946. https://doi.org/10.1021/acsomega.2c07089

Zhu A, Yang MY, Zhang YH, Shao CL, Wang CY, Hu LD, Cao F, Zhu HJ (2018) Absolute configurations of 14,15-Hydroxylated Prenylxanthones from a marine-derived *Aspergillus* sp. fungus by chiroptical methods. Sci Rep 8(1):10621. https://doi.org/10.1038/s41598-018-28996-5

Zuo J, Yin Q, Wang YW, Li Y, Lu LM, Xiao ZG, Wang GD, Luan JJ (2018) Inhibition of NF-κB pathway in fibroblast-like synoviocytes by α-mangostin implicated in protective effects on joints in rats suffering from adjuvant-induced arthritis. Int Immunopharmacol 56:78–89. https://doi.org/10.1016/j.intimp.2018.01.016

# Chapter 14
# Future Directions in Polyphenol Research

**Srinivas Nerella, Bairong Shen, Bey Hing Goh, and Rajeev K. Singla**

**Abstract** Polyphenols are structurally diverse plant secondary metabolites abundant in fruits, vegetables, cereals, tea, cocoa, and many medicinal plants. Accumulating evidence reframes their biological roles beyond direct radical scavenging, emphasising regulation of cellular signalling, gene expression, and metabolic homeostasis at nutritionally achievable exposures. Major advances include recognition of hormetic activation of adaptive stress pathways, appreciation of gut microbiota as a key determinant of biotransformation into more bioactive and absorbable phenolic metabolites, and growing evidence for epigenetic modulation via effects on DNA methylation, histone marks, and microRNA. Parallel progress in chromatographic/mass-coupled approaches, high-resolution mass spectrometry, metabolomics, and system biology has improved structural annotation and enabled multi-omics integration to identify biomarkers of intake and response. Despite these gains, translation

S. Nerella
Department of Chemistry, Government Degree College, Kakatiya University, Wardhannapet 506313, Telangana, India
e-mail: nerellasrinivas1@gmail.com

B. Shen · R. K. Singla (✉)
Department of Pharmacy and Institutes for Systems Genetics, Center for High Altitude Medicine, Frontiers Science Center for Disease-Related Molecular Network, West China Hospital, Sichuan University, Chengdu, Sichuan, China
e-mail: rajeevkumar@scu.edu.cn

B. Shen
e-mail: bairong.shen@scu.edu.cn

B. H. Goh
Sunway Biofunctional Molecules Discovery Centre, Sunway University, Selangor Darul Ehsan, Bandar Sunway, Petaling Jaya, Malaysia

Faculty of Health, Australian Research Centre in Complementary and Integrative Medicine, University of Technology Sydney, Ultimo, NSW, Australia

B. H. Goh
e-mail: beyhingg@sunway.edu.my

R. K. Singla
School of Pharmaceutical Sciences, Lovely Professional University, Phagwara, Punjab, India

 
R. K. Singla et al. (eds.), *Polyphenols in Health*,
https://doi.org/10.1007/978-3-032-27324-6_14

is constrained by low and variable bioavailability, physiologically irrelevant in vitro dosing, heterogeneity in clinical trial design, limited understanding of structure–function relationships, inadequate standardisation, and quality control of extracts and analytical workflows. This chapter covers current trends, delineates knowledge gaps, and proposes a future roadmap centred on chemically characterised interventions, harmonised and adequately powered randomised trials, validated biomarkers, and stratification by genetic and microbiome features to enable precision nutrition. Emerging formulation strategies (nanoencapsulation, biopolymer-based delivery, and co-delivery), expansion of curated databases, coordinated international research networks, and robust regulatory and safety assessment frameworks are highlighted as essential steps to convert polyphenol science into actionable clinical, industrial, and public health applications.

**Keyword** Bioavailability · Gut Microbiota · Metabolomics · Precision nutrition · Polyphenol · Safety · Study design

## 14.1 Introduction

### *14.1.1 Overview of Polyphenol Research and Progress*

Polyphenols are a diverse group of plant-derived secondary metabolites widely distributed in fruits, vegetables, cereals, tea, cocoa, and medicinal plants. They are broadly classified into flavonoids, phenolic acids, stilbenes, lignans, and tannins (Manach et al. 2004). Extensive research shows that polyphenols exert biological effects mainly by modulating oxidative stress, inflammation, and cell signalling pathways such as nuclear factor kappa B (NF-κB) and nuclear factor erythroid 2-related factor 2 (Nrf2), rather than acting solely as direct antioxidants (Scalbert et al. 2005; Williamson 2017). Epidemiological and clinical studies associate polyphenol-rich diets with reduced risk of cardiovascular diseases (CVD), metabolic disorders, neurodegeneration, and certain cancers, although evidence from large randomised controlled trials is still limited (Del Rio et al. 2012). Recent progress includes advanced analytical techniques such as liquid chromatography coupled to mass spectrometry (LC–MS/MS), green extraction technologies, and novel delivery systems such as nanoencapsulation to improve stability and bioavailability (Cory et al. 2018). Future research should emphasise standardised methodologies, metabolite-driven mechanisms, and robust clinical validation. Despite promising findings, challenges persist regarding low bioavailability, variability in biological response, and lack of standardised methodologies. Future research should focus on well-designed clinical studies, metabolite-driven mechanisms, and sustainable production strategies to fully translate polyphenol research into health and industrial applications.

## 14.2 Recent Advances and Current Trends in Polyphenol Research

### *14.2.1 Breakthrough Discoveries in Polyphenol Bioactivity*

Polyphenols were initially studied mainly for their antioxidant capacity; however, breakthrough discoveries over the past two decades have profoundly transformed our understanding of their bioactivity and physiological relevance. Current evidence indicates that the biological effects of polyphenols arise predominantly from their ability to modulate cellular signalling pathways, gene expression, and metabolic homeostasis, rather than from direct free-radical scavenging alone (Scalbert et al. 2005; Williamson 2017).

A major conceptual advance is the discovery of hormetic mechanisms underlying polyphenol action. At nutritionally achievable concentrations, polyphenols activate adaptive stress-response pathways such as Nrf2/antioxidant response elements, AMP-activated protein kinase (AMPK), and sirtuins, enhancing endogenous antioxidant defences, mitochondrial biogenesis, and cellular stress resistance (Calabrese et al. 2011). This paradigm explains their pleiotropic benefits in cardiovascular health, metabolic regulation, neuroprotection, and healthy aging (Del Rio et al. 2012). Figure 14.1 schematically illustrates the pharmacokinetics and pharmacodynamics of dietary polyphenols.

Another landmark discovery is the central role of the gut microbiota in polyphenol bioactivity. Although many polyphenols display low systemic bioavailability, they are extensively transformed by intestinal microbiota into smaller phenolic metabolites with higher absorption and biological potency (Selma et al. 2009). These microbial metabolites have been shown to modulate inflammation, glucose metabolism, vascular function, and immune responses, positioning the microbiome as a critical determinant and limiting factor inducing inter-individual variability while assessing polyphenol efficacy (Williamson and Clifford 2010).

Breakthroughs in epigenetic research further demonstrate that polyphenols can influence DNA methylation, histone modifications, and microRNA expression, thereby exerting long-term regulatory effects on genes involved in cancer, neurodegeneration, and chronic inflammatory diseases (Link et al. 2010). Advances in metabolomics and systems biology have enabled precise tracking of polyphenol metabolites in human tissues, strengthening mechanistic links between intake and health outcomes (Manach et al. 2017).

Collectively, these discoveries redefine polyphenols as multifunctional bioactive molecules with significant implications for preventive nutrition, functional food development, and personalised health strategies.

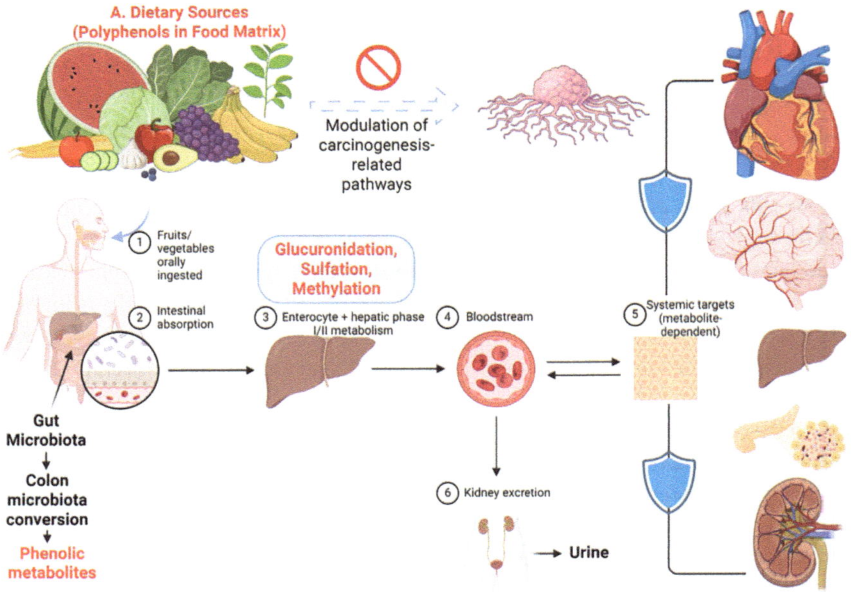


**Fig. 14.1** Systemic fate and biological actions of dietary polyphenols following oral intake. The figure schematically illustrates the absorption and distribution of polyphenols from plant-based foods, their intestinal uptake, and hepatic phase II-biotransformation (glucuronidation, sulphation, and methylation), subsequent entry into bloodstream, and delivery to peripheral target tissues (e.g., heart, brain, liver, pancreas, and kidney). The figure also highlights the contribution of the gut microbiota to polyphenol transformation into phenolic metabolites that can subsequently absorb through intestinal absorption route. This picture has been created by the authors

### 14.2.2 Novel Analytical Techniques and Omics Integration

Recent advances in analytical methodologies have profoundly enhanced polyphenol-based research, enabling accurate characterisation, quantification, and biological interpretation of these structurally diverse compounds. High-resolution platforms such as ultra-high-performance liquid chromatography coupled with tandem mass spectrometry (UHPLC-MS/MS) and high-resolution mass spectrometry (HRMS) allow comprehensive profiling of complex polyphenol matrices, including phase-II conjugates and low-abundance metabolites (Manach et al. 2017). These techniques provide superior sensitivity, selectivity, and structural elucidation when compared with traditional spectrophotometric assays (Scalbert et al. 2005).

The integration of omics technologies has emerged as a transformative approach in polyphenol-based research. Metabolomics enables systematic tracking of dietary polyphenols and their gut microbiota-derived metabolites, offering critical insights into bioavailability and metabolic fate (Wishart 2019). Epigenomic studies further demonstrate that polyphenols can influence DNA methylation, histone modifications, and microRNA expression, linking dietary exposure to long-term regulatory effects (Link et al. 2010).

Systems biology approaches that integrate multi-omics datasets with bioinformatics and machine learning tools are increasingly used to identify biomarkers of intake and response, address inter-individual variability, and strengthen the associated links between polyphenol consumption and health outcomes (Manach et al. 2017). Collectively, these innovations are redefining polyphenol research as a mechanism-driven and translational discipline.

### 14.2.3 Emerging Formulations and Delivery Strategies

Despite strong evidence for their biological potential, the practical application of polyphenols is limited by poor aqueous solubility, chemical instability, rapid metabolism, and low oral bioavailability (Manach et al. 2004; Williamson, 2017). In recent years, significant progress has been made in developing novel formulation and delivery strategies to overcome these constraints and enhance polyphenol efficacy.

Nanoformulation approaches, including polymeric nanoparticles, solid lipid nanoparticles, nanoemulsions, and liposomes, have emerged as leading strategies to protect polyphenols from degradation and improve intestinal absorption. These systems enhance solubility, prolong circulation time, and enable controlled or targeted release. Encapsulation using biopolymers such as chitosan, alginate, proteins, and cyclodextrins has also gained attention for food-grade and nutraceutical applications due to their safety and scalability (McClements 2020).

Another promising strategy involves co-delivery systems, where polyphenols are combined with proteins and complex carbohydrates to improve permeability and metabolic stability (Zhang et al. 2020). Additionally, fermentation-based formulations and microbiota-targeted delivery systems are being explored to exploit gut microbial metabolism and generate more bioactive polyphenol-derived metabolites (Selma et al. 2009).

Advances in formulation science, coupled with pharmacokinetic and metabolomic tools, are accelerating the translation of polyphenols into functional foods, nutraceuticals, and therapeutic adjuncts. Continued research on safety, scalability, and regulatory acceptance will be essential for widespread application.

### 14.2.4 Interdisciplinary Approaches in Polyphenol Science

Polyphenol science has become a strongly interdisciplinary field integrating analytical chemistry, food science, nutrition, pharmacology, microbiology, and data science. Advanced chromatographic and mass spectrometric techniques enable detailed structural and metabolic profiling of polyphenols (Scalbert et al. 2005).

Gut microbiome research explains inter-individual variability in polyphenol bioactivity through microbial biotransformation (Selma et al. 2009). Increasingly, metabolomics, bioinformatics, and machine learning are used to integrate complex datasets and identify biomarkers of intake and response (Manach et al. 2017). This interdisciplinary convergence is driving translational advances in functional foods, nutraceuticals, and personalised nutrition.

## 14.3 Gaps in Current Knowledge

### *14.3.1 Bioavailability and Metabolism Uncertainties*

Polyphenols are widely recognised for their potential health benefits. However, major uncertainties remain regarding their bioavailability, metabolism, and true biological relevance in humans. Unlike essential nutrients, polyphenols are not required for normal physiology, and their health effects depend largely on the extent to which they are absorbed, metabolised, and reach target tissues in biologically active forms (Manach et al. 2004).

A central challenge is that most dietary polyphenols exhibit low intrinsic bioavailability. In plant foods, polyphenols are often present as glycosides, esters, or polymers with high molecular weight that are poorly absorbed in the small intestine (Scalbert et al. 2005). As a result, only a small fraction of ingested polyphenols enters systemic circulation in their native form. Instead, they undergo extensive phase I- and phase II-metabolism in the intestinal epithelium and liver, leading to conjugated metabolites such as glucuronides, sulphates, and methylated derivatives, whose biological activities may differ markedly from those of parent compounds (Williamson 2017).

Another major source of uncertainty is the role of the gut microbiota. Polyphenols that escape absorption in the upper gastrointestinal tract are extensively transformed by colonic microbiota into smaller phenolic acids and other metabolites with higher absorption potential (Selma et al. 2009). These microbial metabolites are increasingly believed to be key mediators of polyphenol bioactivity. However, their formation varies substantially between individuals due to differences in gut microbial composition, diet, age, and health status (Manach et al. 2017). This inter-individual variability complicates the interpretation of clinical studies and limits the development of generalised dietary recommendations.

Furthermore, many in vitro studies employ polyphenol concentrations far exceeding those achievable in vivo, raising concerns about physiological relevance (Del Rio et al. 2012). The lack of validated biomarkers of exposure and effect further hampers efforts to establish clear dose–response relationships. Although advances in metabolomics have improved detection of circulating and tissue-specific metabolites, the correlation between these metabolites and specific molecular targets and health outcomes remains a challenging problem.

Overall, uncertainties surrounding absorption pathways, metabolic transformations, microbiota interactions, and active molecular species continue to limit translation of polyphenol-based research into clinical and nutritional practice. Future progress will depend on integrated pharmacokinetic-metabolomic studies, standardised intervention designs, and a shift towards metabolite-centred mechanisms to more accurately define the role of polyphenols in human health.

### 14.3.2 Inconsistencies in Human Clinical Data

Polyphenols have been extensively studied for their potential health benefits, including cardiometabolic protection, neuroprotection, and anti-inflammatory effects. While pre-clinical studies and observational data are generally supportive, human clinical trials often yield inconsistent or inconclusive results, complicating evidence-based dietary recommendations (Del Rio et al. 2012).

One major source of variability is heterogeneity in study design. Differences in polyphenol source, dosage, form (whole foods vs. extracts), duration of intervention, and subject populations (age, sex, and baseline health status) can all influence outcomes (Manach et al. 2017). Many trials also employ small sample sizes and short intervention periods, limiting statistical power and the ability to detect modest effects.

Bioavailability and metabolism contribute further to inconsistencies. Polyphenols undergo extensive intestinal and hepatic metabolism, producing conjugated and microbial-derived metabolites with variable absorption and biological activities (Williamson 2017). Inter-individual differences in gut microbiota composition can significantly affect metabolite profiles and thus physiological responses (Selma et al. 2009).

Additionally, outcome measures and biomarkers differ widely between studies. Some rely on surrogate markers, such as plasma antioxidant capacity, whereas others evaluate functional endpoints like endothelial function, cognitive performance, or inflammatory cytokines. Variability in analytical methods and lack of standardised biomarkers reduce comparability across trials.

Finally, lifestyle and dietary factors, including background intake of polyphenols and nutrient interactions, can mask or confound observed effects. These factors collectively contribute to inconsistent clinical findings despite robust pre-clinical evidence.

Addressing these challenges requires harmonised study designs, standardised polyphenol interventions, validated biomarkers, and integration of pharmacokinetic and metabolomic data. Such approaches will improve reproducibility, clarify dose–response relationships, and strengthen the translation of polyphenol research into public health recommendations.

### *14.3.3 Limitations in Structure–Function Relationship Understanding*

Polyphenols exhibit diverse chemical structures, including flavonoids, phenolic acids, stilbenes, lignans, and tannins (Chap. 1), which underpin their wide-ranging biological activities. Despite extensive research, establishing clear structure–function relationships remains challenging due to several intrinsic and extrinsic factors (Scalbert et al. 2005).

One key limitation is structural complexity and variability. Polyphenols exist as glycosides, esters, and polymers in foods, and their chemical modifications during digestion or food processing alter biological activity. Minor differences in hydroxylation patterns, conjugation, or polymerisation can dramatically influence antioxidant potential, receptor interactions, and enzyme modulation, making generalisations difficult (Manach et al. 2004).

Metabolic transformations further complicate structure–function analysis. Polyphenols are extensively metabolised in the gut and liver, yielding glucuronides, sulphates, methylated derivatives, and microbial catabolites. These metabolites, rather than parent compounds, often exert the observed biological effects, but their structure–activity relationships are poorly characterised (Williamson 2017).

Additionally, context-dependent effects such as tissue distribution, concentration, and interactions with macronutrients, proteins, or other phytochemicals obscure direct correlations between chemical structure and biological outcome.

Collectively, these factors limit mechanistic understanding and hinder rational design of polyphenol-based functional foods or therapeutics. Advances in metabolomics, computational modelling, and high-resolution analytical techniques are needed to elucidate precise structure–function relationships.

### *14.3.4 Inadequate Standardisation and Quality Control*

Polyphenol research faces significant challenges due to inadequate standardisation and quality control of both experimental materials and analytical methods. Variability in polyphenol content, composition, and stability across plant sources, cultivars, harvest conditions, and processing methods complicates reproducibility and interpretation of biological effects (Manach et al. 2004). Many studies employ crude extracts or commercial preparations with poorly defined composition, making it difficult to compare results across laboratories or to establish dose–response relationships.

Analytical inconsistencies further exacerbate this problem. Diverse extraction procedures, chromatographic techniques, and antioxidant assays differ in sensitivity, selectivity, and quantification accuracy, leading to inconsistent reporting of polyphenol content and activity (Scalbert et al. 2005). In particular, total phenolic content assays often overestimate concentrations due to interference from non-polyphenolic compounds.

Moreover, limited use of validated biomarkers for polyphenol intake and bioactivity hinders translation from pre-clinical studies to clinical and epidemiological research. Inter-laboratory standard reference materials and harmonised protocols remain scarce, restricting the development of evidence-based dietary recommendations and functional foods.

Addressing these challenges requires rigorous characterisation of polyphenol sources, adoption of standardised extraction and analytical methods, and validation of biomarkers to improve reproducibility, reliability, and translational relevance of polyphenol research.

### 14.3.5 Lack of Long-Term and Population-Based Studies

Despite the existence of extensive pre-clinical and short-term clinical studies that demonstrate the potential health benefits of polyphenols, robust long-term and population-based investigations remain scarce. Most evidence for cardiovascular, metabolic, neuroprotective, and anti-inflammatory effects derives from small-scale trials with limited duration, often relying on surrogate biomarkers rather than hard clinical endpoints (Del Rio et al. 2012). This limitation restricts the ability to draw definitive conclusions about the preventive or therapeutic impact of dietary polyphenols.

Population-based observational studies are further constrained by variability in polyphenol content of foods, reliance on dietary questionnaires, and heterogeneity in participant diets, lifestyles, and genetic backgrounds. Such factors introduce significant confounding and inter-individual variability, complicating the establishment of causal relationships between polyphenol intake and health outcomes (Manach et al. 2017).

Additionally, the long-term safety, optimal dosage, and interactions of polyphenol-rich foods or supplements remain poorly characterised. Without large, well-designed cohort studies or randomised controlled trials with sufficient duration, translating mechanistic insights into public health recommendations or functional food development is challenging (Williamson 2017).

Addressing these gaps requires coordinated population-based studies with standardised dietary assessment tools, validated biomarkers of polyphenol intake and metabolism, and long-term follow-up to reliably evaluate efficacy and safety, ultimately strengthening evidence-based recommendations for polyphenol consumption.

### 14.3.6 Underexplored Polyphenol Classes and Sources

While flavonoids and phenolic acids dominate polyphenol research, numerous other polyphenol classes remain underexplored, limiting our understanding of their health potential. Stilbenes, lignans, and polymeric tannins exhibit unique therapeutic effects but have received comparatively less attention due to low abundance in common diets, structural complexity, and analytical challenges (Manach et al. 2004). Resveratrol, a stilbene, has been extensively studied in pre-clinical models, yet other stilbenes remain poorly characterised in humans. Similarly, lignans from seeds, cereals, and vegetables show promising cardiometabolic and estrogen-modulating effects, but robust clinical data are limited (Scalbert et al. 2005).

Emerging dietary sources such as exotic fruits, medicinal herbs, and food processing by-products represent a largely untapped reservoir of polyphenols with potential functional benefits. For example, polyphenols in pomegranate peel, tea residues, and grape pomace contain high levels of tannins and flavonoid polymers, yet systematic studies on bioavailability and biological effects are sparse (Williamson 2017). Additionally, variations in polyphenol composition due to cultivar, geographic origin, and processing complicate their characterisation and functional assessment.

Expanding research to underexplored polyphenol classes and alternative dietary sources, combined with advanced analytical and metabolomic approaches, could reveal novel bioactive compounds, improve mechanistic understanding, and diversify functional food and nutraceutical development.

### *14.3.7 Challenges in Assessing Synergistic Effects*

Polyphenols are consumed as complex mixtures in foods and beverages, raising the possibility of synergistic or antagonistic interactions that influence biological activity. While in vitro studies often suggest additive or synergistic effects, translating these findings to in vivo or clinical settings remains challenging due to several factors (Del Rio et al. 2012).

First, chemical complexity and variability of polyphenol mixtures hinder reproducible assessment. Different polyphenol classes, glycosylation patterns, polymerisation, and matrix interactions can alter bioavailability and metabolic fate, making it difficult to predict combined effects (Manach et al. 2004). Second, metabolic transformations in the gut and liver further complicate interactions. Conjugated and microbial-derived metabolites may differ in activity from parent compounds, and the presence of multiple polyphenols can influence absorption, metabolism, or transport, producing non-linear effects (Williamson 2017).

Third, limitations in study design restrict detection of synergy in human trials. Short-term interventions, small sample sizes, and reliance on surrogate biomarkers reduce sensitivity to subtle interactions. Additionally, standard statistical approaches often do not account for complex multi-component interactions.

Advances in systems biology, metabolomics, artificial intelligence, and computational modelling offer promising tools to deconvolute polyphenol interactions, identify synergistic combinations, and guide rational design of functional foods and nutraceuticals. Addressing these challenges is essential to harnessing the full therapeutic potential of dietary polyphenol mixtures.

### *14.3.8 Research Gaps in Polyphenol-Microbiome Interactions*

Dietary polyphenols interact with the gut microbiome to produce bioactive metabolites that mediate systemic health effects. However, research gaps limit understanding of this interplay. Inter-individual variability in microbiome composition leads to differential metabolite profiles and responses (Selma et al. 2009). Mechanisms of microbial transformation remain poorly defined for many polyphenol classes, and functional consequences on inflammation, metabolism, and gut barrier integrity are incompletely characterised (Del Rio et al. 2012; Manach et al. 2017). Most studies concern short-term, in vitro, or animal-based research, restricting translation. Integrated human studies with metabolomic and metagenomic approaches are essential to informing personalised nutrition and microbiome-targeted interventions.

## 14.4 Recommendations for Future Studies and Clinical Applications

### *14.4.1 Designing Robust and Standardised Human Trials*

Human trials are critical to translating the mechanistic and pre-clinical evidence of polyphenol bioactivity into concrete health recommendations. However, inconsistencies in study design, intervention, and outcome measures have limited reproducibility and interpretability (Del Rio et al. 2012; Williamson 2017). Future studies should prioritise standardisation of polyphenol sources and dosages. Detailed chemical characterisation of interventions, including identification of major polyphenol classes, conjugates, and potential contaminants, is essential. Use of standardised extracts or well-characterised whole foods will improve comparability across studies.

Rigorous study designs are needed, including adequately powered randomised controlled trials (RCT) with appropriate control groups, sufficient sample sizes, and longer durations to assess chronic effects. Stratification based on age, sex, genetic polymorphisms, and gut microbiome composition can help account for inter-individual variability in bioavailability and response (Manach et al. 2017). Incorporation of validated biomarkers of polyphenol intake and metabolism will enhance the precision of exposure assessment and link observed effects to circulating or tissue-specific metabolites.

Harmonised outcome measures across trials are essential. Clinical endpoints, such as cardiovascular events, glucose regulation, and cognitive function, should be prioritised over surrogate biomarkers whenever possible. Integration of multi-omics approaches, including metabolomics, transcriptomics, and microbiome profiling, can elucidate mechanisms of action and identify responder phenotypes (Selma et al. 2009).

Finally, transparent reporting, adherence to Consolidated Standards of Reporting Trials (CONSORT) guidelines (Turner et al. 2012), and inter-laboratory collaboration will improve reproducibility and facilitate meta-analyses. These strategies collectively will strengthen the evidence base, reduce variability, and accelerate the translation of polyphenol research into public health recommendations and functional food development.

### 14.4.2 Advancing Analytical Methods and Databases

Accurate characterisation and quantification of polyphenols are essential for understanding their bioactivity, metabolism, and health effects. Traditional spectrophotometric and colorimetric assays, while useful for total polyphenol estimation, lack specificity and fail to differentiate between structurally diverse compounds (Scalbert et al. 2005). Recent advances in high-resolution analytical techniques, such as UHPLC-MS/MS, HRMS, and nuclear magnetic resonance, now enable precise identification of individual polyphenols, their conjugates, and microbial metabolites (Manach et al. 2017). These methods provide high sensitivity, structural elucidation, and reproducibility, essential for mechanistic and clinical studies.

Complementing analytical advancements, comprehensive polyphenol databases have emerged as critical tools for research. Databases such as Phenol-Explorer, PhytoHub, and FooDB integrate chemical structures, food sources, metabolite profiles, and bioactivity data, facilitating cross-study comparisons, metabolomics analyses, and dietary intake assessments (Crozier et al. 2009). Integration with metabolomics and systems biology platforms allows tracking of polyphenol-derived metabolites in biological matrices and linking them to molecular targets and health outcomes.

Future research should focus on harmonising analytical protocols, expanding databases to include underexplored polyphenols and their microbial metabolites, and integrating machine learning for predictive modelling of structure–activity relationships. These approaches will enhance reproducibility, enable precise dietary recommendations, and accelerate translation of polyphenol research into functional foods and nutraceutical applications.

### 14.4.3 Optimising Formulation and Delivery for Better Bioavailability

Polyphenols are renowned for their health-promoting properties, but their therapeutic potential is limited by poor water solubility, chemical instability, rapid metabolism, and low oral bioavailability (Manach et al. 2004; Williamson 2017). Optimising formulation and delivery strategies is therefore crucial to enhance absorption, protect bioactive compounds, and maximise biological effects.

Polymeric nanoparticles, solid lipid nanoparticles, nanoemulsions, and liposomes can encapsulate polyphenols, improving solubility, protecting against degradation, and enabling controlled release. Biopolymer-based encapsulation using chitosan, alginate, or cyclodextrins offers food-grade and biocompatible solutions suitable for nutraceutical applications (McClements 2020). Co-delivery systems combining polyphenols with bioenhancers or lipids can further improve permeability, metabolic stability, and tissue targeting.

Fermentation and microbiota-targeted delivery strategies represent emerging approaches to exploit gut microbial metabolism, generating more bioactive metabolites and enhancing systemic effects (Selma et al. 2009). Integration of pharmacokinetic studies with advanced analytical and metabolomic tools enables optimisation of dosage forms and personalised interventions.

Future research should focus on standardised evaluation of bioavailability, safety, and efficacy of novel formulations, along with scalable manufacturing techniques for functional foods and supplements. Such strategies will enhance the translational potential of polyphenols, bridging the gap between mechanistic evidence and clinical application. Nanotechnology-based delivery systems have emerged as effective approaches.

### 14.4.4 Personalising Polyphenol-Based Nutrition and Medicine

Polyphenols offer antioxidant, anti-inflammatory, and cardiometabolic benefits, yet inter-individual variability limits their universal efficacy. Factors including genetics, gut microbiome composition, age, sex, and dietary habits significantly influence polyphenol absorption, metabolism, and biological activity (Manach et al. 2017).

Personalised nutrition strategies aim to tailor polyphenol interventions to individual characteristics. Metabolomics, genomics, and microbiome profiling can identify responders and specific metabolic phenotypes, enabling targeted dietary recommendations or supplementation (Selma et al. 2009). For instance, individuals harbouring microbial taxa capable of converting polyphenols into bioactive metabolites may experience enhanced benefits, while genetic variations affecting phase II-metabolism or transport proteins can modulate circulating metabolite levels and therapeutic outcomes.

Integration of multi-omics data with computational tools, including systems biology and machine learning, can predict efficacy, optimise dosage, and identify synergistic polyphenol combinations. Personalised polyphenol-based interventions have potential applications in preventive nutrition, functional foods, and targeted therapy, but require robust clinical studies and validated biomarkers to ensure reproducibility, safety, and effectiveness (Williamson 2017).

Tailoring polyphenol intake to individual metabolic and microbial profiles represents a critical step towards precision nutrition and medicine, maximising health benefits while minimising variability in response.

### 14.4.5 Integrating Polyphenols into Public Health Guidelines

Polyphenols are recognised for their potential to reduce the risk of chronic diseases, including CVD, diabetes, and neurodegenerative conditions. Despite robust preclinical and epidemiological evidence, their integration into public health guidelines remains limited due to variability in bioavailability, metabolism, and inconsistent clinical outcomes (Del Rio et al. 2012).

To inform evidence-based recommendations, standardised methods for assessing polyphenol intake and circulating biomarkers are essential. Food composition databases such as Phenol-Explorer and FooDB provide valuable tools for estimating dietary polyphenol exposure, yet gaps remain for underexplored polyphenols and processed foods (Crozier et al. 2009). Population-level studies and long-term clinical trials are required to establish dose–response relationships, safety, and efficacy across diverse demographics.

Incorporating polyphenols into dietary guidelines should focus on whole-food consumption patterns rather than isolated supplements, promoting intake of fruits, vegetables, tea, cocoa, and legumes. Harmonised recommendations can complement existing nutrient-based guidelines while encouraging dietary diversity.

Future public health strategies should leverage metabolomics, microbiome profiling, and systems biology to better predict individual responses, enabling personalised yet population-relevant guidance. Such integrative approaches will facilitate translation of polyphenol research into actionable public health policies.

### 14.4.6 Regulatory and Safety Assessment Recommendations

Polyphenols are widely consumed as part of diets and as supplements, yet regulatory frameworks and safety assessments remain inconsistent due to chemical diversity, variable bioavailability, and complex metabolic profiles (Williamson 2017). Although generally regarded as safe, high doses of certain polyphenols may cause adverse effects, including gastrointestinal discomfort, interactions with co-administered medications (drug-drug interactions), or pro-oxidant activity under specific conditions (Manach et al. 2004).

To support safe and effective use, regulatory guidelines should require comprehensive chemical characterisation of polyphenol-rich products, including identification of major classes, conjugates, and potential contaminants. Standardised analytical methods, validated biomarkers of exposure, and batch-to-batch quality control are essential to ensuring reproducibility and reliability in research and commercial products (Scalbert et al. 2005).

Pre-clinical toxicology studies and controlled human trials should assess dose–response relationships, metabolic pathways, and potential interactions with drugs or nutrients. Post-market surveillance and adverse event reporting can further safeguard public health. Harmonisation of safety thresholds and labelling standards across countries will facilitate global regulation and consumer confidence.

Future recommendations should integrate pharmacokinetic, metabolomic, and microbiome data to refine safety assessments, particularly for high-dose supplements or novel polyphenol formulations. These approaches will provide a scientific basis for regulatory policies, enabling evidence-based guidelines for polyphenol use in nutrition and medicine.

### *14.4.7 Collaborative International Research Networks*

The complexity and diversity of polyphenols necessitate collaborative international research networks to advance understanding of their bioactivity, metabolism, and health impacts. Such networks enable sharing of standardised methodologies, chemical databases, analytical platforms, and multi-omics data, improving reproducibility and comparability across studies (Manach et al. 2017). Global collaboration facilitates large-scale, population-based studies and long-term clinical trials, accounting for inter-individual variability in genetics, gut microbiota, and dietary patterns (Selma et al. 2009). It also supports exploration of underrepresented polyphenol classes and emerging food sources.

Integration of harmonised protocols, open-access metabolomic and microbiome datasets, and bioinformatics tools accelerates mechanistic insights, identification of biomarkers, and translation into functional foods, nutraceuticals, and personalised nutrition strategies (Del Rio et al. 2012). International consortia foster capacity building, cross-disciplinary training, and coordinated regulatory efforts, ultimately advancing evidence-based recommendations and public health policies related to polyphenol consumption.

### *14.4.8 Roadmap for Translating Research into Clinical Practice*

Despite extensive pre-clinical and epidemiological evidence supporting the health benefits of polyphenols, their translation into clinical practice remains limited due to variability in bioavailability, metabolism, and inconsistent human trial outcomes (Del Rio et al. 2012). A structured roadmap is essential to bridging this gap and enable evidence-based implementation.

##### 14.4.8.1 Step 1: Standardisation of Interventions

Polyphenol sources must be chemically characterised, including major compounds, conjugates, and potential contaminants. Standardised extraction, formulation, and dosing protocols will enhance reproducibility across studies (Manach et al. 2004).

##### 14.4.8.2 Step 2: Robust Clinical Trials

RCT with appropriate controls, longer durations, and diverse populations are needed. Stratification based on genetics, age, sex, and gut microbiome composition can reduce inter-individual variability in response (Selma et al. 2009). Validated biomarkers of intake and metabolism should be incorporated to link exposure with biological effects.

##### 14.4.8.3 Step 3: Multi-Omics and Systems Biology Integration

Incorporating metabolomics, transcriptomics, and microbiome profiling will clarify mechanisms of action, identify responder phenotypes, and guide personalised interventions (Manach et al. 2017).

##### 14.4.8.4 Step 4: Regulatory and Translational Framework

Safety, efficacy, and pharmacokinetics must be rigorously evaluated, and standardised guidelines for functional foods or nutraceuticals should be established. Post-market surveillance and public health integration are critical for real-world impact.

By following this roadmap, polyphenol-based studies can move from mechanistic studies to actionable clinical recommendations, enhancing preventive nutrition and personalised health care.

## 14.5 Translational Potential and Opportunities

### *14.5.1 Polyphenols in Clinical and Therapeutic Settings*

Polyphenols are bioactive plant compounds with demonstrated antioxidant, anti-inflammatory, cardiometabolic, neuroprotective, and anticancer properties, making them promising candidates for clinical and therapeutic applications. Pre-clinical studies and epidemiological data suggest benefits in preventing CVD, type 2-diabetes, obesity, cognitive decline, and certain kinds of cancers (Del Rio et al. 2012). However, translating these findings into clinical practice faces several challenges.

One major limitation is bioavailability. Polyphenols are extensively metabolised in the intestine and liver, producing conjugates and microbial metabolites whose biological activity may differ from parent compounds (Manach et al. 2004; Williamson 2017). Inter-individual variability in absorption, metabolism, and gut microbiota composition further complicates predictable outcomes in human populations.

Clinical trials evaluating polyphenol interventions are often small, short-term, and heterogeneous in terms of dosage, source, and formulation. While some trials demonstrate improvements in endothelial function, blood pressure, insulin sensitivity, or cognitive performance, others report minimal effects, highlighting the need for standardised protocols, validated biomarkers, and extended-term studies (Manach et al. 2017).

Advanced formulation strategies, such as nanoencapsulation, liposomes, and co-delivery with bioenhancers, are being investigated to improve stability, absorption, and targeted delivery. Integration of metabolomics, systems biology, and personalised nutrition approaches can help identify responder phenotypes and optimise therapeutic efficacy.

Overall, while polyphenols hold substantial clinical promise, rigorous, well-designed trials, standardised interventions, and mechanistic insights are essential to fully realising their potential in preventive and therapeutic medicine.

### *14.5.2 Industrial and Nutraceutical Applications*

As a natural phytomaterial, polyphenols are already present abundantly in our regular foods, including fruits and vegetables, which we consume on a regular basis. Further, the food, pharmaceutical, and nutraceutical industries increasingly utilise polyphenols due to their antioxidant, anti-inflammatory, antimicrobial, and health-promoting properties. In the food industry, they serve as natural preservatives, colorants, and functional ingredients, extending shelf life and enhancing nutritional quality (Crozier et al. 2009).

In nutraceuticals, polyphenol-rich extracts from tea, cocoa, berries, grapes, and medicinal herbs are formulated into capsules, tablets, and beverages targeting CVD, metabolic disorders, and cognitive function. Advanced delivery systems, including nanoencapsulation and liposomal formulations, improve stability, solubility, and bioavailability, enhancing therapeutic efficacy (McClements 2020).

Industrial applications also include the valorisation of food by-products, such as grape pomace, pomegranate peel, and tea residues, as cost-effective sources of polyphenols for functional ingredients and supplements. Standardisation, quality control, and regulatory compliance are critical to ensuring safety, efficacy, and reproducibility in industrial production.

The integration of polyphenols into functional foods, nutraceuticals, and industrial formulations underscores their commercial and therapeutic potential, promoting health benefits while supporting sustainable utilisation of plant-based resources.

### 14.5.3 Consumer Awareness and Integration into Diet

Polyphenols are key bioactive compounds with demonstrated health benefits, yet consumer awareness of their dietary sources and functional roles remains limited. Fruits, vegetables, tea, cocoa, and legumes are rich in polyphenols, and their regular consumption is associated with reduced CVD risk, metabolic disorders, and cognitive decline (Del Rio et al. 2012).

Effective integration into the diet requires education on food sources, portion sizes, and preparation methods that preserve polyphenol content. Functional foods and fortified products offer additional avenues to increase intake, but evidence-based labelling and clear communication are essential to guiding consumers (Williamson 2017). Personalised dietary strategies considering individual metabolic and microbiome profiles may further optimise polyphenol benefits.

Increasing consumer knowledge through public health campaigns, nutritional guidelines, and digital tools can enhance dietary adherence, promote health, and translate scientific research into practical and sustainable lifestyle changes.

## 14.6 Ethical, Economic, and Social Considerations

### 14.6.1 Ethical Issues in Research and Application

Ethical considerations are critical in polyphenol research and application, particularly in human studies and commercial translation. Clinical trials must ensure informed consent, appropriate risk–benefit assessment, and transparency in reporting outcomes, especially given uncertainties in bioavailability and long-term safety (Manach et al. 2017). Overstated health claims in nutraceuticals and functional foods raise ethical concerns related to consumer misinformation and regulatory compliance (Williamson 2017). Sustainable sourcing of polyphenol-rich plant materials and fair use of traditional knowledge are additional ethical priorities. Addressing these issues through rigorous study design, responsible communication, and regulatory oversight is essential for maintaining scientific integrity and public trust.

### 14.6.2 Socio-Economic Impacts of Polyphenol Innovations

Innovations in polyphenol research have significant socio-economic implications, contributing to public health promotion, agricultural sustainability, and industrial growth. Development of polyphenol-rich functional foods and nutraceuticals can reduce healthcare costs by supporting prevention of chronic diseases (Del Rio et al. 2012). Valorisation of plant resources and agro-industrial by-products generates new income streams for farmers and rural communities, fostering sustainable bioeconomies (Crozier et al. 2009). However, equitable access, affordability, and evidence-based health claims remain critical to ensuring societal benefit. Integrating polyphenol innovations into food systems and health policies can stimulate economic development while improving population health outcomes.

### 14.6.3 Ensuring Global and Equitable Access

Ensuring global and equitable access to polyphenol research is essential for maximising its public health and socio-economic benefits. Open-access databases, shared analytical platforms, and international research consortia can reduce disparities between high- and low-resource regions (Manach et al. 2017). Capacity building through training, technology transfer, and inclusion of diverse populations in clinical studies improves relevance and equity of findings (Del Rio et al. 2012). Ethical sharing of data, sustainable sourcing, and fair intellectual property practices further support inclusivity. Such strategies promote balanced global participation and enable translation of polyphenol research into accessible nutritional and therapeutic solutions worldwide.

## 14.7 Conclusion

### 14.7.1 Summary of Gaps and Future Prospects

Despite extensive research highlighting the health-promoting potential of polyphenols, several critical gaps limit their effective translation into clinical and public health applications. Major challenges include poor and variable bioavailability, extensive metabolism, and significant inter-individual differences driven by genetics, gut microbiota, and lifestyle factors (Manach et al. 2004–2017). Inconsistencies in human clinical data, short intervention durations, small sample sizes, and lack of standardised dosing and formulations further complicate interpretation of outcomes (Del Rio et al. 2012). Additionally, limited understanding of structure–function relationships, synergistic interactions among polyphenols, and long-term safety restrict evidence-based recommendations.

Future research should prioritise standardised analytical and clinical protocols, integration of multi-omics approaches, and identification of validated biomarkers of intake and efficacy. Expanding studies to include underexplored polyphenol classes, diverse food sources, and population-based cohorts will enhance generalisability. Advances in formulation science, personalised nutrition, and microbiome-targeted strategies hold promise for improving bioavailability and therapeutic efficacy. Strengthening international collaboration and open-access data sharing will accelerate progress, enabling translation of polyphenol research into functional foods, nutraceuticals, and precision health interventions.

### 14.7.2 *Vision for Polyphenol Research and Application*

The future of polyphenol research lies in transforming abundant mechanistic and epidemiological evidence into precise, safe, and effective nutritional and therapeutic applications. While polyphenols are widely recognised for their roles in chronic disease prevention, progress is constrained by variability in bioavailability, metabolism, and individual responsiveness (Del Rio et al. 2012). A clear research vision should therefore prioritise integration of chemistry, nutrition, pharmacology, and systems biology to achieve translational impact.

Advances in high-resolution analytical techniques, metabolomics, and microbiome science will enable accurate tracking of polyphenol-derived metabolites and elucidation of their molecular targets (Manach et al. 2017). Coupling these tools with robust, standardised human trials will strengthen causal evidence and support regulatory acceptance. Personalised nutrition strategies, guided by genetic and microbial profiling, offer opportunities to tailor polyphenol interventions for maximal efficacy (Williamson 2017).

From an application perspective, innovation in formulation, sustainable sourcing, and food system integration will expand access to polyphenol-rich products. International collaboration, open data platforms, and ethical governance are essential to ensuring equitable global benefits. Collectively, this vision positions polyphenols as key components of precision nutrition, preventive medicine, and sustainable health solutions.

## 14.8 Funding Source:

This work was supported by the Sichuan Science and Technology Program (Grant No. 2025YFHZ0213).

**Acknowledgements** We thank Caproslaxy Media (https://www.caproslaxymedia.com/) for linguistic assistance during the preparation of this manuscript.

## Abbreviations

| | |
|---|---|
| AMPK | AMP-activated protein kinase |
| UHPLC-MS/MS | Ultra-high-performance liquid chromatography coupled with tandem mass spectrometry |

| | |
|---|---|
| HRMS | High-resolution mass spectrometry |
| CONSORT | Consolidated Standards of Reporting Trials |
| LC-MS/MS | Liquid chromatography coupled to mass spectrometry |
| NF-κB | Nuclear factor kappa B |
| Nrf2 | Nuclear factor erythroid 2-related factor 2 |
| RCT | Randomised controlled trial |

## References

Calabrese EJ, Mattson MP, Calabrese V (2011) Hormesis, cellular stress response and vitagenes as critical determinants in aging and longevity. Mol Aspects Med 32(4–6):279–304. https://doi.org/10.1016/j.mam.2011.10.007

Cory H, Passarelli S, Szeto J, Tamez M, Mattei J (2018) The role of polyphenols in human health and food systems: a mini-review. Front Nutr 5:87. https://doi.org/10.3389/fnut.2018.00087

Crozier A, Jaganath IB, Clifford MN (2009) Dietary phenolics: chemistry, bioavailability, and effects on health. Nat Prod Rep 26(8):1001–1043. https://doi.org/10.1039/b802662a

Del Rio D, Rodriguez-Mateos A, Spencer JPE, Tognolini M, Borges G, Crozier A (2012) Dietary polyphenols in human health: structures, bioavailability, and evidence of protective effects. Antioxid Redox Signal 18(14):1818–1892. https://doi.org/10.1089/ars.2012.4581

Link A, Balaguer F, Goel A (2010) Cancer chemoprevention by dietary polyphenols: promising role for epigenetics. Biochem Pharmacol 80(12):1771–1792. https://doi.org/10.1016/j.bcp.2010.06.036

Manach C, Scalbert A, Morand C, Rémésy C, Jiménez L (2004) Polyphenols: food sources and bioavailability. Am J Clin Nutr 79(5):727–747. https://doi.org/10.1093/ajcn/79.5.727

Manach C, Hubert J, Llorach R, Scalbert A (2009) The complex links between dietary phytochemicals and human health deciphered by metabolomics. Mol Nutr Food Res 53(10):1303–1315. https://doi.org/10.1002/mnfr.200800516

Manach C, Milenkovic D, Van de Wiele T, Rodriguez-Mateos A, de Roos B, Garcia-Conesa MT, Landberg R, Gibney ER, Heinonen M, Tomás-Barberán F, Morand C (2017) Addressing the inter-individual variation in response to consumption of plant food bioactives: towards a better understanding of their role in healthy aging and cardiometabolic risk reduction. Mol Nutr Food Res 61(6):1600557. https://doi.org/10.1002/mnfr.201600557

McClements DJ (2020) Advances in edible nanoemulsions: digestion, bioavailability, and potential toxicity. Prog Lipid Res 81:101081. https://doi.org/10.1016/j.plipres.2020.101081

Scalbert A, Johnson IT, Saltmarsh M (2005) Polyphenols: antioxidants and beyond. American Am J Clin Nutr 81(1 Suppl):215S–217S. https://doi.org/10.1093/ajcn/81.1.215S

Selma MV, Espín JC, Tomás-Barberán FA (2009) Interaction between phenolics and gut microbiota: role in human health. J Agric Food Chem 57(15):6485–6501. https://doi.org/10.1021/jf902107d

Turner L, Shamseer L, Altman DG, Weeks L, Peters J, Kober T, Dias S, Schulz KF, Plint AC, Moher D (2012) Consolidated standards of reporting trials (CONSORT) and the completeness of reporting of randomised controlled trials (RCTs) published in medical journals. Cochrane Database Syst Rev 11(11):MR000030. https://doi.org/10.1002/14651858.MR000030.pub2

Williamson G (2017) The role of polyphenols in modern nutrition. Nutr Bull 42(3):226–235. https://doi.org/10.1111/nbu.12278

Williamson G, Clifford MN (2010) Colonic metabolites of berry polyphenols: the missing link to biological activity? Br J Nutr 104(Suppl 3):S48–S66. https://doi.org/10.1017/S0007114510003946

Wishart DS (2019) Metabolomics for investigating physiological and pathophysiological processes. Physiol Rev 99(4):1819–1875. https://doi.org/10.1152/physrev.00035.2018

Zhang L, McClements DJ, Wei Z, Wang G, Liu X, Liu F (2020) Delivery of synergistic polyphenol combinations using biopolymer-based systems: advances in physicochemical properties, stability and bioavailability. Crit Rev Food Sci Nutr 60(12):2083–2097. https://doi.org/10.1080/10408398.2019.1630358

FSC
www.fsc.org
MIX
Papier aus verantwortungsvollen Quellen
Paper from responsible sources
FSC® C105338

If you have any concerns about our products,
you can contact us on
**ProductSafety@springernature.com**

In case Publisher is established outside the EU,
the EU authorized representative is:
**Springer Nature Customer Service Center GmbH**
**Europaplatz 3, 69115 Heidelberg, Germany**

Printed by Libri Plureos GmbH
in Hamburg, Germany